World Savannas
Ecology and Human Use

Jayalaxshmi Mistry, BSc, PhD
Department of Geography, Royal Holloway, University of London

Routledge
Taylor & Francis Group

LONDON AND NEW YORK

First published 2000 by Pearson Education Limited

Published 2014 by Routledge
2 Park Square, Milton Park, Abingdon, Oxon OX14 4RN
711 Third Avenue, New York, NY 10017, USA

Routledge is an imprint of the Taylor & Francis Group, an informa business

ISBN 13: 978-0-582-35659-7 (pbk)

British Library Cataloguing in Publication data
A CIP catalogue record for this book can be obtained from the British Library.

Library of Congress Cataloging-in-Publication data
A catalog record for this book can be obtained from the Library of Congress.

Typeset by Mathematical Composition Setters Ltd, Salisbury, Wiltshire.

Contents

Preface

Images of large herds of mammals roaming the African landscape of grassy plains with scattered umbrella-shaped trees have been plaguing me for some time. I'm not denying that they exist, but it seems that this is the only picture people envisage when they think of the word 'savanna'. Many of my students were incredulous at the thought of South America having savannas. 'Brazil – savannas. Really?' was a typical comment! This was the main incentive for me to write this book. It is a way of showing that in fact savannas are found not only in Africa, but also in Asia, Australia and, yes, South America.

I was also unhappy with the distinction made in many texts between the ecology and human influence on ecosystems. What I have tried to do in this book is show that humans are actually inherent to the ecology of savannas. The material I have sought to cover in the book is in no way comprehensive, but is intended to give a taster for different savannas and the basic concepts and issues surrounding them. If more detailed information is required, this can be obtained from the key readings at the end of each chapter, the bibliography and the savanna Internet sites of interest.

I would also like to apologise to the animal lovers out there. Being inclined towards plants, I admit that I have concentrated more on the vegetation. However, in defence, I should say that the aim of this book is to look at savanna ecology, and so only when animals are intrinsic to the functioning of savannas are they discussed. It is hoped that more can be added in the future.

I begin the book with a basic introduction to savannas, looking at what determines their functioning, and some of the key theories behind savanna ecology at present. This is followed by more detailed accounts of various savanna formations around the world. The chapters all follow a similar format, beginning with the general ecology of the savanna, followed by the main ecological and human determinants – the latter described in historical terms – and ending with an analysis of current management problems. The final chapter is a form of conclusion, and looks forward into the future of savannas.

I would like to thank the following people for helping me with the book. Justin Jacyno, Kevin Dobbyn and Nigel Page of the Department of Geography, Royal Holloway, kindly agreed to help put together the various maps presented in the book. I would also like to thank the contributors of photographs and figures, but particularly Steve Archer and Juan Silva for their enthusiastic responses to my e-mails. Finally, a massive hug to my partner, Andrea Berardi, who gave me the confidence and support to see it through to the end.

Jay Mistry, London

Acknowledgements

I am grateful to the following for permission to reproduce copyright material: Table 1.2 Springer-Verlag GmbH & Co. KG, Biodiversity and savanna ecosystem processes: a global perspective, *Ecological Studies* (Berlin) 121: 1–27 (Solbrigg *et al.*, 1996). Figure 1.4 from International Union of Biological Sciences, Paris. A hierarchy model of savannas. Solbrig, 1991. *Biology International.* Figure 1.6 from the Ecological Society of America, *Ecological Monographs*, 58: 111–127 (Archer *et al.*, 1988). Figure 1.7 from *Ecoscience*, 2: 82–99, p. 87 (Archer, 1995). Figure 1.8 from Cambridge University Press, *Journal of Tropical Ecology*, 11: 651–669 Figure 6 (previously published by Gautier-Villars) (Farji Brener and Silva, 1995). Figure 2.6 Universidade de Brasilia, Fig 3 (Castro Neves and Miranda, 1996, in Miranda, Saito and Dias (eds). Figure 3.3 Elsevier Science, Amsterdam. Adaptation of woody species in savannas, *Biological Reviews*, 60: 315–355 (Sarmiento *et al.*, 1985). Figure 3.5 from Editions Elsevier, Changes in a protected savanna, *Acta Oecologica*, 12(2): 237–247 (San José and Farinas, 1991). Figure 3.6 from Editions Elsevier, *Acta Oecologica*, 11(6): 783–800 (Silva *et al.*, 1990). Table 3.2 from *Ecotropicos*, 10(2): 51–64 (Sarmiento and Silva, 1997). Figure 3.7 from Blackwell Science Ltd, Oxford, Large ranches as conservation tools in the Venezuelan *llanos. Oryx*, 31(4): 274–284 (Hoogesteijn and Chapman, 1997). Table 3.3 from Springer-Verlag GmbH & Co. KG, Photosynthetic responses of native and introduced C_4 grasses from Venezuelan grasses. *Oecologia* (Berlin), 67: 388–399 (Baruch *et al.*, 1985). Figure 4.3 from Stockholm Environment Institute, Stockholm, *Miombo ecology and management: An introduction* (Chidumayo, 1997). Table 4.7 from International African Institute, London. A subsistence society under pressure: the Bemba of northern Africa, *Africa*, 55: 40–59 (Stromgaard, 1985). Figure 4.5 from UNESCO, Paris and Parthenon Publishers, Carnforth. *Man and the Biosphere*, 12: 139–178. Changes in land use in Zimbabwe from 1911 to 2000 (Murphree and Cumming, 1993, in Young and Solbrigg (eds) *The World's Savannas*). Figure 5.3 from Blackwell Science Ltd, Australia. De Bie *et al.*, Table 3 in Woody plant phenology in savanna, 1998, *Journal of Biogeography*, 25: 883–900. Figure 5.6 OPULUS Press, The Netherlands, Tree and grass routing patterns in a humid savanna, *Journal of Vegetation Science*, 8: 65–70, p 67 (Mordelet *et al.*, 1997). Figure 5.8 Wageningen Agricultural University, The Netherlands, Wildlife resources of the West Africa savanna. De Bie, 1991. Wageningen Agricultural University Papers, Nos 91–92. Figure 6.5 from Cambridge University Press, *Journal of Tropical Ecology*, 14: 565–576 (Van de Koppel and Prins, 1998). Figure 6.6 from UNESCO, Paris and Parthenon Publishing, Carnforth. *Man and the Biosphere*, 12: 93–120 (Lane and Scoones, 1993, in Young and Solbrigg (eds) *The World's Savannas*). Figure 6.8 from John Wiley & Sons Limited, Chichester. After Perkins and Thomas, 1993. Spreading deserts or spatially confined environmental impacts? Land degradation and cattle ranching in the Kalahari desert of Botswana. *Land Degradation and Rehabilitation*, 4:179–124. Copyright John Wiley & Sons Limited. Reproduced with permission. Table 6.5 from UNESCO, Paris and Parthenon Publishing, Carnforth. *Man and the Biosphere*, 12: 93–120 (Lane and Scoones, 1993, in Young and Solbrigg (eds) *The World's Savannas*). Table 6.6 University of Wisconsin, Madison, West Bend, WI, Canada. Bruce *et al.*, 1998, Synthesis of trends and issues raised by land tenure, in Bruce, J.W. (ed.) *Country profiles of land tenure: Africa*, 1966 Research Paper No. 130. Table 7.2 from Cambridge University Press, *Vegetation of Southern Africa*, pp. 158–277 (Scholes, 1997). Table 7.3 from Cambridge University Press, *An African savanna. Synthesis of the Nylsvley study*, Table 12.2 (Scholes and Walker, 1993). Figure 8.5 from Green World Foundation, *Aspects of the temporal pattern of dry season fires in the dry dipterocarp forests of Thailand*, p. 239 (Kanjanavanit, 1992). Figure 8.8 from Blackwell Science Limited, *Case studies of human–forest interactions in Northeast Thailand*, KU/KKU/Ford Foundation, 850–0391, Final Report 2, Northeast Thailand Upland Social Forestry Project, Khon Kaen (Subhadira *et al.*, 1987). Table 8.8 from Silkworm Books, Bangkok, Thailand, *Seeing forests for trees. Environment and environmentalism in Thailand* (Ganjanapan, 1996). Figure 9.3 from Blackwell Science, Australia, *Australian Journal of Ecology*, 22: 279–287. The impact of experimental fire regimes on seed production in two tropical eucalypt species in northern Australia, Fig 2, p 281 (Setterfield, 1997). Figure 9.5 Australian Academy of Science, Canberra. Fire and native grasses, p 236 (Mott and Andrew, 1985, in Tothill and Mott (eds) *Ecology and management of the world's savannas*, pp 56–82). Figure 9.6 reprinted from: Monsoonal Australia – Landscape, ecology and man in northern lowlands. Haynes, C.D., Ridpath, M. and Williams, M.A.J. (eds). 90 6191 638 0, 1991, 28 cm, 242pp, EUR 104.50/US$123.00/GBP74. A.A. Balkema, P.O. Box 675, Rotterdam, Netherlands. Figure 10.2 from Blackwell Science Limited, Oxford, *Global Environmental Change*, Figure 7.4 Photosynthesis of the crops maize and wheat under increasing CO_2 concentrations at constant light conditions. (Moore *et al.*, 1996). Figure 10.6 from R. Duffy, The Environmental Challenge to the Nation State: Super Parks and National Parks Policy in Zimbabwe, *Journal of South African Studies*, Volume 23, Issue 3, 1997, pp. 441–451.

Every effort has been made to trace and acknowledge ownership of copyright. The publishers will be pleased to hear from any copyright holders whom it has not been possible to contact.

Chapter 1

The savanna ecosystem

1.1 Introduction

Savannas are the most common vegetation type in the tropics and subtropics (Solbrig, 1991) (Figure 1.1). They can be loosely defined as ecosystems with a continuous and important grass/herbaceous stratum, a discontinuous layer of trees and shrubs of variable height and density, and where growth patterns are closely associated with alternating wet and dry seasons (Bourlière and Hadley, 1983). Rainfall in savannas is highly seasonal, and the dry season can last from 2 to 9 months. Not only does this affect plants and animals, but it is also a major limitation to over one-fifth of the world's population, who live in or around savanna areas (Frost *et al.*, 1986). Many of these communities rely on subsistence agriculture or pastoralism. Productivity of crops and pastures is therefore governed by the uneven and often unpredictable nature of rainfall distribution. This is exacerbated by the generally nutrient-poor status of many savanna soils. Nevertheless, over time, plants, animals and humans have adapted to the savanna environment, and today savannas support a rich diversity of species and human cultures.

1.2 The key ecological determinants

Savannas are extremely heterogeneous ecosystems at different spatial and temporal scales (Figure 1.2). Over a landscape or at the patch scale, from a week to a century, savanna boundaries fluctuate through a variety of causal factors including the physical environment and human land use. They are also found intermingled with other vegetation forms, such as gallery forest and grassland. This heterogeneity has led to much confusion over the definition of savannas. In addition, there was a common misconception that all savannas were derived, i.e. they were previously forest, but through disturbance such as human activity, they became savannas. This may be the case for some areas (see Section 1.6.5), but palaeoevidence indicates that the majority of savannas are natural and ancient.

The actual word 'savanna' is thought to originate from an Amerindian word which, in a work on the Indies published in 1535, was used by Oviedo y Valdes to describe 'land which is without trees but with much grass either tall or short' (Bourlière and Hadley, 1983). Subsequently, its use was extended to include trees, as by Schimper in 1903. The term, however, has undergone constant

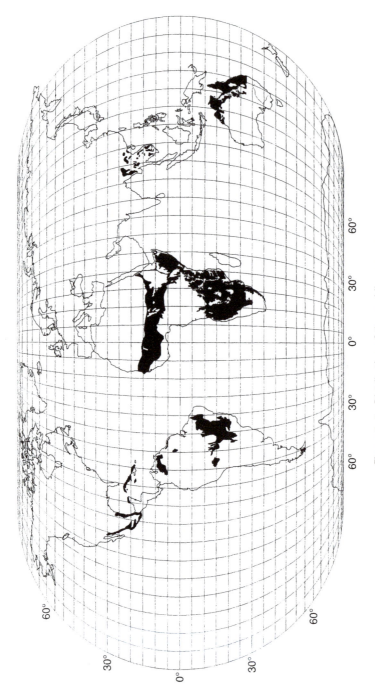

Figure 1.1 Distribution of the world's savannas.
(Reproduced by kind permission of Justin Jacyno)

(a)

(b)

Figure 1.2 The large range of savanna types has led to confusion over its definition. (a) An East African savanna; (b) a Brazilian savanna.
(Photographs by the author)

metamorphoses, and has been used as either a climatic or a vegetation concept (Table 1.1).

The huge range of definitions has been a major obstacle in attempts to relate research results and observed responses to disturbance and management in one savanna to those in another. These disparities brought together a group of savanna ecologists from around the world, in an attempt to identify a classificatory framework from which global as well as local comparisons could be made between savanna types, and between different components of savannas, e.g. tree or grasses. The group of savanna ecologists discussed the problem at a Responses of Savannas to Stress and Disturbance (RSSD) workshop sponsored by the International Union of Biological Sciences, in Harvard Forest, Massachusetts, USA. A conceptual model was developed, based on four key ecological determinants recognised as controlling the structure and function of savannas (Frost et al., 1986; Goldstein et al., 1988; Medina, 1993). These determinants are water availability, nutrient availability, fire and herbivory (Stott, 1991a). The model, however, evolves around the two factors considered to have primary control: plant-available moisture (PAM) and plant-available nutrients (PAN). Termed the PAM/PAN plane (Figure 1.3), this model allows the substitution of biological meaningful measures, i.e. PAM and PAN, for purely physical variables, thus enabling ready comparisons at various levels of savanna sites and plants. For example, PAM could be measured using factors such as the number

Table 1.1 Some definitions of savannas

Definitions of savanna	Authors
Climatic	
Ecosystems which lie in the tropical savanna (Aw) and monsoon (Am) climatic zones, largely between the latitudes of Cancer and Capricorn, where annual precipitation is between 250 and 2000 mm, most of which falls in the wet season	Köppen (1884, 1900)
Any formation or landscape within the region experiencing a winter dry season and summer rains is a savanna.	Jaeger (1945), Troll (1950) and Lauer (1952)
Ecosystems bound by dry forests at higher rainfall (> 1000 mm), by thorn forests at lower rainfall (< 500 mm), and by thorn steppe and temperate savannas at lower temperatures (< 18 °C)	Holdridge (1947)
Ecosystems with low to moderate rainfall (500–1300 mm) and high mean annual temperatures (18–30 °C)	Whittaker (1975)
Vegetation	
A mixed physiognomy of grasses and woody plants in any geographical area	Dansereau (1957)
A mixed tropical formation of grasses and woody plants, excluding pure grasslands	Walter (1973)
Open formations dominated by grasses, in the lowland tropics, where trees and shrubs, if present, are of little physiognomic significance	Beard (1953) and Whittaker (1975)

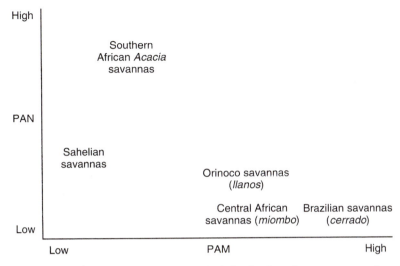

Figure 1.3 The PAM/PAN plane, and hypothetical distribution of some savanna formations (after Frost *et al.*, 1986).

of ecologically humid days (i.e. the period when growth is not limited by water availability) and/or the number of days in which rainfall exceeds evapotranspiration, and PAN could be based on the sum of exchangeable bases (K^+, Ca^{2+}, Mg^{2+} and Na^+).

Scale is fundamental to the model as both PAM and PAN will function differently over discrete levels of space and time. The four geographical scale divisions considered in the model are *patch*, a small area that is homogeneous in relation to some characteristic such as vegetation or topography; *catena*, a topographically determined unit in which a series of patches may be linked through a continuum of processes; *landscape*, a contiguous set of catenas; and *region*, a set of landscapes (Solbrig, 1991).

1.2.1 Plant–available moisture (PAM)

Precipitation in savannas varies from about 300 mm year^{-1} to over 1600 mm year^{-1}, and there is a dry season of between 2 and 10 months (Frost *et al.*, 1986) (Table 1.2). Savannas are characterised by a positive water regime (precipitation greater than evapotranspiration) during the rainy season, and a negative one during the dry season. The spatial and temporal distribution of rainfall is often highly variable, within a wet season, during a year, and between years.

Temperatures depend on the altitude and the latitudinal position of the savanna. Savannas located at higher elevations, such as some Brazilian savanna (*cerrado*) localities, and at the margins of the tropical zone, such as those in southern Africa and southern South America, can experience extremely low temperatures, and differences between mean January and mean July temperatures can be more than 10 °C (Nix, 1983). The latitudinal position also determines the length of the day and distribution of solar radiation.

Table 1.2 Climatic variables of some savanna locations (after Solbrig, 1996)

Locality	Country	Latitude	Longitude	Altitude (m above sea level)	AMT	TAP	Month < 20 mm	APE	Month ET > P
Corrientes	Argentina	−27.28	−58.50	29	22	1268	0	1130	3
Pretoria	South Africa	−25.45	28.14	1460	17	746	3	921	9
Campo Grande	Brazil	−20.27	−54.37	566	22	1444	0	1085	5
Ft. Jameson	Zimbabwe	−19.39	32.41	1256	22	1050	6	1071	8
Townsville	Australia	−19.15	146.46	4	24	1333	2	1161	10
Goiana	Brazil	−16.41	49.17	729	22	1487	3	1064	6
Cuiaba	Brazil	−15.36	−56.00	172	26	1375	3	1187	6
Broken Hill	Zambia	−14.24	28.24	1300	21	946	5	1028	9
Lilongwe	Malawi	−13.59	33.45	1136	20	849	6	992	9
Porto Amelia	Mozambique	−12.58	40.30	50	26	865	5	1199	9
Lobito	Angola	−12.22	13.12	3	24	221	8	1136	12
Lindi	Tanzania	−10.00	39.42	41	26	897	5	1199	8
Daru	Sierra Leone	−9.04	143.12	8	27	2098	0	1222	5
Barumba	Democratic Republic of Congo	1.15	23.29	24	25	1798	0	1158	3
Tabou	Ivory Coast	4.25	−7.22	6	26	2383	0	1186	4
Gambela	Ethiopia	8.15	34.35	1345	27	1240	3	1227	7
Barinás	Venezuela	8.38	−70.12	180	26	1461	2	1194	6
Calabozo	Venezuela	8.56	−67.20	119	28	1303	4	1235	7
Jos	Nigeria	9.54	8.53	1330	22	1404	4	1075	8
Menaka	Mali	15.52	2.30	280	30	263	8	1404	12
Bombay	India	18.54	72.49	11	27	2078	6	1252	9

AMT = annual mean temperature (°C); TAP = total annual precipitation (mm); Month < 20 mm = number of months in which rainfall is less than 20 mm; APE = annual potential evapotranspiration (mm); Month ET > P = number of months in which evapotranspiration is greater than precipitation.

1.2.2 Plant-available nutrients (PAN)

Savanna soils vary widely, depending on a variety of soil-forming factors including climate, geology, geomorphology, topography, the vegetation cover and animal activity (Montgomery and Askew, 1983). However, most tropical savannas occur on old and weathered surfaces, the result of geomorphological processes over millions of years (Cole, 1986). As a result, many savanna soils have low levels of nutrients (soil types: oxisols and ultisols) since they have been subject to soil-forming processes for prolonged periods of time. Generally, the higher moisture regimes of American, Australian and West African savannas make them extremely nutrient-poor compared to drier savannas, where nutrient levels are higher. This is, of course, subject to smaller-scale variations in soil-forming factors outlined above. Overall, savannas have small amounts of exchangeable calcium, magnesium and phosphorus, and a low cation exchange capacity; highly weathered soils also have high levels of exchangeable aluminium, which can reach toxic levels (Frost *et al.*, 1986).

1.2.3 Measuring the PAM/PAN plane

The PAM/PAN plane is a useful concept, in that if a savanna site can be positioned on the plane within an acceptable level of precision, it can be used to make predictions about the site's inherent structure and about various functional attributes that are difficult to measure, such as primary productivity, the seasonal course of evapotranspiration, nitrogen dynamics, and responses to fire and herbivory. It can also enable predictions to be made about possible changes in savannas in response to phenomena such as climate change. These inferences can then be extended to other sites with similar locations on the PAM/PAN plane.

However, Belsky (1991) notes that until the axes of the PAM/PAN plane are associated with quantifiable variables, the model provides little improvement over the earlier models. Few studies have as yet attempted to characterise PAM and PAN. One such work was by Walker and Langridge (1997), in which their objective was to find the optimal values (a trade-off between increasing complexity and the likelihood of obtaining the necessary data) for PAM and PAN, so as to be able to develop indices for comparing the savanna structure of 20 different sites in Australia. Their findings showed that general predictions of woody leaf, grass and total biomass did not need a detailed water budget analysis, but could be estimated from simple values of total rainfall, mean number of rainy days, potential evapotranspiration and soil texture. The best overall predictor of PAM was mean annual rainfall divided by potential evapotranspiration, and for PAN, the available phosphorus in the subsoil (Walker and Langridge, 1997). Williams *et al.* (1996), working in the Northern Territory, Australia, also found that savanna composition and structure could be related to the relatively simple variables of mean annual rainfall and soil texture, representing PAM and PAN respectively.

The data used in the above studies had several shortcomings, predominantly the weakness in soil data, estimates of woody leaf biomass derived from other site data, and most of the sites used having been disturbed in the past, therefore having less overall biomass (Walker and Langridge, 1997; Williams *et al.*, 1996). Nevertheless, they do indicate the potential for developing indices for PAM and PAN. But what are the problems in characterising PAM (or PAN) using a single value? First, in savannas the total amount of soil water and the length of time during which growth is possible are both important PAM factors. Secondly, any index would have to account for the predominant characteristic of savannas, i.e. the co-dominance of trees and grasses, and the factors that determine this coexistence (see Section 1.3.3). Lastly, PAM is not a meaningful site descriptor for all structural and functional variables of interest. For example, grass biomass is annually very variable and driven by short-term measures of PAM, and does not correlate well with a long-term average PAM.

1.2.4 Fire and herbivory

PAM and PAN are the primary determinants of savannas, with fire and herbivory considered 'modifiers', both partly controlled by PAM (Stott, 1991a). Fire is an endemic force in all the world's savannas, and before the human use of fire, lightning during thunderstorms probably caused the ignition of highly combustible fuel at the end of the dry season. With the discovery of fire by hominids, some 1.4 million years ago, the frequency of fires increased and fires were set at different times during the dry season (Schüle, 1990). These were used mainly for hunting and tribal warfare activities. Fire occurrence has steadily increased up to the present day, and today, some areas of savanna burn every year. Burning is used to clear areas for agriculture, during deforestation, to promote grass regrowth in pastures and for hunting, although accidental and criminal fires are also common.

Herbivores in savannas comprise vertebrates and invertebrates. The vertebrates that are most well known are the large native mammal fauna of browsers and grazers, a conspicuous feature of many Africa savannas. However, there are many smaller herbivorous animals in savannas including armadillos and porcupines. Although invertebrates, such as grass-eating harvester termites, are an important feature of savannas, they have been neglected as components of savanna functioning. Yet termites, ants and grasshoppers, to name a few of the insect herbivores, play a vital role in many savanna processes including consuming grass material and seed predation.

1.2.5 New paradigms in savanna dynamics

Initial theories developed to interpret functioning and dynamics of ecosystems were based on the equilibrium paradigm, where nature was assumed to maintain a permanence of structure and function if left undisturbed (Clements, 1916). Equilibrium dynamics assumes that an ecosystem is a closed, biotic system, and the stable state of the ecosystem is maintained through inter- and intra-specific interactions gradually 'equilibrating' to constant, external conditions. If there is a disturbance, the ecosystem may return to equilibrium (resilience) or retain equilibrium (resistance). This paradigm has had an overwhelming influence on the natural sciences, and has dominated ecological thinking, guiding the focus of ecological research, management policy and expectations of environmental productivity throughout this century (Sullivan, 1996).

The equilibrium paradigm states that historical effects, spatial heterogeneity, stochastic factors and environmental perturbations play a negligible role in ecosystem functioning and dynamics, while research has repeatedly shown that all of these factors are fundamental in ecological systems (e.g. Holling, 1973; Behnke et al., 1993). As a result, there has been increasing support for alternative paradigms to interpret ecosystem functioning and dynamics, including the non-equilibrium paradigms and the multiple-equilibria paradigms (DeAngelis and Waterhouse, 1987). This is particularly true for arid and semi-arid ecosystems, in which many savannas lie. Here, biological processes are moisture or PAM limited, and rainfall is inherently variable and unpredictable. The biotic system is therefore driven by infrequent rainfall events which vary in space and time. Thus, a savanna ecosystem rarely exhibits a stable state. Consequently, studies need to explore how these non-equilibrium ecosystems are maintained in the face of continual change and the strategies used by plants, animals and humans to cope with resource fluctuations (Sullivan, 1996). Unfortunately, this has not been the case, and dominant thinking about savannas remains focused on imposing 'stability', the issue of desertification being a case example (see Section 10.2.4).

Whether savannas are in equilibrium or non-equilibrium depends on the scale, both spatial and temporal, of observation, and on the kind and degree of disturbance. For example, on a large or intermediate scale, savannas may show stability, but at a patch scale they may be in non-equilibrium. A new conceptual framework has been proposed to accommodate variations in scale, dispensing with the attempt to categorise ecosystems as in 'equilibrium' or 'non-equilibrium' or something in-between. The paradigm of hierarchical patch dynamics (Wu and Loucks, 1995) provides a framework for explicitly incorporating scale while integrating aspects of equilibrium, non-equilibrium and multiple-equilibria systems.

Hierarchy theory has already been used as a conceptual tool in savanna ecology (Solbrig, 1991). PAM and PAN are prime controls on savanna form and functioning, whereas fire and herbivory are seen as modifiers. As such, the determinants are placed on different levels of the hierarchy, and within different spatial scales (Figure 1.4). The hierarchical patch dynamics paradigm is different in that it conceptualises ecological systems as nested hierarchies of patch mosaics. A patch is defined as a spatial unit differing from its surroundings in nature or appearance (Kotliar and Wiens, 1990). These patches can be characterised in terms of size, shape, content, duration, structural complexity and boundary characteristics, and can be organism-dependent (Wu and Loucks, 1995). For example, environmental heterogeneity within a spatial unit will create different types of patches depending on whether the organism under investigation is a mobile animal or a static plant.

Vegetation patterns are the most evident form of biotic patchiness and can provide the basis for patches at higher trophic levels (Wiens, 1976). Different

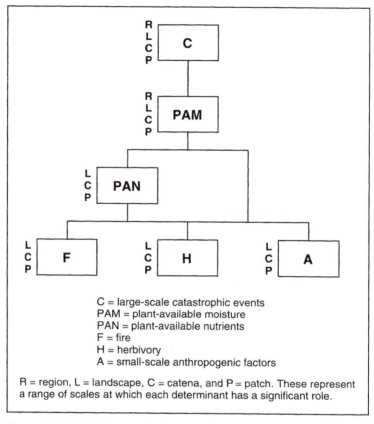

C = large-scale catastrophic events
PAM = plant-available moisture
PAN = plant-available nutrients
F = fire
H = herbivory
A = small-scale anthropogenic factors

R = region, L = landscape, C = catena, and P = patch. These represent a range of scales at which each determinant has a significant role.

Figure 1.4 A hierarchy model of savannas (after Solbrig, 1991).

causes and mechanisms operate on different spatial, temporal and organisational scales, thus creating a hierarchical framework of patch determinants (Figure 1.5). For example, in savannas, climatic variables are the most significant factors determining vegetation pattern at the continental scale, whereas disturbance such as fires can create patches from the local to the landscape scale (Delcourt and Delcourt, 1988; Wickham *et al.*, 1995). At an even higher resolution, it is species interactions such as competition and predation, that can create patchiness.

The 'hierarchical' part of the hierarchical patch dynamics paradigm has been developed so as to order the spatial and temporal scales within which these patch-forming mechanisms are functioning. The hierarchy patch dynamics paradigm may be a useful conceptual tool for savannas and their management because it considers disturbance as a force that shapes and maintains ecosystems, and analysis at different spatial scales as fundamental. Additionally, resiliency is seen as imperative in the systems, as is consideration of past land-use/environmental change legacies and future land-use/environmental variations, as opposed to focusing only on the present state of the site.

Type of patchiness   *Examples/causes of patchiness*

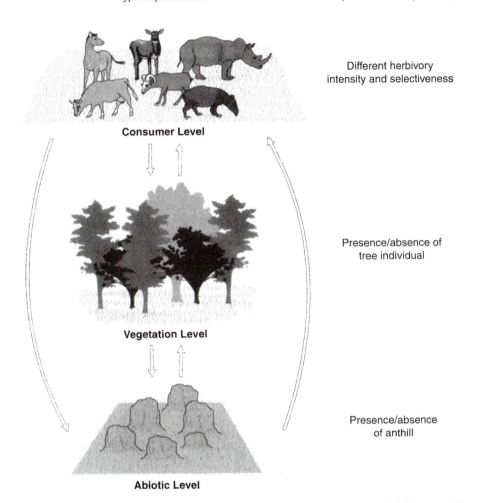

Different herbivory
intensity and selectiveness

Consumer Level

Presence/absence of
tree individual

Vegetation Level

Presence/absence
of anthill

Abiotic Level

Figure 1.5   A diagram of patch hierarchy in savannas. This shows examples of different patch-iness at the consumer, vegetation and abiotic levels, which although occurring at different spatial and temporal scales, interact to create the savanna ecosystem. Depending on the nature of interactions and the scale of investigation, the patches could maintain a constancy, i.e. equi-librium, or experience random changes, i.e. non-equilibrium.

1.3 Vegetation

1.3.1 Plant characteristics

The significant plant life forms in savannas are grasses and trees. Many of these have particular characteristics and strategies in respect of the four main ecolog-ical determinants. Savanna grasses, for example, have a C_4 photosynthetic pathway (described below) as an adaptation to the limitations of PAM. Attributes of savanna trees for reducing evapotranspiration include sclerophylly, i.e. hard-leaved, and being leafless during the dry season. Many plants also store water in underground organs, and some may have extended root systems to allow greater foraging for moisture.

Plant adaptations to PAN include internal recycling of limiting nutrients, especially just before leaf fall, and root mycorrhizal associations. The latter may be particularly important. There are two types: ectomycorrhizae (ECM) and vesicular–arbuscular mycorrhizae (VAM). ECM are more common in savannas, and have an advantage over VAM in infertile soils as they enable plants to take up nutrients direct from the litter. Savanna trees generally have thick corky bark to protect internal organs from fire, and both trees and grasses protect their dormant buds using dense, hairy sheaths. In response to herbivory, savanna plants have physical, chemical and biological deterrents. These include thorns, the presence of allelochemicals which are either bad-tasting or indigestible, and the presence of mutualisms, e.g. ants living on savanna trees to protect them from herbivores.

Phenology of savanna plants is also crucial. Trees being leafless, and annual grasses surviving as buried seeds, during the dry season, is important in terms of PAM, but also for avoiding the effects of fire. Perennial grasses also survive the dry season by dying back, and remaining dormant as underground structures. On the other hand, some plants require fire to complete their life cycles, as it can stimulate germination, leaf flush, flowering and fruiting.

The C_4 photosynthetic pathway in savannas

Savanna grasses mainly fix CO_2 by way of the C_4 photosynthetic pathway. In 'normal' photosynthesis, found in the majority of high-latitude plants and in savanna trees, CO_2 is first fixed to produce a 3-carbon compound, phosphoglyceric acid (PGA). This is termed C_3 photosynthesis or the Calvin cycle. In C_4 photosynthesis, on the other hand, CO_2 is initially fixed as a 4-carbon compound, oxaloacetic acid, and then transported to specialised cells where it is broken down and normal C_3 photosynthesis takes place.

For savanna grasses, the advantage of the C_4 mechanism is that the enzyme that fixes the CO_2, phosphoenolpyruvate (PEP) carboxylase, has a much higher affinity for CO_2 than the enzyme used in the C_3 pathway, rubisco. This means that PEP carboxylase can accumulate CO_2 more effectively from low concentrations. PEP carboxylase is also not inhibited by oxygen (photoinhibition) as rubisco is, so less CO_2 is wasted. This efficiency means that when plants are under water stress, as is the case in savannas, they can spend less time with their stomata open for gaseous exchange, and lose less water vapour in transpiration. In addition, C_4 photosynthesis has a higher temperature tolerance, so is able to take place at the higher temperatures and irradiation conditions found in savannas. The precise conditions under which C_4 or C_3 photosynthesis is advantageous result from interactions between light, temperature and moisture, and are not fully understood yet. It is hypothesised that most tropical trees, including savanna trees and shrubs, are C_3 photosynthesisers because of growing as seedlings and saplings in shade beneath canopies.

1.3.2 Floristics

Savannas share very few Linnaean species between continents, especially among the woody element. Herbaceous species, on the other hand, are wide-ranging and two families dominate throughout all savannas: the grasses (Gramineae) and sedges (Cyperaceae) (Solbrig, 1996). There are some woody floral affinities between continents which result from passive transport, range extension of individual species, but particularly because of old pantropical land connections. For example, families such as Proteaceae, Bombacaceae and Combretaceae are well represented among woody plants in savannas of all the continents of the old Gondwanaland. Australia has a scarcer affinity with other continents, although some genera, such as *Acacia*, *Adansonia* and *Cochlospermum*, are also found in Africa. Some families, including Leguminosae and Myrtaceae, are widespread, whereas others, such as Velloziaceae and Vochyziaceae in South America, are endemic to one continent (Solbrig, 1996).

Savannas share more species with other local vegetation types, and particularly with other savanna types in the same continent. For example, a number of woody species, such as *Acacia* spp., can be found throughout the savanna belt in Africa, from west to east and southern Africa. The Neotropics also have woody species, including *Byrsonima crassifolia* and *Byrsonima coccolobifolia*, which are distributed throughout the savanna range. Similarities with other vegetation types are varied, but include, for example, similarities between the *cerrado* and Amazonian/Atlantic rain forest flora (Table 1.3). Southeast Asian savanna woodland tree species show affinities with dense semi-deciduous forest (e.g. *Dipterocarpus obtusifolius*, *D. tuberculatus*, *Shorea obtusa*), and with dense rain forest (e.g. *Dipterocarpus intricatus*, *Pterocarpus* sp., *Syzygium* sp.).

From botanical surveys, it is clear that savannas are very rich in their woody component, although less so than other tropical vegetation types such as tropical rain forests. The most diverse is undoubtedly the Brazilian *cerrado*, which

Table 1.3 Some examples of pairs of closely related species from Brazilian rain forest (Amazonian and Atlantic) and from *cerrado* (after Sarmiento, 1983a)

Tropical rain forest species	*Cerrado*
Andira retusa	*A. humilis*
Aspidosperma duckei and *A. pallidiflorum*	*A. tomentosum*
Caryocar villosum	*C. brasiliense*
Connarus cymosus	*C. suberosus*
Copaifera lucens	*C. langdorffii*
Dalbergia nigra	*D. violacea*
Dimorphandra parviflora	*D. mollis*
Emmotum glabrum	*E. nitens*
Hymenaea altissima and *H. stilbocarpa*	*H. stigonocarpa*
Machaerium villosum	*M. opacum*
Qualea jundiahy	*Q. multiflora*
Sclerolobium rugosum	*S. aureum*
Vochysia tucanorum	*V. thyrsoidea*

has an estimated 10 000 vascular species (Ratter, 1986). The poorest is the Australian savanna (Solbrig, 1996). The number of grass species in any savanna numbers between 30 and 60 (Medina and Huber, 1994), and the number of sedges between 15 and 30, although normally 6–10 species are dominant.

1.3.3 Tree–grass interactions

The key characteristic of savannas is the long-term coexistence of trees and grasses. Both these functional types essentially use the same resources, i.e. sunlight, water and nutrients, and there is little evidence for niche separation in either the depth of rooting or the season of moisture uptake (e.g. Akpo, 1993; Scholes and Walker, 1993; Belsky, 1994; Mordelet *et al.*, 1997) (Box 1.1). It would therefore be expected that competition between the two would lead to the dominance of one. However, this is not so, and co-dominance of trees and grasses is prevalent. The paradigm shift from equilibrium dynamics (Walker *et al.*, 1981; Walker and Noy-Meir, 1982; Eagleson and Segarra, 1985) to a

---

Box 1.1 The equilibrium theories for tree–grass interactions based on niche separation

The classical concept of tree–grass interactions in savannas was based on the 'Walter hypothesis' (Walter, 1971). This proposes that trees have roots in both the surface and deeper soil layers, while grasses only have surface layer roots, and that to coexist with trees, grasses must have more efficient water-use capabilities. However, empirical observations indicate that although there is a rooting depth difference between trees and grasses, it is small. Trees and grasses both have the majority of their roots in the first 40 cm of the soil (Jackson *et al.*, 1996), and grass root density exceeds tree root density to a depth of 1 m (Scholes and Walker, 1993). Some savanna trees do have long tap roots, but these are generally few and of a small diameter (Scholes and Archer, 1997). This may give some trees an advantage during drought periods, but is insufficient to supply the transpiration needs of a complete canopy, and explains why many savanna trees are deciduous.

Niche separation with regards to soil moisture may not be soil depth, but season of use. Deciduous woody plants are able to store water and nutrients, and achieve full leaf expansion within weeks of the onset of the rainy season (Scholes and Walker, 1993). In contrast, the peak leaf area of grasses is generally several months later. Deciduous trees may also retain leaves for several weeks after the grasses have died. An early start for trees may also give them access to an early flush of nutrients, which in some cases may be more fundamental (Scholes, 1997). The water-use efficiency of grasses, both in depth and time, is not substantially or consistently higher than trees, so mature trees should always outcompete grasses. However, this does not happen, and so leads to the idea that tree : grass ratios are not deterministic equilibria, but dynamic and determined by disturbance.

non-equilibrium view suggests that disturbance allows coexistence, that trees and grasses are resilient, and that past environmental changes are important with regard to the present state of the vegetation (Skarpe, 1992; Scholes and Walker, 1993). Where tree growth is only limited by competition with other trees, disturbances have the potential to prevent savannas from developing into a pure woodland. As a result, disturbances of different 'patch sizes' (Wu and Loucks, 1995), such as fire, drought and grazing, over time and space, combined with rainfall factors may generate and sustain the coexistence of trees and grasses (Coughenour and Ellis, 1993; Jeltsch et al., 1996).

The effect of disturbance on tree–grass ratios: some examples

Grazing in the Texan savannas

Studies in the savannas of the Rio Grande Plains of southern Texas and northern Mexico have been investigating the development and stability of tree–grass mosaics. The dominant vegetation here is termed a savanna parkland. It is characterised by discrete patches of woody vegetation beneath the legume tree, *Prosopis glandulosa* (honey mesquite), in a matrix of C_4 grasses (Archer, 1991). Over time, woody encroachment has increased in these savannas, and they are gradually being replaced with subtropical thorn woodlands, dominated by *P. glandulosa*. For example, aerial photographs from 1941 to 1983 showed that woody cover had increased from 13 to 36%, the average size of woody patches had increased from 494 to 717 m^2, and convergence of discrete patches was evident (Archer et al., 1988) (Figure 1.6).

The dynamics of these patches show that they typically consist of one *P. glandulosa* individual in the centre of the patch, it being the largest plant in terms of diameter, height and canopy area (Archer et al., 1988) (Figure 1.7). The number of subordinate woody species in the patch ranges from 1 to 15 and is positively correlated with the size of *P. glandulosa*, the composition being similar to nearby closed-canopy woodlands. With the exception of *P. glandulosa*, all the woody species are bird-dispersed, and most woody saplings and seedlings in the herbaceous zone are of *P. glandulosa*. Models of patch growth and precipitation showed that establishment of woody species under *P. glandulosa* occurred within 10–15 years (Archer, 1989).

Aerial photographs show that small patches established and persisted for many years, but then there was a shift in favour of *P. glandulosa* (Archer, 1989). *P. glandulosa* has several attributes for successful invasion of grasslands: the production of abundant, long-lived seeds; germination and establishment occur over a wide range of soil, water and light regimes; root development is rapid; it fixes nitrogen early in life; the seedlings reproduce vegetatively following top removal within two weeks of germination; and 2–3 year old individuals can survive hot fires and severe droughts (Archer et al., 1988; Brown and Archer, 1990). Nevertheless, evidence shows that *P. glandulosa* has changed little in geographical range in the last 300–500 years (Johnston, 1983).

It seems that the increase in *P. glandulosa* coincides with the development of domestic grazing, especially cattle, in the region. This has facilitated the spread

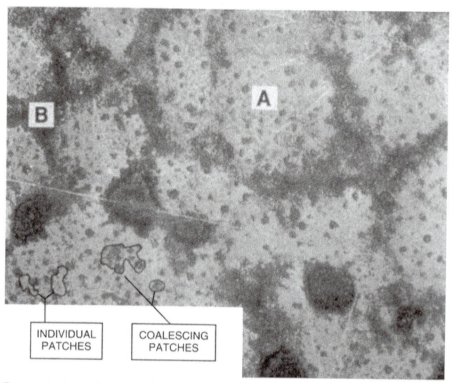

Figure 1.6 Aerial photograph showing the formation and coalescence of *P. glandulosa* patches in subtropical savanna parklands of southern Texas. A = tree/shrub patches within a matrix of C$_4$ grasses; B = closed canopy woodlands (after Archer *et al.*, 1988).

of *P. glandulosa* by increasing dispersal into new habitats and enhancing seed germination and seedling establishment (Brown and Archer, 1987). *P. glandulosa* pods are nutritious and heavily used by livestock, especially during drought periods when grass productivity is low. A high percentage of the hard seeds that are ingested escape mastication and are scarified in the digestive tract and deposited in moist, nutrient-rich dung away from the parent plant, which may harbour host-specific predators (Brown and Archer, 1987).

Germination in this way can be high and establishment is facilitated because herbaceous interference, and fire frequency and intensity, are reduced by grazing. In the absence of fire, *P. glandulosa* individuals can develop in stature, and provide seed for additional dispersal, a vertical structure attractive to avifaunal dispersal of other woody plant species, and a microhabitat conducive to germination and/or the establishment of other woody species (Brown and Archer, 1989). Therefore, the development of woody assemblages in heavily grazed areas in the Rio Grande Plains may reflect better dispersal and survival of *P. glandulosa*.

Ants in the Venezuelan savannas
Disturbance caused through nesting of leaf-cutting ants (*Atta* spp.) is considered the primary cause or accelerator of invasion of some grasslands by trees in

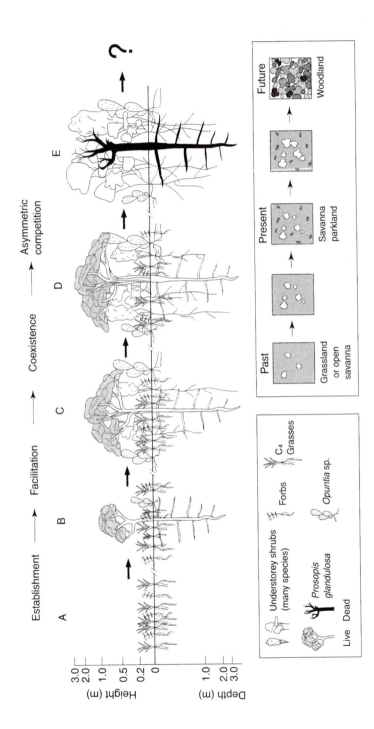

Figure 1.7 The development of subtropical thorn woodlands in subtropical savanna parklands of southern Texas. A–B = *P. glandulosa* establishment; B–C = *P. glandulosa* facilitation; C–D = overstorey, understorey co-existence; D–E = asymmetric competition (after Archer, 1995).

the Gran Chaco region of South America (Jonkman, 1976, 1978). Trees have been found to grow preferentially on abandoned nests of *Atta* spp. (Bucher, 1982), and similar processes have been observed in termite mounds (Glover *et al.*, 1964; Oliveira-Filho, 1992; Ponce and Da Cunha, 1993). Some savanna areas in the Orinoco *llanos* region in Venezuela are characterised by a higher frequency of small forest groves, resulting in a parkland landscape (Sarmiento, 1983a). This is particularly apparent between deciduous forests and savannas. Previously thought of as locally declining forest, these groves are now seen as an advancing front of forest trees onto grassland.

Farji Brener and Silva (1995a) suggest that initial colonisation of savanna may be by woodland tree species, as in their studies, smaller groves were entirely composed of woodland species and in most groves there is a central taller woodland tree. These woodland trees tend to improve site conditions, increasing soil organic matter, creating a more mesic microclimate and decreasing fire frequency by shading out savanna grasses (Kellman, 1979; Hobbs and Mooney, 1986; Belsky *et al.*, 1989). The aggregation of woodland species would therefore increase the chances of successful colonisation by forest species, relegating further woodland species colonisation to the periphery of the groves, where conditions such as increased radiation and fire are minor threats to their survival (Farji Brener and Silva, 1995a).

However, further site improvement may take place through the action of leaf-cutter ants. Farji Brener and Silva (1995b) found that *Atta laevigata* nests are richer in nitrogen, magnesium, organic carbon and calcium than similar soils in open grassland and in groves without nests. The leaf-cutter ants were found to increase in frequency and size of nests with the area of the grove, concurrent with the increasing presence of forest trees (Farji Brener and Silva, 1995a) (Figure 1.8). The above study did not show any clear sequence of arrival of the ants or the forest trees, but it seemed that the two may act synergistically, improving conditions for each other. In contrast to the case studied by Archer *et al.* (1988), the Orinoco *llanos* are not turning into shrubland or forest. This may be because the small-scale spatial heterogeneity of variables, such as soil, water-table depth and other PAM and PAN factors, may be significant in determining the tree–grass ratio (Medina and Silva, 1991).

Small patch disturbances

The coexistence of trees and grasses in the form of tree clumps may be common in some savanna types, such as the Texan savannas and Venezuelan *Trachypogon* savannas (San José *et al.*, 1991a), but in many others, including some East African savannas, the distribution of trees is more scattered. It seems that in these areas, the distribution of trees may be determined by small-scale disturbances in the environment, causing differential probabilities of tree establishment and survival. For example, a recent study using modelling (see Section 10.3.2) found that fire and grazing brought about coexistence in the form of tree clumps in a semi-arid southern Kalahari savanna (Jeltsch *et al.*, 1996). Then by introducing various small-scale disturbances into their model, Jeltsch *et al.* (1998) found that increased seed availability in localised clumps

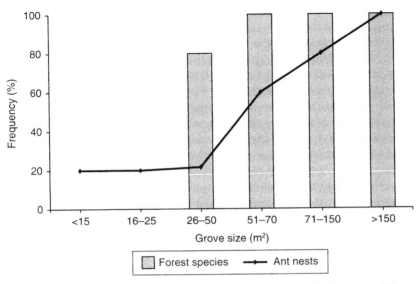

Figure 1.8 The frequency of *Atta laevigata* nests increases with the area of the grove and frequency of forest species in the savannas of Venezuela (after Farji Brener and Silva, 1995a).

(deposited in herbivore dung or collected by rodents), disturbance patch size, and the spatio-temporal correlation among individual disturbance patches, had the strongest impact on long-term tree–grass coexistence.

These results indicate that small-scale disturbances, favouring the establishment of woody seedlings by either increasing opportunities for establishment as above, or by reducing competition with grasses (i.e. vegetation clearing), are probably crucial during periods of tree decline. For example, during drought periods competition for moisture reduces the probability of tree establishment, and the patches of additional establishment may help to facilitate tree persistence, albeit at low levels. Therefore, small-scale disturbances may favour tree establishment by functioning as ecological 'buffer mechanisms' that prevent the extinction of the tree populations in critical situations (Jeltsch *et al.*, 1998).

1.4 Fauna

Knowledge about savanna fauna is limited and scattered. It consists mainly of taxonomic studies describing animal behaviour such as feeding and reproduction. Unfortunately, few ecological studies relating fauna to savanna functioning are available. Nevertheless, in certain groups, such as birds and vertebrates, there are some adequate data. For example, the large herbivores which characterise African savannas have been subject to substantial research (see particularly Section 6.2.3). These large mammals are absent from American, Asian and Australian savannas, although they may have been present in these savannas until the end of the Pleistocene. Debate as to the cause of their extinction, whether it be environmental change, human-induced or both, continues (e.g. Bowman, 1998). Asian savannas, notably the dry deciduous savanna forests, were occupied by large numbers of kouprey (*Bos sauveli*), the world's largest wild cattle species, but unfortunately the majority have been killed through hunting and war (Stott, 1991b).

Hunting and land use (see below) have been major contributors to fluctuations in the population of savanna fauna. For example, during the late nineteenth and twentieth centuries, the ivory trade decimated the elephant populations in many areas of Africa. Poaching is still widespread, whether it be for subsistence or commercial purposes, and many savanna species are on the IUCN World Conservation Union endangered list. Animal numbers and diversity have also been affected by land-use patterns in savannas. The continued conversion of savanna into arable land, for example, has fragmented habitats and restricted species ranges. Although conservation areas have been established in savanna regions, in many cases for the preservation of fauna, the majority of animals live outside these protected areas, and are therefore subject to considerable human impact.

1.5 Land use

Savannas are probably the oldest ecosystems used by people. *Homo sapiens* evolved in a savanna environment in East Africa about one million years ago (e.g. Abbate *et al.*, 1998). Human populations probably existed by hunting wildlife and gathering plant materials, and some of these cultures, such as the Aborigines of Australia, still survive today. Fire was an important tool in many activities including hunting, stimulating fruit production on trees, controlling weeds, and in tribal wars.

Where agriculture (about 10 000 years ago) and herding animals (6000 years ago) evolved, a great diversity of agriculture and agropastoral societies developed. This was particularly apparent in Africa and Asia, where shifting cultivation and extensive grazing were practised. This long period of land use in the Old World savanna regions, together with increased population densities, wars and other forms of conflict, has put greater pressure on savanna lands for production and resources. On the other hand, in South America it was not until the sixteenth century, and in Australia, the middle of the nineteenth century, that livestock grazing was introduced. Because of this, population and land use in savanna regions of these two continents is comparatively low.

Present-day savanna land use by both traditional and non-traditional cultures mainly involves grazing livestock, wood harvesting and growing crops (Figure 1.9). Related to this is the increase in domestic herbivory, and the use of fire for clearing areas and pasture regeneration. Commercial agriculture and more intensive modern ranching is growing in savanna regions, both needing large inputs such as fertilisers, pesticides and irrigation. The long season without available plant moisture is a severe limitation to plant growth, as are the low nutrient levels and high acidity of the soil. Consequently, a large proportion of agricultural research in savanna regions is aimed at improving the tolerance and resistance of crop varieties to these limitations.

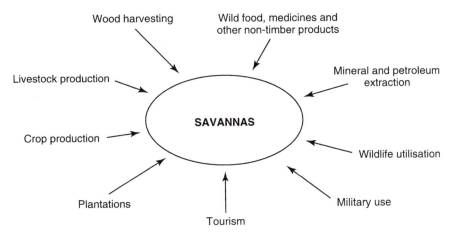

Figure 1.9   The different forms of savanna use.

1.6 The different savannas of the world

A mosaic of savanna types are found throughout the tropics and subtropics. They range from grasslands with scattered woody plants, to woodlands with a heterogeneous grass cover. Most of these savannas are 'natural', i.e. they have historically been savannas, although their range may have fluctuated with past environmental change. Aside from these natural savannas, in many parts of the world human disturbance has induced the conversion of other ecosystems into 'savannas', so-called 'derived' savannas (see below). For example, there has been particular focus on moist tropical forests, and their gradual conversion to 'savanna' through clearance and burning (e.g. Cavelier *et al.*, 1998). In this book, however, the interest focuses on natural savannas which cover large areas of the Americas, Africa, Asia and Australia.

1.6.1 Americas

Savannas cover over 2 million km^2 in the Neotropics (Figure 1.10). In some regions, such as the Brazilian *cerrado* (Eiten, 1972, 1978) or the Colombian and Venezuelan *llanos* (Sarmiento, 1984), they form a continuous landscape, interrupted by other vegetation formations through small-scale environmental discontinuities. However, in other areas they appear more as isolated patches amid other major formations. Along the middle Magdalena Valley in northern Colombia are savannas closely related to the seasonal savannas of the piedmont region of the *llanos* (Sarmiento, 1983a). In the Guiana region of south-eastern Venezuela, tablelands from 800 to 1200 m above sea level house the *Gran Sabana* in Venezuela (Sarmiento, 1983a), and the Rio Branco–Rupununi savannas in Guyana and Brazil (Eden, 1964, 1974). These are mainly composed of a *Trachypogon/Curatella* tree savanna. Several types of savanna occur in the coastal savanna region of the Guianas. Jansma (1994) gives a full description of these.

Figure 1.10 The savannas of the Americas.
(Reproduced by kind permission of Justin Jacyno and Nigel Page)

The Amazonian region has a range of seasonal and hyperseasonal savanna forms, but also has, in areas of pure white sand with podzolised soils, savannas that range from low sclerophyllous woodland to scrub and grass formations (Sarmiento, 1983a). The large lowland regions of the seasonally flooded Gran Pantanal (Eiten, 1975) and Llanos de Mojos (Hasse, 1990) are mosaics of forests, woodlands, woody savannas and grasslands. The western coast of southern Mexico, Central America and the Caribbean, most notably Cuba, have patches of seasonal, hyperseasonal and woodland savannas (Sarmiento, 1983a). Savannas dominated by the genera *Prosopis, Acacia* and *Andropogon* are also found in the Rio Grande Plains of southern Texas and northern Mexico (Küchler, 1964; Johnston, 1983). Mixed woodland, savanna parkland and flooded savanna type are extensive in the Chaco region of western Paraguay, eastern Bolivia, northern Argentina and part of southeastern Brazil (Adámoli *et al.*, 1991).

1.6.2 Africa

The most detailed classification of African vegetation is that of White (1983) in which the continent is divided into a number of zones. Savannas fall into the Zambezian zone, the Sudanian zone, the Somalia–Masai zone and various transition zones (see Figures 4.1, 5.1, 6.1 and 7.1). The Zambezian zone of central and southern Africa comprises various savanna woodlands, but principally *miombo*, which is dominated by *Brachystegia, Isoberlinia* and *Julbernardia* tree species. As one moves south and south-west from the main *miombo* region, arid–eutrophic savannas such as *Acacia* savanna, *mopane* (*Colophospermum mopane*) woodland and other dry savanna woodlands, are found. The Sudanian woodlands (where *Isoberlinia doka* and *Acacia* tree species are common) are bordered on the north by drier Sahelian types and on the south by wetter Guinean types. The Somalia–Masai zone consists of a range of savanna types from *Acacia–Commiphora* woodlands to the grassy plains dominated by various grass communities.

1.6.3 Asia

Savannas extend throughout Asia, and are the predominant formation across mainland Southeast Asia and the Indian subcontinent. The key savanna forest type of mainland Southeast Asia is the dry dipterocarp forest, starting in Manipur State, India, and widespread in Burma, Thailand, Cambodia, Laos and Vietnam (Blasco, 1983; Stott, 1984; Mistry and Stott, 1993) (see Figure 8.1). On the Indian subcontinent, the prevalence of *sal* (*Shorea robusta*) savanna forests is apparent (Champion, 1936; Puri, 1960). These two types, the dry dipterocarp and *sal*, have close taxonomic parallels, and the various structural, phenological and physiological similarities between the two have long been recognised (Kurz, 1875).

1.6.4 Australia and Oceania

Northern Australia and a small proportion of southern Papua New Guinea have a woodland savanna which becomes both taller and denser towards the northerly tips of the continent (Gillison, 1983). These savannas are dominated by numerous species of *Eucalyptus* although *Acacia* species such as *mulga* (*A. aneura*), *gidgee* (*A. georginae*), lancewood (*A. shirleyi*) and ironwood (*A. estrophiolata*) are present in drier areas (see Figure 9.1). Eucalypts are also found in the savannas of Timor, whereas in most of the other Southwest Pacific islands, savanna vegetation is probably anthropogenically derived (Gillison, 1983).

1.6.5 Derived savannas

Although savanna boundaries may change over time and space in relation to the main four ecological determinants, in some regions of the world, prolonged or intense use of particular ecosystems have bought about a process of 'savannisation'. A case example would be the Indian subcontinent. Much of the Indian subcontinent has undergone drastic land use transformation over the millennia, and what was dry deciduous and subhumid deciduous forest has now become derived savanna (Misra, 1983).

Dabadghao and Shankarnarayan (1973) classified these Indian 'savannas' according to the dominant grass species present. The *Sehima–Dichanthium* type covers the dry subhumid zone of peninsular India, and associated thorny bushes include *Acacia catechu*, *Mimosa rubicaulis* and *Zizyphus* species. The

Dichanthium–Cenchrus–Lasiurus type is found in the semi-arid zone extending from the northern part of Gujarat and eastwards towards Punjab. Woody species such as *Acacia senegal, Calotropis gigantea, Cassia auriculata* and *Prosopis spicigera* make it seem shrubby. The *Phragmites–Saccharum–Imperata* type of the moist subhumid zone covers the Ganga alluvial plain in northern India, and associated tree and shrub species include *Acacia arabica, Anogeissus latifolia* and *Butea monosperma*. The *Themeda–Arundinella* type extends over the moist sub-humid zones of the most eastern Indian states and north of Delhi State. These savannas are derived from various humid forests and therefore their woody elements are diverse.

1.6.6 Research in savannas

The research that has been carried out in different savanna formations has been highly focused on certain subjects, largely reflecting perceptions of the importance of the different determinants. All savannas are controlled primarily by PAM and PAN. However, fire and herbivory tend to play a greater or lesser role in particular savanna formations. The American, Asian and Australian savannas do not have large herbivores so fire is considered more important in these areas, although invertebrate herbivores may be significant. This is illustrated by the quantity of research into savanna fire ecology in these savanna regions. On the other hand, mammalian herbivory dominates African savannas, and as a consequence, much research on the continent has been on herbivores and their effects.

The interest in savanna research is also a reflection of the history of occupation and land use. The *cerrado* and *llanos* of South America were relatively unstudied until the beginning of the twentieth century because of their geographical isolation in the perceived 'backwaters' or interior of the country. However, as land began to be converted to other forms of use in these areas, greater interest in the ecology and management of these savannas became apparent. In Australia, the failure of agriculture in the savannas has prompted researchers to focus on other forms of land use, especially tourism. It is recognised that this will mean gaining increased knowledge about the savannas and their management, as well as about the indigenous Aboriginal culture, and as a consequence, there is substantial research in these areas.

The African and Asian savannas have the longest history of land use, and colonialism in these regions has particularly influenced research. In Africa, studies have been concentrated on wildlife and its conservation, agriculture and pastoralism. In Asia, the focus has been on wood extraction, deforestation, resettlement, and interrelated problems. However, it is also in these long-exploited savannas of Africa and Asia that there is innovative research on savanna management strategies involving local people and traditional knowledge.

1.7 Concluding remarks

Savannas are dynamic ecosystems, determined by plant-available moisture, plant-available nutrients, fire and herbivory, at different spatial and temporal scales. They cover about 40% of the tropics (Solbrig, 1991), or nearly a third of the world's land surface (Werner *et al.*, 1991). Savannas have a long history of human use, and presently support over a fifth of the world's population, many of whom live at a subsistence level. There is much bias in the information available between savannas, and also the various subjects of savanna ecology and management. However, despite the large gaps in knowledge, there is currently much active research going on in savanna regions. There is, therefore, a lot to look forward to in the future!

The following chapters will describe the ecology and management aspects of selected savanna formations in different continents. In areas where information is lacking, regional material is presented, and inferences made about the savannas. Chapters 2 and 3 deal with the two most extensive savanna formations in the Americas, namely the *cerrado* and the *llanos*. Chapters 4–7 tackle the African savannas, Chapter 8, the dry dipterocarp savannas of Southeast Asia, focusing principally on Thailand, and Chapter 9, the *Eucalyptus* savannas of northern Australia. The book ends (Chapter 10) by examining issues currently facing savannas, which may become increasingly significant in the next century. Chapter 10 also looks at the different sorts of technology that may be used in the future for managing savannas, and for helping advance our knowledge of savanna functioning.

Key reading

Bourlière, F. (ed.) (1983). *Tropical Savannas. Ecosystems of the World 13*. Elsevier Scientific Publishing Company, Amsterdam.

Huntley, B.J. and Walker, B.H. (eds) (1982). *Ecology of Tropical Savannas*. Springer-Verlag, Berlin.

Tothill, J.C. and Mott, J.J. (eds) (1985). *Ecology and Management of the World's Savannas*. Australian Academy of Science, Canberra.

Walker, B.H. (ed.) (1987). *Determinants of tropical savannas*. IUBS, Paris.

Werner, P.A. (ed.) (1991). *Savanna Ecology and Management: Australian Perspectives and Intercontinental Comparisons*. RSSD Publication 5. Blackwell Scientific, Oxford.

Young, M.D. and Solbrig, O.T. (eds) (1993). *The World's Savannas: Economic Driving Forces, Ecological Constraints and Policy Options for Sustainable Land Use* (MAB 12). UNESCO and Parthenon Press, Paris and Carnforth.

Chapter 2

The *cerrado* of Brazil

2.1 Introduction

2.1.1 Distribution, climate and soils

The Brazilian savanna, commonly called *cerrado* (meaning 'closed' in Portuguese) is a complex vegetation form, characterised by a mosaic of physiognomies ranging from pure grasslands through open scrubland to dense woodlands (Eiten, 1972, 1978). Situated between latitude 3°S and 24°S and longitude 41°W and 63°W, it occupies over 1.8 million km^2, 22% of the Brazilian territory (Goodland, 1971a; Coutinho, 1990) (Figure 2.1), and in terms of area, is second only to the Amazon rain forest (Furley and Ratter, 1988). The *cerrado* is centred on the Brazilian *Planalto*, characterised by a realm of plateaux and high tablelands (termed *chapadas*) ranging in altitude from approximately 300 m to 1000 m above sea level (Ab'Saber, 1971). From here *cerrado* descends southwards to the lowlands of the Mato Grosso *Gran Pantanal*, and northwards to the Amazon rain forests. To the east and south, the transition with the humid forest landscapes of the Atlantic region is gradual, while to the north-east a rather steep climatic gradient leads to the large depressions of the dry *caatinga* region (Eiten, 1994). The tablelands covered in savannas are occasionally dissected by wide valleys penetrated by gallery forest.

The whole *cerrado* region has a tropical seasonal climate (Aw type of Köppen, 1931) with average annual rainfall in the order of 1500 mm. The dry season lasts from three to five consecutive months during the winter of the Southern Hemisphere (May to September) (Ab'Saber, 1983). The soils of the *cerrado* are mostly oxisols, and shales are the predominant parent material (Parada and Andrade, 1977). They are characteristically deep (> 3 m), well drained, red or yellow, clay-rich, acidic, structurally strong but nutrient-poor (Furley and Ratter, 1988). Because of the low cation exchange capacities and the high levels of aluminium saturation of the soil, the amounts of exchangeable calcium, magnesium and phosphorus are low, with aluminium reaching toxic levels in many areas (Goedert, 1983).

2.1.2 Vegetation structure and composition

The *cerrado* encompasses several structural types of open vegetation, from grass-

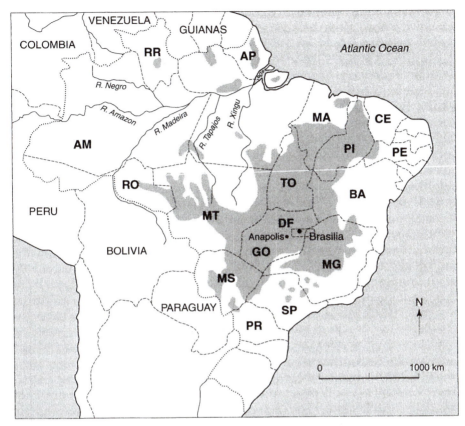

Figure 2.1 The extent of *cerrado* in Brazil. State abbreviations are as follows: AM, Amazonas; AP, Amapa; BA, Bahia; CE, Ceará; DF, Distrito Federal; GO, Goiás; MA, Maranhão; MG, Minas Gerais; MS, Mato Grosso do Sul; MT, Mato Grosso; PA, Pará; PE, Pernambuco; PI, Piauí; RO, Rondônia; RR, Roraima; SP, São Paulo; TO, Tocantins.
(Reproduced by kind permission of Justin Jacyno)

lands to dense woodlands. With such variation in form, the delimitation of physiognomic types is fairly arbitrary. A generally accepted classification based on the presence of woody elements recognises five structural types of *cerrado* vegetation (Eiten, 1972) (Figures 2.2 and 2.3):

1. *campo limpo* ('clean field') – a pure or almost pure grassland;
2. *campo sujo* ('dirty field') – a tree and shrub savanna, with widely scattered woody species;
3. *campo cerrado* – a wooded savanna, where the scattered low trees have a total crown cover of about 3%, but the herbaceous species still appear as a conspicuous part of the landscape;
4. *cerrado sensu stricto* – a savanna woodland where the total woody cover is about 20%;
5. *cerradão* – a woodland or open low forest, with a tree canopy cover above 50%.

Campo limpo>Campo sujo>Campo cerrado>Cerrado sensu stricto>Cerradão

Figure 2.2 The physiognomic gradient of *cerrado* vegetation.

Throughout the *cerrado* region, these physiognomic forms can be found mixed and intergraded with one another, forming complex vegetation mosaics. The 'forest–ecotone–grassland concept' of Coutinho (1978a) describes this intricacy as a continuum of grassland formations (*campos*), savanna intermediary formations (*cerrado sensu stricto* – the most common *cerrado* formation) and forest formations (*cerradão* – the rarest *cerrado* formation). Goodland (1971a) quantitatively analysed 110 stands of *cerrado* vegetation in a region of Minas Gerais in central Brazil. He found a continuous variation in physiognomy and species composition. For example, from *campo sujo* to *cerradão*, canopy cover

Figure 2.3 Typical *cerrado sensu stricto* vegetation from the Distrito Federal, central Brazil. (Photograph by the author)

ranged from 0% to 85%, ground cover from 30% to 2%, number of tree species from 19 to 72, and number of herb species from 79 to 21. This gradual change along a physiognomic gradient suggests the ecotonal nature of the *cerrado*, especially in the intermediary formations.

Other savanna formations present within the *cerrado* include hyperseasonal savannas and *veredas*, generally occurring as treeless grasslands or sometimes as palm savannas with *Mauritia vinifera* (Askew *et al.*, 1970; Eiten, 1975). They occupy wet sites on valley sides or tableland margins throughout the area (Eiten, 1978), though they become more important in Mato Grosso, towards the transition to the *Gran Pantanal* formation (Eiten, 1975, 1978).

The flora of the *cerrado* is notably rich and diverse (Sarmiento, 1983a; Eiten, 1994). Heringer (1971) recorded more than 300 species in one hectare of protected *cerrado* near Brasilia. Filgueiras and Pereira (1994) found 2366 species of phanerogams in a small area of *cerrado* in the Distrito Federal. Ratter (1986) estimates that the entire *cerrado* has over 10 000 vascular plant species. This high diversity is thought to have originated due to two factors: the antiquity of the *cerrado*; and the dynamic effects on *cerrado* distribution during the Pleistocene.

It has been suggested that the *cerrado* may have existed in a prototypic form in the Cretaceous before the separation of the American and African continents (Ratter and Ribeiro, 1996), although there is little fossil evidence for this. However, data for the late Quaternary do show alternating dry/wet climatic conditions that would have caused the extension and contraction of *cerrado* vegetation (e.g. Behling, 1998; Ledru *et al.*, 1998; Salgado-Labouriau *et al.*, 1998). For example, prior to 20 000 years BP, humidity levels were much higher than today. Following this, there was a dry phase lasting until about 7000 years BP, after which a more seasonal climate is recorded. From about 4000 years BP to today, modern climatic conditions became established. These recurrent climatic changes probably led to complicated patterns in biota and fragmentation, but also high speciation. A large amount of work has been devoted to this subject in the Amazon, and to whether high diversity arose through islands of vegetation forming isolated 'refugia' (Haffer, 1969) or through biotic interchange at ecotones (Bush, 1994). However, little attention has been given to the *cerrado*.

Though the study of *cerrado* phytogeography is still at an early stage, it has been shown that marked differences occur in floristic composition at a large scale between various regions of Brazil (Ratter *et al.*, 1973, 1988, 1996; Gibbs *et al.*, 1983; Ratter, 1986, 1987; Felfili and Silva Jr, 1988; Oliveira-Filho *et al.*, 1989; Oliveira-Filho and Martins, 1991; Ratter and Dargie, 1992; Castro, 1994a,b), and at a smaller scale between areas close to each other, such as the protected areas within the Federal District (Felfili and Silva Jr, 1993). Biodiversity 'hot-spots' have been recorded in Mato Grosso, Goiás (Alto Araguaia region), Tocantins and in the Federal District (Ratter *et al.*, 1997).

Plant life strategies

The ground layer of the *cerrado* is dominated by perennial herbs and contains almost no annual herbs (Coutinho, 1990). The woody layer is made up of some trees and shrubs which can remain alive at relatively short heights for many years, producing new branches and increasing in girth (Eiten, 1994). These have rather large, usually stiff leaves, or large compound leaves with tiny to large leaflets; only a few species have soft, hairy or mesomorphic leaves (Eiten, 1982). Generally, *cerrado* trees have fewer branches for their size than tropical meso-phytic forest trees or temperate zone trees, and many have rather open crowns (Eiten, 1982).

Phenological studies have only been undertaken for grasses. De Almeida (1995) has identified three phenological groups of the *cerrado* perennial grass community from the Distrito Federal. According to the beginning of flower-ing during the growing wet season, there are short-cycle precocious species (PCC), long-cycle precocious species (PCL), and late species (TAR) (Table 2.1). PCC species have the shortest reproductive cycle, lasting from one to three months, November to January, i.e. entirely within the rainy season. PCL species have the longest cycle, lasting from five to nine months. Flowering begins between November and December, and seeds are dispersed between May and August. TAR species were the most common in the area studied. The reproductive cycle in these species lasted from four to six months, with flower-ing taking place in the second half of the rainy season, and dispersal completed by mid dry season. PCL and TAR species flowered regularly during the growing season, but PCC species showed variations. Similar to the *llanos* perennial phenological groups (see Section 3.1.2), PCC species are short with basal leaves, whereas PCL and TAR species are tall with both basal and cauline leaves.

2.1.3 Fauna

The *cerrado* has a rich variety of fauna, with over 400 species of birds (one-fourth of all Brazilian species) and a diverse range of rodents, carnivores,

Table 2.1 Examples of the different phenological groups (based on De Almeida, 1995)

Phenological groups	Species
Short-cycle precocious species (PCC)	*Digitaria matogrossensis* *Paspalum erianthum* *Elyonurus muticus*
Long-cycle precocious species (PCL)	*Echinolaena inflexa* *Panicum cervicatum* *Axonopus barbigerus*
Late species (TAR)	*Agenium goyazense* *Axonopus marginatus* *Schizachyrium condensatum* *S. hirtiflorum* *S. tenerum* *Trachypogon* spp.

marsupials, bats, primates and edentates (Alho and Martins, 1995). However, many of these vertebrate species are not endemic to the *cerrado*, and are widely distributed, the rodents having the highest endemicity of 37%. Many of these vertebrates also rely on other vegetation forms in the *cerrado* for their survival, particularly gallery forest, indicating that the mosaic nature of the *cerrado* is vital for animal diversity. The largest species of carnivore in the *cerrado* is the manned wolf or *lobo guará* (*Chrysocyon brachyurus*), and the destruction of its habitat has placed it on the endangered list (see Section 2.4.1). Other animals are also being threatened by habitat fragmentation, including deer, monkeys, sloths, armadillos and anteaters (Alho and Martins, 1995). The giant anteater, *Myrmecophaga tridactyla tridactyla*, is practically extinct. Invertebrate fauna is not very well known, but it is estimated that there is a high diversity and degree of endemicity, represented mainly in the Mollusca, Annelida and Arthropoda families. The latter is the most diverse, with large numbers of insects, termites, butterflies, ants, bees and wasps.

2.2 The main savanna determinants

The origins and the principal determinants of the *cerrado* vegetation have been discussed since the last century (Saint-Hilaire, 1824; Warming, 1892). In his classic work *Lagoa Santa*, Warming considered the *cerrado* to be a climatic climax determined by dry conditions during the winter months. Other authors, such as Rawitscher (1942) and Ferri (1944), gave greater relevance to fire, although their work was restricted to the southern *cerrado* physiognomies. Some, such as Goodland (1969, 1971a,b), Lopes and Cox (1977) and Queiröz Neto (1982), have emphasised the role of soil dystrophy and aluminium toxicity. But, there is no one single cause that governs *cerrado* formation. It is an interplay of the main determinants at different scales.

2.2.1 Plant-available moisture (PAM)

The geomorphology of the *cerrado* region is one of the main determinants of the hydrology and therefore PAM at the landscape scale. The Distrito Federal, for example, is characterised by relief forms of *chapadas*, plateaux whose surfaces are level or slightly rolling, and are dissected by elongated valleys (Novaes-Pinto, 1994). With 57% of the *chapadas* above 1000 m in altitude, they act as water divides for Araguaia–Tocantins (Amazon), Paraná and São Francisco drainages (fluvial valleys) (Novaes-Pinto, 1994).

These *chapadas* are covered with metamorphic rocks of low porosity and permeability, with a heterogeneous distribution of laterite, quartzites, metasiltites and latosols (Barros, 1994). At the subsoil level, the prevalence of metamorphic rocks and various depths of hard laterite limit the replenishment of subterranean water (Barros, 1994; Haridasan, 1994). Characteristics of the latosol soils, which have a high clay content, are their high water-holding capacity, low infiltration ability and problems with aeration after intense rains (Haridasan, 1994). Where quartzites predominate, the *chapadas* are covered in lithosolic soils of a shallow, stony and humus-rich composition, freely-drained and with good aeration.

Gradients of groundwater level along topographic catenas are closely matched by gradients in vegetation, from *cerrado* to gallery forest (Askew *et al.*, 1970; Oliveira-Filho and Martins, 1986). Since *cerrado* species cannot tolerate soil waterlogging, even for a relatively short period (Eiten, 1972, 1975; Joly and Crawford, 1982), the occurrence of woody *cerrado* vegetation on more elevated, level sites reflects its requirement for soils that are well drained throughout the year (Ratter *et al.*, 1973; Furley and Ratter, 1988). Where the water-table is permanently high, swampy gallery forest presides, and in areas where the soil is inundated for part of the year, but dries up during the dry season, wet *campo* grassland is prominent (Freire, 1979; Furley, 1985).

This intolerance to waterlogging is most apparent in a distinct community of tree species formed on raised islands of ground within wet *campos* (Furley and Ratter, 1988). The island formations, called *campos de murundu* (Eiten, 1982; Diniz de Araújo Neto *et al.*, 1986), are particularly common in inundated areas such as the Pantanal do Mato Grosso and the Ilha do Bananal. They consist of raised earthmounds bearing shrubs, termitaria, and *cerrado* trees, most commonly *Curatella americana* (Dilleniaceae) and *Byrsonima crassifolia* (Malpighiaceae) (Ratter *et al.*, 1973; Prance and Schaller, 1982), two species which can tolerate periodic inundation of the soil. During this seasonal rise of the water-table, the deeper roots of these species die, causing chlorosis of the leaves, and subsequent annual leaf fall (Foldats and Rutkis, 1975).

Adaptations to PAM

The evolutionary significance of PAM is evident in the many examples of adaptations to water stress present in *cerrado* plant species. Studies of transpiration and water balance in a number of tree, shrub and grass species have shown that deep-rooted woody species do not curtail their transpiration during the dry season, and some grasses, such as *Echinolaena inflexa*, simply dry out completely (Ferri, 1944; Rachid, 1947; Maitelli, 1987). This led to the conclusion that the deep roots of *cerrado* woody species enable them to reach enough disposable water in the lower levels of the soil during the dry season to survive (Rawitscher *et al.*, 1943; Rawitscher, 1948; Miranda and Miranda, 1993). Many species also use xylopodia to store water (Coutinho *et al.*, 1978). Sclerophylly is a common adaptation to water stress, and many species have leaves with thick cuticles, sunken stomata and greatly lignified tissues (Furley and Ratter, 1988).

2.2.2 Plant-available nutrients (PAN)

Higher levels of the *chapadas*, steep slopes, and low-lying areas all have different soil types of varying nutrient content and availability (Camargo and Bennema, 1966; Ranzani, 1971; Freitas and Silveira, 1977). For example, the dark red latosols found at higher levels of tablelands in central Brazil have low nutrient reserves compared to other red latosols found on gently sloping topography. Also related to soil fertility is the presence of aluminium, which in the *cerrado*, can reach high, almost toxic levels (Box 2.1). Differences in nutrient levels of the soil are reflected in the vegetation present.

Box 2.1 Aluminium in *cerrado* soils

Goodland (1971a) described aluminium as an ecological factor with a strong negative influence over *cerrado* vegetation. At high soil concentrations, this element impedes the growth of roots by inhibiting the mechanisms of phosphorylation in the cells, and thereby interfering with normal growth. Aluminium also makes the essential plant nutrients phosphorus and calcium insoluble, thus reducing soil fertility (Haridasan, 1992). The array of factors associated with high soil aluminium concentrations have also been attributed to characteristics other than growth responses, such as the high degree of scleromorphy. It should be noted though that the more serious toxic effects attributed to aluminium may be overrated, as 40–60% of the Amazon rain forest (non-sclerophyllous) has equally high levels of aluminium (Tothill, 1985).

Aluminium levels are directly related to soil type and will therefore vary accordingly. For example, Ratter *et al.* (1977) recorded aluminium levels in mesotrophic soils in Mato Grosso and Goiás of 0 and 0.62 me 100 g^{-1} soil respectively, whereas Ribeiro (1983) found 2.36 me 100 g^{-1} soil in dystrophic soils in the Distrito Federal.

Although aluminium can reach toxic levels harmful to most plants, a taxonomically unrelated group of plants can actually accumulate aluminium in their tissues (Goodland, 1971b). These aluminium-tolerant plants include, most notably, the Vochysiaceae, e.g. *Qualea grandiflora*, *Qualea parviflora* and *Vochysia thyrsoidea* (accumulating an extraordinary 14 120 mg kg^{-1} in its leaves), *Palicourea* spp. (Rubiaceae) and *Miconia* spp. (Melastomataceae) (Haridasan, 1982; Haridasan *et al.*, 1987; Haridasan and Araújo, 1988). Aluminium-tolerant species are just as abundant amongst the larger trees of the *cerradão* as they are in the smaller species of more open forms of *cerrado*, suggesting that this characteristic does not seem to be associated with differences in the stature of the community (Haridasan, 1992). Nevertheless, the amount of aluminium in the soil at smaller scales will determine which species will most successfully establish there, and in turn delimit the composition of the *cerrado* form present.

An example is the occurrence of two types of *cerradão*. One occurs on soils with a high calcium content (between 3.25 and 7.58 me 100 g^{-1} soil), and is identified by a number of indicator species such as *Magonia pubescens* and *Callisthene fasciculata*, allowing it to be recognised as a mesotrophic facies *cerradão* (Ratter, 1971; Ratter *et al.*, 1973, 1977; Araújo, 1984). It is widespread in the *cerrado* region and has been recorded from many localities in Goiás, Minas Gerais, Mato Grosso and Mato Grosso do Sul, some of these more than

1500 km apart. Often associated with the margins of deciduous or semi-decid-uous forest (the climax vegetation of better soils in the *cerrado* region), this *cerradão* indicates an intermediate in soil fertility between the dystrophic forms of *cerrado* and the deciduous forest (Furley and Ratter, 1988). A floristically dif-ferent type of *cerradão*, dystrophic facies *cerradão* (Ratter *et al.*, 1977; Araújo, 1984), with indicator species such as *Hirtella glandulosa* and *Emmotum nitens*, is found on dystrophic soils. Here, the soil calcium content falls to a range of 0.08–0.2 me 100 g^{-1} soil. This *cerradão* is widespread in central Brazil (in the absence of disturbance), especially in the Distrito Federal.

Oliveira-Filho *et al.* (1989) observed small-scale variations in soil nutrient status, and suggest they may be due to localised soil patchiness. This patchiness could be accounted for by a number of factors including nutrient enhancement from soil erosion, leaf-cutter ants (*Atta* spp.), which accumulate nutrients in their chambers (Egler and Haridasan, 1987; Constantino, 1988; Haridasan, 1994), and through the recycling of nutrients after patchy fires.

Adaptations to PAN

Research on physiological adaptations of *cerrado* species to poor nutrient levels is in its infancy. It is possible that many *cerrado* species have similar nutrient adaptations to other savanna plants, which include mycorrhizal symbiotic asso-ciations between plant roots and certain fungi, and storage of nutrients within living tissues (Högberg, 1992; Chidumayo, 1994a) (see Section 4.2.2). The only known examples of plant adaptations to poor nutrient availability in the *cerrado* are related to fire, and include the presence of xylopodia, which absorb mineral nutrients particularly after the occurrence of fire (Coutinho *et al.*, 1978).

2.2.3 Herbivory

Herbivory by native fauna, a major determinant of other savanna formations, appears to play a less significant role in determining the *cerrado*, principally due to the low abundance of large herbivores present (Ojasti, 1978; Nascimento, 1987; Moura *et al.*, 1989; Prado, 1989; Nascimento *et al.*, 1990; Nascimento and Lewinsohn, 1992). This is the result of the megafaunal extinctions which took place at the end of the Pleistocene, and beginning of the Holocene. The leaves of many *cerrado* plant species also contain high levels of aluminium and silica, an effective anti-herbivory defence strategy (Nascimento *et al.*, 1990).

Although the effects of native mammals may be negligible, insects, such as leaf-cutter ants, termites and grasshoppers, are probably extremely important herbivores in the *cerrado*, and have been observed to remove large amounts of leaf biomass from *cerrado* plants (Helena Castanheira, pers. comm., 1997). Further evidence for herbivory occurring comes from mutualistic relationships in the *cerrado* (Del-Claro *et al.*, 1996; Oliveira, 1997). For example, De Assis Dansa and Duarte Rocha (1992) found that the presence of honeydew-produc-ing homopteran *Aconophora teligera* and associated ants probably reduces her-bivory on apical meristems of the *cerrado* shrub *Didymopanax vinosum*.

2.2.4 Fire

Fire is undoubtedly extremely important in the *cerrado* (Figure 2.4). Frequent fires favour the more open forms, whereas protection from fire allows the woody vegetation to establish, and succession continues to the closed *cerradão* when other factors are not limiting (Ferri, 1973; Ratter *et al.*, 1973, 1978; Ratter, 1991; Henriques, 1993). Several authors, like Rizzini and Heringer (1962) and Rizzini (1963) have considered the *cerradão* as the original forest type in the whole *cerrado* area, the other structural types having been derived by human activities, particularly burning. Though this hypothesis may be applicable to some restricted areas, it is improbable that the whole *cerrado* region is determined by ancient anthropogenic management (Sarmiento, 1983a).

There is a large body of literature on fire in the *cerrado*, which has recently been reviewed in detail in Mistry (1998a). The following sections give a summary of some of the causes and effects of fire, and the adaptations of plants and animals to it.

The causes of fire

Wildfires have been significant in the *cerrado* since the late Quaternary, some 20 000 years BP (Vernet *et al.*, 1994; Salgado-Labouriau *et al.*, 1997), and even today, natural causes of fire, such as lightning, are reported by farmers and park rangers. However, their importance has not been recognised due to the sheer lack of research on the subject (Coutinho, 1990). Even so, humans are unquestionably the principal cause of fire in the *cerrado*.

Figure 2.4 Fire in the *cerrado*.
(Photograph by the author)

People have been using fire for more than 32 000 years in central Brazil (Guidon and Delibrias, 1986). Anthropological research shows that even before the colonisation of Brazil by the Portuguese, Indians were using fire by burning vegetation for hunting and tribal wars (Lukesch, 1969; Villas-Boas and Villas-Boas, 1976). Today, however, the burning of the more open forms of the *cerrado*, which are frequently used by cattle ranchers as natural pastures, is the major cause of fire (Klink *et al.*, 1993). This is closely followed by the vast areas of *cerrado* cleared and burned at the end of the dry season in order to bring in new agricultural land (Coutinho, 1990). Other burns arise from various causes, such as the control of shrubs in pastures, pest control, carelessness in fire management in intentionally burned areas (such as during the cutting and burning of vegetation while cleaning the edges of highways and railroads), and the falling of balloons with the wicks still alight during the June religious festivals (Coutinho, 1990). Although carelessness with cigarettes does not seem to be relevant in the *cerrado* (Coutinho, 1990), arson is common, whether by hunters hoping to catch wild animals escaping from the flames, or local people who burn the vegetation for the aesthetic value fire has.

According to Eiten and Goodland (1979), to sustain a *cerrado* physiognomy, burning should occur at a frequency of every 3 years. This allows enough time for both the herbaceous and woody layers to recover sufficiently. However, population increases and agricultural expansion have led to an increase in burning rates, and large areas still covered in natural vegetation are now burned almost every year (Klink *et al.*, 1993). Though fire frequency is important, the nature of a fire is equally decisive in determining how an area of *cerrado* may recover from, or degrade, as a result of fire. Since the *cerrado* consists of various physiognomies, fire behaviour will vary from one form to another.

Factors affecting fire
Fires in the *cerrado* are characteristically surface-level and fast-moving (reaching speeds of 30 m min^{-1}: Kauffman *et al.*, 1994), consuming the herbaceous layer, but rarely igniting the taller woody plants (Miranda and Miranda, 1993). The main factors controlling the nature of fire in the *cerrado* are climate, topography and fuels.

Climate
Fires usually begin with the onset of the dry season (May–June), increase in frequency and intensity during June and July, and attain a maximum peak in August (Coutinho, 1990). As the wet season starts (September–October), the occurrence of fires drops markedly, and although prescribed burning does not take place, the vegetation is still susceptible to burning, particularly in areas where there have been no burn-offs for several years, and/or after a sequence of hot days in the absence of rain (*veranico*) (Cochrane *et al.*, 1988). The period between August and the beginning of September is particularly favourable for the propagation of fire, as relative air humidity during the hottest hours of the day (25–30 °C) can reach below 20%, and the days are very windy (Coutinho, 1982a). In years of frosts, a great part of the epigeous phytomass in

the herbaceous/undershrub stratum dies. This, and the accompanying fall of leaves from many trees and shrubs accumulate on the soil as an easily dehydratable and highly combustible material, greatly increasing the risk of fire (Coutinho, 1990).

As a result of variations in climate over the dry season, generally, fires during the early dry season (May to June) are patchy and of a low intensity, due to the high moisture still present within the vegetation from the rainy season. The mid-season fires (July to August) are of a higher intensity and more homogeneous, since most of the combustible fuel may be dry, and the climatic conditions may be ideal for fire propagation. In the late dry season, two scenarios are possible: the build-up of dry combustible fuel and peak air temperatures may cause very intense fires; or the onset of the rains may result in patchy, low-intensity burns (Miranda and Miranda, 1993).

Topography
The *cerrado* landscape is characterised by gently sloping hills, and in view of this fact, topography probably plays an important role in fire spread, which increases with degree of upslope due to fuel preheating. Unfortunately, research has yet to be carried out on this aspect of fire behaviour in the *cerrado*. The types of fire that occur in the *cerrado* vary in behaviour most notably in relation to fuels, which is intrinsically related to physiognomy (Kauffman *et al.*, 1994).

Fuels
Although total fuel biomass is significantly greater in *cerrado sensu stricto* and *cerradão*, the biomass of grasses is considerably lower in these than in the *campo limpo* or *campo sujo* (Coutinho, 1982a; Pivello and Coutinho, 1992; Ward *et al.*, 1992; Miranda and Miranda, 1993; Kauffman *et al.*, 1994). Because of their high degree of flammability, grasses and other ground-layer vegetation are a major influence on fire behaviour in different *cerrado* physiognomies (Figure 2.5). In a *campo sujo*, for example, where grasses can represent up to 91% of the combustible fuel (Ward *et al.*, 1992), fires can reach temperatures higher than 800 °C (Cesar, 1980; Berardi, 1994), whereas *cerrado* types with prevailing woody elements, e.g. *cerrado sensu stricto*, yield significantly lower temperatures (Miranda *et al.*, 1993). There is virtually no smouldering combustion following flaming combustion in either *campo limpo* or *campo sujo* (because of the low woody biomass), whereas in *campo cerrado* and *cerrado sensu stricto*, smouldering combustion is prevalent (Kauffman *et al.*, 1994). This suggests that fire has a greater influence in the open *cerrado* forms, and more woody plants will be killed, rather than scorched, in the closed physiognomies.

It is also possible to explain the difference in fire behaviour along the gradient of *cerrado* physiognomies according to fuel moisture content. Kauffman *et al.* (1994) found that fuel moisture content was very low in dry, non-green grass (22–29% dry weight basis), comprising 78% of the fuel mass in *campo limpo* and 40% in *campo cerrado*. In contrast, fuel moisture content of the woody component ranged from 118 to 140% (dry weight basis) – 28% of the

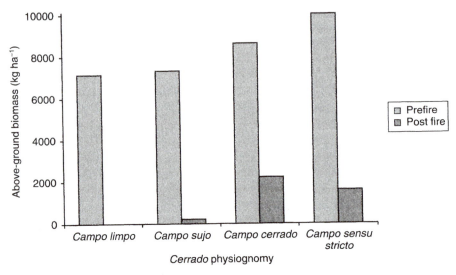

Figure 2.5 Above-ground biomass (the fuel load) (kg ha⁻¹) before and after fires in four *cerrado* physiognomies. The dominance of grasses in *campo limpo* and *campo sujo* means that more fuel is consumed by fires compared to *campo cerrado* and *cerrado sensu stricto* (based on Kauffman *et al.*, 1994).

fuel mass in *campo cerrado* yet less than 6% in *campo limpo*. Total mass of water in fuels was calculated to be < 3100 kg ha⁻¹ in *campo limpo* and *campo sujo*, and $\geqslant 4600$ kg ha⁻¹ in *campo cerrado* and *cerrado sensu stricto*. The differences in moisture content affect the ignitability of fuels, thus possibly contributing to the lower impact of fire on the tree-dominated communities, in comparison to the grasslands.

The different physiognomies of *cerrado* vegetation, in terms of their densities and corresponding microclimates, also affect fire behaviour. Miranda *et al.* (1993) recorded a maximum fire temperature of 260 °C in an area of *cerrado sensu stricto* which had been protected for 15 years and burned three days after rain. Many patches of the vegetation remained unburned. In contrast, *campo sujo* burned on the same day under the same conditions, left only the woody vegetation unburned, and attained temperatures in the region of 650 °C. Shading of the fine fuel on the ground by trees and shrubs affects the rate of fuel moisture loss, causing mosaic-like burns in the more closed *cerrado* types. In the open *campo sujo*, most of the dry matter is not in close contact with the wet soil surface, and is well exposed to wind and solar radiation, so moisture is quickly lost to the environment (Miranda *et al.*, 1993).

'Time since last fire' also determines the characteristics of *cerrado* fires. Areas protected from fire for long periods of time will burn at higher temperatures than those areas burned regularly, regardless of their physiognomic type, due to the build-up of combustible fuels (Miranda and Miranda, 1993; Miranda *et al.*, 1993).

Vertical distribution of fire

As fires vary horizontally, they also have a vertical pattern of distribution in terms of temperature. At the soil level, temperature increases during *cerrado* fires are relatively small, e.g. around 50 °C, decreasing exponentially with depth, and becoming more or less negligible at and below 5 cm depth (Coutinho, 1976, 1978b; Cesar, 1980; Miranda *et al.*, 1993) (Figure 2.6). These insignificant soil temperature changes are irrespective of the physiognomic form being burned, and the maxima observed are unlikely to have any direct effect on soil organic matter, microbial populations, or buried seeds (Miranda *et al.*, 1993).

Miranda *et al.* (1993) found that regardless of *cerrado* physiognomy, at 1 cm above the ground, 60 cm height and 160 cm height, maximum temperatures ranged from 85 to 326 °C, 180 to 840 °C and 107 to 650 °C respectively. Also, the residence time (duration of temperatures) above 60 °C varied from 90 to 270 seconds at 1 cm above the ground, from 90 to 200 seconds at a 60 cm height, and from 20 to 70 seconds at a 160 cm height. Other data confirm these results, which suggest that the highest temperatures occur between 1 and 60 cm above the ground, and that the residence time above 60 °C also peaks at this height range (Miranda and Miranda, 1993). This could be related to the height of combustible fuels within the herbaceous layer. Above 60 cm, fire temperatures decrease, with temperatures reaching peaks of up to 700 °C for short periods of time.

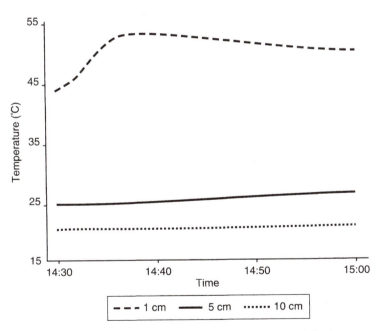

Figure 2.6 Temperatures at different soil depths during a prescribed fire in *campo sujo* (after Castro Neves and Miranda, 1996).

Fire and mineral nutrient cycling

Generally, the soils underlying the *cerrado* are rather poor in mineral nutrients, distinctly acidic, and with high levels of aluminium (Goodland, 1971b; Lopes and Cox, 1977). Fire is intimately related to this nutritional status since it is involved with the cycling of mineral nutrients (Coutinho, 1990). Through the action of fire, most of the epigeous biomass is rapidly mineralised, with nitrogen, carbon, sulphur and to a lesser extent, phosphorus and potassium, volatilised and lost to the atmosphere, and the remaining material either deposited on the soil surface as ash, or removed as particulate matter in smoke (Frost and Robertson, 1987).

In the *cerrado*, the immediate effects of burning result in an increment at the soil surface (0 to 5 cm) in concentrations of calcium, magnesium, phosphorus and potassium, and the complete disappearance of aluminium, which can remain at zero levels for up to 40 days (Cavalcanti, 1978; Coutinho, 1982a). Deeper down in the soil there is no change in these nutrients or in aluminium levels, suggesting that the ash deposited on the topsoil is highly beneficial to the growth of herbaceous/undershrub plants with superficial root systems compared to the trees and shrubs, since the former are provided with a large quantity of mineral nutrients and a significant reduction in aluminium toxicity (Coutinho, 1990). Coutinho (1982b) suggests that compensation to the tree/shrub vegetation may be brought about through the small-scale action of leaf-cutter ants (*Atta* spp.), which are frequently encountered in the *cerrado*, and transport nutrients to their nests at depths of 6–7 m in the soil.

Kauffman *et al.* (1994) looked at the relationship between fire and nutrient dynamics along the physiognomic gradient in the *cerrado*. They found that though the pool size of nitrogen, carbon and sulphur (within the fuel load) increased along the gradient from *campo limpo* to *cerrado sensu stricto*, the percentage of those nutrients lost by fire decreased. For example, along the gradient, the total mass of nitrogen increased from 24 to 55 kg ha^{-1}, but nitrogen lost by fire was greater or equal to 90% of the pool in the grasslands, in comparison to less than 56% in the tree-dominated communities. This is probably because of the significantly greater amounts of readily accessible combustible fuel in the grasslands, while nutrients remain locked up in the woody plants, which rarely burn. Other data showed that in *cerrado sensu stricto*, which has a lower combustion efficiency, greater quantities of nutrients were lost as particulates, whereas in *campo limpo*, most of the nutrients were volatilised during fire. These levels of losses through volatilisation are especially important because they are likely to be ecosystem losses in contrast to particulate losses, which may be redistributed within the ecosystem. However, the amount of nutrients lost in the *cerrado* due to fire represents a minor proportion of the total pool (in roots, soils and above-ground woody vegetation), and is likely to be replaced through natural inputs, particularly through rainfall (Coutinho, 1979; Schiavini, 1984), in one to three years (Pivello-Pompêia, 1985; Kauffman *et al.*, 1994).

Plant adaptations to fire

The closed *cerradão* comprises many species that are extremely fire-sensitive, e.g. *Emmotum nitens* and *Ocotea pomaderroides* (Moreira, 1992). The flora of the open *cerrado* forms, on the other hand, and especially the herbaceous stratum, is typically pyrophytic (Coutinho, 1990). A small proportion of species are annuals, growing and developing during the rainy months, thus escaping the dangers of dry season fires as seeds buried in the soil. Many perennial species possess subterranean organs such as bulbs, underground shoots, rhizomes and xylopodia, which avoid damage from fire. The densely imbricated sheaths of some grasses provide protection by limiting combustion due to inadequate aeration, e.g. *Aristida pallens* (Rachid-Edwards, 1956). Some woody species present in the herbaceous/undershrub stratum develop their entire system of trunks and branches subterraneously, with only the small vegetative branches and yearly reproductive sprouts protruding above the soil. This example of cryptophytism can be found in trees such as *Anacardium pumilum* and *Andira humilis* and among palms such as *Acanthococos emensis* and *Attalea exigua* (Rawitscher *et al.*, 1943; Rawitscher and Rachid, 1946; Lopez-Naranjo, 1975).

In the woody layer, the most conspicuous pyrophytic characteristic is the strong suberisation of the trunk and branches of the trees, permitting thermal isolation of the living internal tissues (Eiten, 1994). Other species maintain the capacity to produce vigorous sprouts from subterranean roots following the total carbonisation of the aerial branches (Rachid-Edwards, 1956). Even the seedlings of certain tree species may present this type of adaptation (Dionello, 1978). Some trees protect their dormant apical buds using dense, hairy cataphylls, e.g. *Anemia anthriscifolia* (Rachid-Edwards, 1956).

Burning induces flowering and fruit dehiscence in many *cerrado* species, and some seeds of the *Mimosa* genus require a thermal shock in order to germinate (Coutinho and Jurkewics, 1978; Coutinho, 1982a; Almeida and Silva, 1989). Coutinho (1977) postulates that fire has a beneficial effect on these species by cleaning out obstructing vegetation, thus facilitating pollination and seed dispersion.

Effects of fire on animals

Although there is abundant information about the death and survival of particular *cerrado* animals from fire, for example their related adaptive strategies, the dependence of many animals on fire 'products' (e.g. deer species licking ash), and the post-fire flush of new grass, flowers, fruits and seeds during a period of food scarcity in the dry season (Sick, 1965; Alho, 1981; Coutinho, 1990; Rodrigues, 1996), little is known about the effects of fire on animal populations.

Studies on lizard populations show that periodic heterogeneous burns probably help to maintain populations and species diversity, which will fall with an increase in fire frequency and homogeneity (Araujo *et al.*, 1996). Periodic fires have also been shown to be beneficial for insect herbivores, such as leaf-galling midges, which increase substantially in burned areas (Prado, 1989; Prada *et al.*, 1995; Vieira *et al.*, 1996). Arboreal ant populations were found to fall by 69% due to fire, whereas underground colonies were hardly affected (Naves, 1996). Other studies conducted include the effects of fire on guilds of spiders

(Dall'Aglio, 1992), on plant–insect interactions (Righetti, 1992; Seyffarth *et al.*, 1996), and on the construction of ant nests (Dias, 1993). Of bird communities, most studies have concentrated on the effects of fire in gallery forest, a formation common along waterways in the *cerrado* region, rather than actual *cerrado* vegetation (e.g. Figueiredo and Cavalcanti, 1992; Marini and Cavalcanti, 1996). These studies, however, stress the importance of habitat diversity in the *cerrado* for animals during and after fires (Mares *et al.*, 1986; Marini and Cavalcanti, 1996). Nevertheless, for managing fire in natural areas, data on other wildlife, such as anteaters, armadillos, the manned wolf and deer, is necessary, but as yet, is lacking.

2.3 Human determinant of the savanna

Before the arrival of Europeans, the *cerrado* was inhabited by groups of tribal people, relying on the natural resources and small-scale agriculture for survival. After colonisation, many of these indigenous people were exterminated, and large portions of land, sometimes the size of small European countries, were allocated to families of the Portuguese aristocracy. However, *cerrado* land use was mainly extensive cattle-raising, and it was not really until the establishment of Brasilia in 1959/1960, that development within the *cerrado* region expanded from peripheral areas of the coastal states of Rio de Janeiro and São Paulo, to more western and northern *cerrado* in Mato Grosso, Mato Grosso do Sul and Tocantins (Alencar, 1979). The principal advance has been in large-scale commercial arable cultivation (wheat, maize, soybean, sorghum), followed by extensive low production ranching (Furley and Ratter, 1988). Today, agriculture is still the main growth area, but other major factors increasingly affecting the *cerrado* region include urban and industrial land use, the construction of reservoirs and dams, and the extensive planting of eucalypts and pines for pulp and charcoal production (Dias, 1994).

2.3.1 Pre-colonial land use

There are numerous tribes which probably inhabited the *cerrado* in small settlements prior to colonisation (Dias, 1994). These include the Kayapó, Xavante, Karajá, Guajajara, Guarani and Terena, who relied on hunting, gathering and small-scale shifting cultivation for subsistence. The range of crops grown was large and diverse. For example, Kerr and Posey (1984) found that Kayapó people cultivate 22 varieties of sweet potato, 22 varieties of cassava, 12 varieties of maize, various vegetables and beans, and numerous fruit species. They use certain natural signs to indicate periods for different activities. For example, planting is initiated by particular bird calls (usually indicating rain), and the flowering of certain *cerrado* species.

The rich vocabulary of 'fire' words in *cerrado* tribes such as the Xavante (McLeod and Mitchell, 1980; Giaccaria and Heide, 1984; Hall *et al.*, 1987) suggests that fire played an important role in indigenous lives. Anderson and Posey (1985, 1989) observed Kayapó people that burn the *cerrado* during the dry season. They found that the timing of the burns was determined by tribal elders, who used both astrological and ecological indicators. The time of fruiting of *Caryocar brasiliense*, a key food source for the Indians, was probably the most important of these indicators. The burns were effectively controlled: precautionary firebreaks were constructed by removing dry grass and shrubs, and during the fire, branches were employed to avoid fires penetrating other areas. Indigenous tribes also used fire for hunting, and as signals to other tribes (Nimuendajú, 1983; Maybury-Lewis, 1984).

2.3.2 Colonisation and post-colonial land use

After Portuguese colonisation in 1500, land in Brazil was divided up among élite Portuguese families. The *cerrado* region was no different, and huge landholdings, called *sesmarias*, were used for highly extensive cattle-raising. The oldest settlements began from the mid-eighteenth century, and were in the region near Cuiaba (Mato Grosso) to the West of Goiás, following the discovery of gold and precious stones (Cunha *et al.*, 1994). Although this period, known as the Gold Cycle, was short-lived (the mines were quickly exhausted), it was a vital opening for subsequent occupation of the region.

The environmental (poor soils and long dry season) and geographical isolation of the *cerrado* made it inadequate for large-scale commercial agriculture, and so kept it far removed from the exporting activities that were going on along the coast. While the Atlantic Forest was being rapidly cleared for the production of different agricultural products such as sugar cane, coffee and cotton, the *cerrado* region remained in its own particular long 'cattle period' (Aragão, 1994). It was not until the 1930s that a more intensive process of occupation began, starting with the construction of a railway from São Paulo (the main economic centre on the coast) to Anapólis (in the centre of Goiás) (Cunha *et al.*, 1994). This encouraged the economies of southern Goiás and south-east Minas Gerais, but development was based on traditional cattle-raising or agriculture in areas of relatively fertile soils south of the *cerrado*. The rest of the region was still occupied by subsistence crops, grown by small land-holders, and cattle-raising estates (Goncalves, 1995).

2.3.3 A new period of land use in the *cerrado*

With the construction of the new capital, Brasilia, in 1959/60, a transportation infrastructure was developed, facilitating access to the region and encouraging agricultural activities. This period also marked the end of land available for cultivation in the south of the country, which, by the end of the Second World War had transferred most of its public land into private ownership, compared to less than one-third in the *cerrado* region (Mueller *et al.*, 1994). These developments encouraged the clearing and settlement of the *cerrado*, and from the 1970s onwards, together with public policies and programmes (Box 2.2), gave rise to the expansion of commercial agriculture in the region.

Box 2.2 Some public programmes and policies that have influenced the expansion of agriculture in the *cerrado*

POLOCENTRO (Programme for the Development of the *Cerrado*)
In 1975, POLOCENTRO was created with the intention of increasing the economy of the *cerrado* region. The programme was concentrated in 12 areas in the states of Minas Gerais, Goiás, Mato Grosso do Sul and Mato Grosso, selected for their farming potential and basic infrastructure. The objective of the programme was to assimilate 3 million hectares of *cerrado* into farming activities over four years, with 60% of the area to be used for crops and 40% for livestock production. The project actually lasted until 1982, during which 2.4 million hectares of *cerrado* were directly transformed into agricultural land. However, the impact of the programme may have been even greater as many areas were integrated into agricultural production as a result of benefits such as technological innovations and road construction. Contrary to the plans, only 40% of the area was occupied by crops, while 60% was converted to pasture for cattle-raising. POLOCENTRO loans (forms of rural credit) were given at very low, fixed interest rates and with no monetary correction and long repayment times. This meant that when inflation began to increase at rates of 40% per year after 1974, credits from the programme were almost like donations. Mostly large and medium farmers benefited: 81% of the producers in the programme had farms larger than 200 hectares and received 88% of the total amount of credit. As credit concession depended on the size of cultivated land, many farmers ploughed areas larger than they were prepared to cultivate. Also, greater support was given for farming export crops such as soybeans, rice, coffee, wheat, maize and sugarcane, rather than staple food crops such as beans and cassava.

PRODECER (Japanese–Brazilian Programme of Co-operation for the Development of the *Cerrado*)
PRODECER was a programme developed jointly by the Brazilian and Japanese governments and implemented in 1979 (to present) with the aim of settlement and agricultural production with modern techniques in the *cerrado*. The programme was not co-ordinated by the government, but by a private company named Campo, comprising both Brazilian and Japanese executives. Loans were granted at market-comparable interest rates, and could also be given for land purchase. However, as with POLOCENTRO, rural credit was the main instrument of PRODECER, and attracted the more successful, larger farmers, and encouraged the clearing of virgin land.

Minimum price policy
In the 1980s, pressure from the International Monetary Fund and the World Bank led to the substitution of rural credit and agricultural

Box 2.2 *(Continued)*

subsidies with minimum prices. Beginning in 1985, the government began buying large quantities of agricultural products produced with the support of PRODECER, especially soybeans, rice and maize. Because prices paid by the government were uniform throughout the country, these purchases particularly benefited farmers in more remote areas such as the *cerrado*, as transportation costs from these areas were considerably reduced. Purchases by the government followed market trends in that when market prices were higher than the minimum prices, the government tended to buy less, and when market prices fell, the government bought more. The government incurred substantial financial losses during this period, mostly due to price differences, high transportation and storage costs, and due to the deterioration of products. However, the policy kept commercial agriculture in the region artificially profitable and encouraged the expansion of cleared land.

Sources: Alho and Martins (1995), Cunha *et al.* (1994), Klink *et al.* (1993), Mueller (1995).

An essential element of the expansion and success of modern agriculture in the *cerrado* was the development of technology that allowed the natural acidity and aluminium toxicity of the soil to be reduced, fertilisers to be applied and the development of high-yielding cultivars of export crops, particular soybean. EMBRAPA, the Brazilian Agricultural Research Enterprise, was founded in 1973, and in particular, the Soybean National Research Centre and the Cerrado National Research Centre (CPAC) worked on increasing output in the region through technological innovation.

Up until the construction of Brasilia, the *cerrado* region was relatively depopulated (Ajara, 1989). Since then, however, there has been a strong trend of migration into the region, which resulted in a density increase of 0.92 people per km^2 in 1950 to 4.02 people per km^2 in 1980 (Ajara, 1989). During the periods 1970–1980 and 1981–1991, population growth rates in the region were 3.4% and 2.5% respectively – rates higher than the national average (Alho and Martins, 1995). Although these migrators have come from all parts of Brazil, there is a strong influx of subsistence farmers and workers from poorer regions, such as north and north-east Brazil.

Land-ownership in the *cerrado* is highly concentrated. In 1980, 74% of the land area in central Brazil was owned by 7.5% of 'farmers' (Gusmão, 1988). Large areas are under-utilised, and most rural labour is employed on small-holdings (Hees *et al.*, 1987). In some areas, absentee landlords own considerable tracts of land, which are then leased to other people. This has led to social tension, and the invasion of public and private land by the landless (Mesquita, 1989).

Today, it is estimated that 37% of the *cerrado* has been 'transformed' (Dias, 1994) (Figure 2.7). This is through urban and industrial developments, dam-building, degraded areas (due to mining, erosion and farm abandonment), but particularly by agriculture.

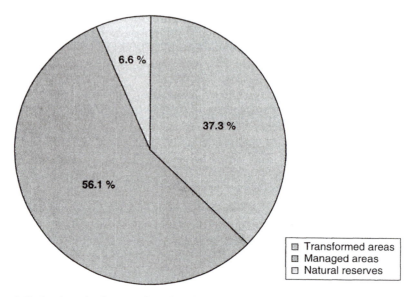

Figure 2.7 Land use in the *cerrado* and their estimated area in 1985. 'Transformed' = planted pastures, planted crops and urban development. 'Managed' = natural pastures, indigenous reserves and managed conservation areas. 'Natural reserves' = unmanaged parks and reserves (based on Dias, 1994).

Agriculture

Agricultural development has grown apace in the *cerrado* (Table 2.2). For example, from 1970 to 1985, the area covered by farms grew by 33.4%, or 27.4 million ha (an area slightly smaller than Belgium) (Alho and Martins, 1995). The area within farms being cleared in the same period has increased by 30.6 million ha, with the most significant growth in the planting of pasture.

The main crops grown in the region are soybean, corn, rice, beans and cassava. Of these, soybean and corn represent 25.4% and 16% of the total Brazilian production, respectively (Alho and Martins, 1995), with soybean showing the most dramatic increase: from being virtually non-existent in 1960 to some 4 million tonnes being grown in 1985 and nearly 9 million tonnes in

Table 2.2 Patterns of agricultural land use in the core areas of *cerrado* in 1970 and 1985. Figures in millions of hectares (after Mueller, 1995)

Patterns of land use	1970	1985
Farm-holdings	81.9	109.4
Area altered	20.2	50.8
Area of crops	4.1	9.5
Area of planted pastures	8.7	30.9
Unused altered area	7.5	10.3

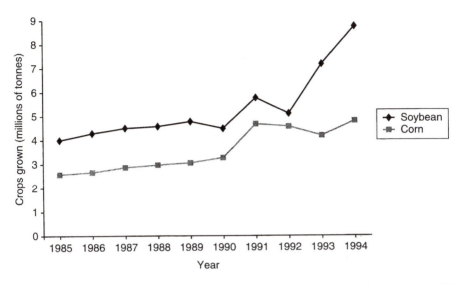

Figure 2.8 The growth of soybean and corn production in the *cerrado* from 1985 to 1994 (based on Alho and Martins, 1995).

1994 (Alho and Martins, 1995) (Figure 2.8). However, this growth in agricultural activities has not been evenly distributed. Some areas, such as Mato Grosso do Sul, Goiás, Distrito Federal and Minas Gerais, are characterised by relatively modern, concentrated farming, whereas the states of Maranhão, Piauí and parts of Bahia and Tocantins have been hardly developed, and contain large areas of virgin land (Alho and Martins, 1995).

The *cerrado* is also an important cattle-raising region, which varies from relatively modern and efficient farms to extensive operations using rudimentary methods and having low productivity (Alho and Martins, 1995). Bovine herds in the *cerrado* have increased from 16.6 million heads in 1970 to almost 38 million heads in 1985; from 20% to 32% of the region's share of the national herd (IBGE, 1960–1987). Part of this increase can be attributed to the change from traditional extensive use of the natural savanna to cultivated pastures, which in central Brazil, have increased by 15.6 million ha since 1970 (Mesquita, 1989) (Figure 2.9). The main forage gramineas are species of *Brachiaria*, but other genera such as *Panicum, Andropogon, Hyparrhenia, Melinis* and *Cynodon* are also common (Macedo, 1995).

Charcoal production

After agriculture, charcoal production for Brazil's steel industry exerts most pressure on the *cerrado*. Most deforestation for charcoal production so far has been associated with farming expansion (Medeiros, 1995), where, after clearing land, farmers sell the wood to charcoal producers to compensate the costs of preparing soils for crops or pasture cultivation. Today, the steel mills of Minas

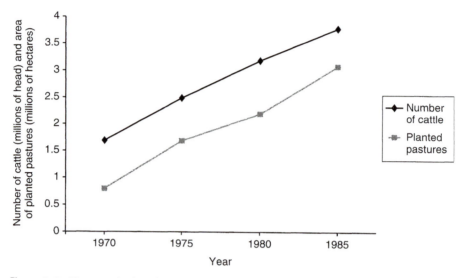

Figure 2.9 The growth of cattle production and planted pastures in the *cerrado* from 1970 to 1985 (based on Alho and Martins, 1995).

Gerais are the largest charcoal users in the world (Alho and Martins, 1995), but as these locations are quickly being exhausted, other *cerrado* states are being used (Medeiros, 1995). Eucalypt plantations are now also being used to supply demand, and it can be predicted that charcoal production from plantations will be a future growing activity in the *cerrado* region.

Mining

Mining has been an important activity in the *cerrado*, which is rich in mineral resources such as copper, iron, manganese, nickel, gold, beryllium, silver and uranium (Verdesio, 1994). Most of these mines are open, and the use of chemicals, such as mercury in gold mining, has caused extensive damage to local river systems as well as to vegetation cover. The other common element mined extensively is limestone, which is used for cement production and for agricultural liming. The explosives used directly destroy the local area. Many areas of limestone in the *cerrado* region support a semi-deciduous and deciduous woodland, rich in plant and animal diversity (Rocha, 1998).

Biological resources

Many rural communities depend on the direct consumption of resources from the *cerrado* (Alho and Martins, 1995). Timber is obviously extremely important, and is used for firewood and building construction. However, many other products are used (Table 2.3). Unfortunately, no assessments have been made of the extent of their use, their sustainability, and to what degree these extractism processes affect the ecosystem.

Table 2.3 Commonly used *cerrado* plants (based on Alho and Martins, 1995)

Use	Plants
Food	Seeds, fruits and palm-hearts of almost 80 species are used. Commonly eaten are 'pequi' (*Caryocar* sp.) and 'guariroba' (*Syagrus* sp.)
Seasoning, spices and food colour	Seasoning plants include 'pimenta de macaco' (*Xylopia aromatica*) and 'canela batalha' (*Cryptocaria aeschersoniana*). Spices used include vanilla, and the most frequently used food colouring is 'açafrão do *cerrado*' (*Escobedia grandiflora*).
Textiles	Seeds, leaves and inner bark used to make fabrics, rope, hammocks, hats, pillows, etc. Genera include *Eriotheca*, *Pseudobombax*, *Mauritia*, *Attalea* and *Xylopia*.
Cork	Almost 20 species of plants with cork in their trunks, including 'pau santo' (*Zollernia illicifolia*), 'mama de porca' (*Fagara cinerea*), 'tamboril do *cerrado*' (*Enterolobium gummiferum*) and 'fruta de papgaio' (*Aegiphila lhotzkiana*).
Tannin	The high tannin content in some *cerrado* species is high enough for industrial use. These include 'angico' (*Anadenanthera* sp.) and 'carvoeiro' (*Sclerolobium* sp.).
Trunk exudates	Resin-producing species include 'jatobá' (*Hymenaea* sp.), 'breu' (*Protium brasiliense*) and 'laranjinha do campo' (*Salacia campestris*). Common gum-producing species are *Vochysia* sp., 'angico vermelho' (*Piptadenia rigida*) and 'aroeira' (*Schinus molle*). Balsam can be extracted from 'bálsamo' (*Copaiba langsdorffii*) and 'cabreúva' (*Myroxylon balsamun*). Latex can be found in 'leiteiro' (*Sapium obovatum*) and *Himatanthus obovatus*.
Oils and fats	Most frequently used are 'pequi' (*Caryocar* sp.), 'babaçu' (*Orbignia martiana*) and 'macaúba' (*Acrocomia sclerocarpa*).
Medicine	More than 100 species are used for various problems.
Ornamental plants	Approximately 200 species can be used for ornamental purposes. They vary from grasses to large trees.
Handicrafts	These plants are used to produce objects such as baskets, hammocks, mats, brooms, wooden cutlery, and bouquets made from some ornamental plants.
Honey	There are around 220 species used by honey-producing bees.

2.4 Current management problems and strategies

Today, management of the *cerrado* is seen as increasingly important: both the population and agricultural activities have grown substantially in the last 35 years; delimited areas of conservation, such as national parks and reserves, are small and constantly invaded by criminal/accidental wildfires and alien species; biodiversity is being threatened; and there is global concern regarding greenhouse warming, and the contribution of vegetation burning in the release of carbon dioxide (Alho and Martins, 1995).

2.4.1 Agricultural expansion

The growth of agriculture in the *cerrado* has, with no doubt, had a major impact on the flora and fauna of the region (see Figure 10.1). The process of land clearing has significantly reduced the natural and diverse *cerrado* habitats for plants and animals. Diseases carried by domestic fauna may have also had an effect on wildlife. Little has been documented, but the reduction in populations of animals such as the manned wolf (*Chrysocyon brachyurus*), pampas deer (*Ozotoceros bezoarticus*) and rodent (*Juscelinomys candango*) may be indications of agricultural impact (Mittermeier *et al.*, 1992; Klink *et al.*, 1993; Rolim and Correia, 1995).

The heavy use of mechanisation and chemical inputs in the *cerrado* has caused serious environmental problems in the region, the most apparent being soil degradation. Soil erosion, compaction and nutrient loss are common consequences of deforestation, and farming practices, such as a mechanised fallow at the end of the dry season, leaving soils bare when thunderstorms are most likely (Dedecek, 1986). Studies also indicate that some crops are more likely to be eroded than others. For example, the architecture and planting intervals of 1 m of maize make it more erosive than soybean planted at intervals of 0.5 m, with no-tillage soybeans showing the least erosion (Dedecek *et al.*, 1986). The frequent use of fire to open land for cultivation, kill weeds and renew pastures, can also over time cause soil degradation, as well as kill any native flora and fauna present in fields and pastures (Mistry, 1998b).

The use of pesticides and the depletion of water reserves by giant rotating irrigators in intensive farming systems in the *cerrado* are amongst the most serious causes of water pollution in the region (Ratter, 1995). Unfortunately, there is a general lack of information about water pollution by farming activities in the *cerrado*, although a few cases have been recorded. For example, nitrogen and pesticide pollution has been detected in almost all underground aquifers around Brasilia (Klink *et al.*, 1993). De Souza (1994) found that crops, occupying only 10% of the Corumba Basin in Goiás, were the source of more than 85% of the solid particles found in water streams of the region. Although no tests for toxic substances were carried out, the author points out that these particles resulting from erosive processes probably carry with them residues of pesticides and fertilisers.

2.4.2 Fire management

Burning is the oldest, cheapest and most widely used management tool in the *cerrado* (Coutinho, 1990). It is a natural force and has many beneficial effects including the stimulation of germination, resprouting, flowering and fruiting, as well as accelerated nutrient recycling (Coutinho, 1990), all of which are important for animal and human populations. Prescribed burning also prevents fuel build-up and the occurrence of uncontrollable wildfires, and is an important determinant of the many fire-adapted species present in the *cerrado* (Dias, 1997). However, its misuse, both intentionally and unintentionally, has led to a strong feeling against fire usage, and has delayed the development and implementation of practical fire policies and management strategies (Pivello and Coutinho, 1996; Pivello and Norton, 1996).

Current legislation makes it illegal for prescribed burning to take place without obtaining permission first from IBAMA (the government environmental agency) (PREVFOGO, 1995; PREVFOGO/IBAMA/SUPES-MT, 1996). However, farmers, the major cause of fires in the *cerrado*, regularly burn their land, and observations and a study by Mistry (1998b) indicate that most burn illegally. Mistry (1998b) also shows that of the farmers who knew that permission was needed to burn, the detailed procedure for obtaining a licence and the lack of fines meant that they never bothered applying for a licence.

The uncertainty about fire prescription is also reflected in practical fire management research for the *cerrado*, which has been scarce to date. However, controlled burns are now being recognised as a suitable tool for *cerrado* management, and increasingly, *cerrado* fire research is being undertaken with a management aim/perspective. For example, Mistry (1996, 1998c,d) investigated corticolous lichens as potential bioindicators of fire history in the *cerrado*. It is important to know what role fire has played in an area of vegetation, i.e. its fire history, in order to apply correct and effective burning regimes. A Lichen Fire History (LFH) Key was developed, which uses lichen community responses on tree bark to detect local fire history, and comprises a simple, illustrated user guide and checksheets, which can be used by both scientific and lay people. The Key can estimate the frequency and behaviour of past fires up to 20 years ago, and can be used in conjunction with other fire knowledge, such as current fuel loads. Still, the LFH Key was developed for a small area in central Brazil, and further testing and validation needs to be undertaken in order to prescribe correct burning regimes to areas of *cerrado*. Other emerging fire management work involves the use of satellite imagery for mapping fires and building fire histories (Prins and Menzel, 1994), and expert systems for planning prescribed fires in conservation areas (Pivello, 1992) (see Section 10.3.1).

2.4.3 Conservation

The *cerrado* has always been undervalued as an ecosystem. For example, in contrast to the Amazon and Atlantic rain forests, the Pantanal, and other coastal ecosystems, the Brazilian Constitution does not recognise the *cerrado* as a National Heritage (Alho and Martins, 1995). Nevertheless, attitudes are changing, and more and more governmental, non-governmental and international institutions are becoming active in *cerrado* research.

Brazilian law requires that 20% of private land in the *cerrado* be maintained as reserve areas. Unfortunately, many of the incentives to clear land have meant that this rule has not been adhered to, if taken at all seriously. As Mistry (1998b) found, many small farmers do not know about the rules and regulations regarding the reservation of land, extracting resources or burning in the *cerrado*. Of the whole *cerrado* biome, only 1.5% is protected as conservation areas, and of these, 90% are less than 50 000 ha in size (Alho and Martins, 1995). Although there are calls to increase these conservation areas (e.g. Alho and Martins, 1995), choosing locations is not an easy task. This is presently being carried out by teams of researchers who are surveying the floristic patterns of *cerrado* vegetation to discover representative areas and biodiversity 'hot-spots' (Ratter *et al.*, 1997). However, once locations are identified, other ecological issues, such as the size of the reserves, but also social and economic factors, including resource use by local communities, will need to be considered.

The major threats to areas of conservation, such as national parks and reserves, are the constant invasion by criminal/accidental wildfires, and the introduction of exotic species. For example, 80 000 ha of Emas National Park burned over seven days in 1988 after an accidental fire, and over 80% of Brasilia National Park burned in 1997 following an extremely severe dry season and criminal ignition. Much of this problem is due to poor resources resulting in inadequate vigilance methods, and the lack of policies on prescribed fires. Fire problems are also exacerbated by the invasion of alien species. In the *cerrado*, invasive grasses not only reduce biodiversity, but also detrimentally alter fire regimes.

2.4.4 Invasive grasses

African grasses are the most common exotics invading *cerrado* areas today, and the most studied. Some, such as *Hyparrhenia rufa*, were accidentally introduced into Brazil, brought in by slave ships during colonial times, whereas others, including *Melinis minutiflora*, were purposely used in the conversion of natural to artificial pastures (Alho and Martins, 1995). The implications of these biological invaders are wide, including reduced productivity in agricultural areas, economic costs related to control methods, as well as ecologically adverse effects such as reduced biodiversity in natural areas as a result of competitive exclusion (Alho and Martins, 1995). It has also been hypothesised that exotic plants may alter the fire regime in the *cerrado*, as they do in other ecosystems (D'Antonio and Vitousek, 1992; Smith and Tunison, 1992).

A study by Berardi (1994) investigated the effects of *Melinis minutiflora* on the biodiversity and fire regime of natural *campo sujo* in the Distrito Federal (Figure 2.10 and Box 2.3). His results indicate that where *M. minutiflora* reaches high cover values, the diversity of the herbaceous flora is considerably lower (6 species) than equivalent areas of native *campo sujo* (18 species). The fire results were also striking. Compared to native *campo sujo* fires, where the highest temperatures are usually in the first 100 cm above the ground (most of the fuel is situated here), and rarely exceed 800 °C (Cesar, 1980; Miranda *et al.*, 1993), fire temperatures in *M. minutiflora* were highest at 60 and 160 cm heights, with peaks of 817 °C and 1006 °C respectively, and flames over 6 m high (Figure 2.11). Also, native fires normally have very short residence times, whereas in the *M. minutiflora* fire, average temperatures of 300 °C were found to last over three minutes.

Figure 2.10 The African invasive grass *Melinis minutiflora*
(Photograph by the author)

These results have important consequences for the conservation of native *cerrado* vegetation. The presence of *M. minutiflora* and other alien species may be aggravating the situation of native species that are already threatened by the reduction of their habitat outside reserves and increasing the chances of extinction by environmental, demographic, genetic stochasticity, and natural catastrophes. Normally, *cerrado* woody species have thick bark insulating inner living tissues from the effects of fire, and higher parts of the trunk and leaves are rarely affected (Coutinho, 1990). This could be ineffective where fires are sustained by *M. minutiflora*.

Guedes (1993) calculated that the critical bark thickness below which the cambium will be killed after a 42-second exposure to a temperature of 180 °C would be 3 mm. This suggests that many woody plants could be killed rather than scorched by fire (Berardi, 1994). Also the higher peak temperatures at higher heights could contribute to adverse effects in the canopy layers. All these factors, could, in turn, reduce species diversity. Berardi (1994) also points outs the effects on nutrient cycling. Fires within *M. minutiflora* would result in a greater loss of nutrients because of the higher temperatures attained, which would volatise a greater proportion of nutrients, and lead to larger ecosystem losses.

Box 2.3 *Melinis minutiflora*: origin, characteristics and distribution

Melinis minutiflora is a C_4 perennial grass from tropical Africa that can spread vegetatively by rhizomes as well as by seeds, has high productivity, and is palatable to cows (Parsons, 1972; Cabrera and Baruch, 1983; Hernandez *et al.*, 1983). The species attains extremely high dead/live biomass ratios (80–90%), and can burn at very high relative humidities (85–90%) as well as high fuel moisture (20–25%) (Smith and Tunison, 1992). After fire, *M. minutiflora* resprouts rapidly and recovers its pre-fire biomass within one growing season (Hughes *et al.*, 1991). *M. minutiflora* is thus able to initiate a grass–fire interaction capable of maintaining cleared forest land as a derived savanna or grassland, preventing succession back to forest (Blydenstein, 1967; Medina, 1987). *M. minutiflora* is also able to displace native species in undisturbed sites, as opposed to the majority of introduced species which need disturbance to enter a new habitat. This may be a result of increased nutrient availability at protected sites allowing *M. minutiflora* to outcompete other species (Bilbao and Medina, 1991). Its ability to develop into dense mats capable of overgrowing and smothering native species in vine-like fashion, has been shown to be a serious threat to native communities (D'Antonio and Vitousek, 1992).

Locally called '*capim gordura*' (fat grass), *M. minutiflora* was, until recently, one of the most economically valuable pasture grasses of central Brazil, and was common throughout the states of Pernambuco, São Paulo, Parana, Goiás, Mato Grosso, Rio de Janeiro and Minas Gerais (Correa, 1984). Although superseded within pastures by more productive introductions, *M. minutiflora* is now found extensively in disturbed areas such as roadside verges, abandoned coffee fincas and abandoned pastures, and has also been observed within nature reserves (Ferri, 1973; Coutinho, 1982a; Medina, 1987; Klink, 1992). The degree and rate of invasion of the Brazilian savanna by *M. minutiflora* and its effects on the ecosystem have not been researched and only passing observations have been made. In the state of São Paulo, *M. minutiflora* has been observed replacing native grasses within an area of *cerrado* protected from fire for more than 30 years (Ferri, 1973). Klink (1992) describes a considerable area of *M. minutiflora* persisting on abandoned pasture which has not been grazed or burnt for the last 18 years within the Reserva Ecologica do Roncador, Distrito Federal. Here, small foci of *M. minutiflora* have been noticed within the native vegetation surrounding the original *M. minutiflora* area but these have not been monitored, so rate of spread is unknown.

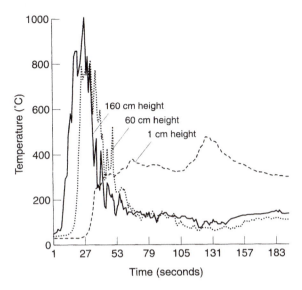

Figure 2.11 Graph of temperature against time during *Melinis minutiflora* fire (after Berardi, 1994).

Unfortunately, as yet, little work has been done on the invasive dynamics of *M. minutiflora* within the *cerrado*. Research on the extent and spread of *M. minutiflora* and viable control techniques need to be addressed.

2.5 Concluding remarks

Research in the *cerrado* is somewhat biased towards plant taxonomic, biodiversity and fire ecology studies. Although these will be essential for future conservation plans, other aspects of *cerrado* management will need to be tackled, especially since the outlook for the *cerrado* is of further, particularly agricultural, expansion (Garcia, 1995). The environmental effects of large-scale land conversion on remaining *cerrado* patches, the effects on fauna, and the social and economic implications on local communities, are just a few of the areas needing research. None the less, with a growing body of scientific knowledge, people are realising the urgency of greater scientific endeavour as well as political intervention, if the *cerrado* is to remain the 'superstar' of savanna diversity.

Key reading

Alho, C.J.R. and Martins, E.S. (1995). *Little by Little, The Cerrado Loses Space.* Discussion paper. WWF/PRO-CER, Brazil.
Furley, P.A. and Ratter, J.A. (1988). Soil resources and plant communities of the central Brazilian *cerrado* and their development. *Journal of Biogeography*, **15**: 97–108.

Klink, C.A., Moreira, A.G. and Solbrig, O.T. (1993). Ecological impact of agricultural development in the Brazilian *cerrados*. In: Young, M.D. and Solbrig, O.T. (eds), *The World's Savannas. Economic Driving Forces, Ecological Constraints and Policy Options for Sustainable Land Use. Man and the Biosphere Vol.12*. UNESCO, Paris, pp. 259–82.

Mistry, J. (1998). Fire in the *cerrado* (savannas) of Brazil: an ecological review. *Progress in Physical Geography*, **22**: 425–448.

Novaes-Pinto, M. (ed.) (1994). *Cerrado, Caracterização, Ocupação e Perspectivas* (2nd edition). Editora Universidade de Brasilia, Brasilia DF.

Ratter, J.A., Ribeiro, J.F. and Bridgewater, S. (1997). The Brazilian *cerrado* vegetation and threats to its biodiversity, *Annals of Botany*, **80**: 223–230.

Chapter 3

The *llanos*

3.1 Introduction

3.1.1 Distribution, climate and soils

The savannas of northern South America (the Orinoco *llanos*) extend from the Guaviare River in Colombia to the eastern coast of Venezuela, and cover approximately 500 000 km² (Beard, 1953; Blydenstein, 1967; Sarmiento, 1983a) (Figure 3.1). They are found on a variety of highly leached soils derived from Tertiary sediments and alluvial deposits, and are generally acid and infertile (Blydenstein, 1967; Sarmiento, 1984). The *llanos* region is characterised by a warm and humid climate during the rainy season (about 80–85% relative humidity) and a warm and dry climate during the dry season (about 65% relative humidity). Total rainfall varies widely, and ranges from 800 to 2500 mm, falling mainly in the wet season of 5–8 months (April to November)

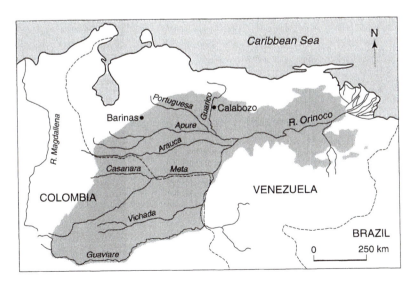

Figure 3.1 The extent of the *llanos* in Venezuela and Colombia.
(Reproduced by kind permission of Justin Jacyno and Nigel Page)

(Monasterio, 1970). The mean annual temperature is 26–27 °C with annual variation less than 3 °C between the monthly mean temperatures, contrasting with daily variations of 10–15 °C (Blydenstein, 1967).

3.1.2 Vegetation structure and composition

The *llanos* are characterised by a combination of an open tree layer and a continuous herbaceous layer (Beard, 1953) (Figure 3.2). These open tree, seasonal savannas are found on the tablelands (*mesas*), and on the alluvial fans and upper river terraces (termed 'piedmont' region), at altitudes of between 100 and 500 m (Sarmiento, 1983a). Common woody species include *Antonia ovata, Bowdichia virgilioides, Byrsonima coccolobifolia, B. crassifolia, B. verbascifolia, Casearia sylvestris, Curatella americana, Genipa americana, Palicourea rigida, Plumeria inodora, Psidium* spp., *Roupala complicata* and *Salvertia convallariodora*. The herbaceous layer is dominated by *Trachypogon plumosus*, with *Aristida setifolia, A. tincta, Axonopus canescens, Bulbostylis paradoxa, Echinolaena inflexa, Leptocoryphium lanatum* and *Thrasya paspaloides* as co-dominants.

However, vegetation physiognomy can vary widely across the *llanos*: from a tree-less savanna grassland to a savanna woodland with up to 80% tree cover (Sarmiento, 1984). For example, savanna woodlands are found on lower mountainous slopes of the piedmont region, where *Curatella americana* dominates, and hyperseasonal savannas and *esteros* occur on the river flats which are either

Figure 3.2 The *llanos* vegetation.
(Reproduced by kind permission of Juan Silva, Universidad de los Andes)

pure grasslands, or areas where small trees occur on termite mounds (Sarmiento, 1983a). Seasonal swamps and *morichales*, grass and sedge *esteros*, with the *moriche* palm (*Mauritia minor*) the only tree, are found along rivulets and on slowly draining lowlands.

Pollen records indicate that the present-day savanna vegetation, dominated by grassland with scattered woody taxa such as *Curatella* and *Byrsonima*, has been present since the last glacial, indicating dry climatic conditions with a marked dry season (Behling and Hooghiemstra, 1998). The greatest expansion of grassland savanna occurred in the early Holocene, from around 9730 to 5260 years BP (dates vary according to sites), suggesting that this period was the driest that has been recorded (Behling and Hooghiemstra, 1998). In the following period, until about 3500 years BP, grassland savanna decreased, suggesting a change to wetter conditions (Wijmstra and Van der Hammen, 1966; Behling and Hooghiemstra, 1998). During the late Holocene, after about 3600 years BP, stands of *Mauritia* increased significantly, interpreted as greater human influence on the savanna, primarily by more frequent burning under wetter conditions (Wijmstra and Van der Hammen, 1966; Rull, 1991, 1992; Behling and Hooghiemstra, 1998). Notwithstanding anthropogenic pressure, forest expansion in the *llanos* is occurring, suggesting that the late Holocene trend to wetter climatic conditions is still probably in progress.

Plant life strategies

The woody component of the *llanos* is generally characterised by a low stature varying from 2 to 6 or 7 m, with few species growing taller than 8 m. The tree species are all evergreen with seasonal growth. Leaf fall normally starts at the beginning of the dry season (around November) and continues throughout the dry period, as does leaf flush, therefore the plants always have some foliage (Sarmiento *et al.*, 1985). At the start of the rainy season (around April), leaf growth ceases altogether, and does not start again until the next dry season (Figure 3.3). Reproduction also takes place almost entirely during the dry season, although the ripening and dispersal of seeds may continue for some time afterwards. On the other hand, radial growth in stem and branches appears to occur almost exclusively during the rainy season.

The herbaceous stratum is normally 50–100 cm in height, and the majority of species are perennial medium-tall tussock (bunch) grasses and sedges (Monasterio and Sarmiento, 1976; Sarmiento and Monasterio, 1983). They can be divided into two main architectural types. Some species, such as *Trachypogon plumosus* and *Andropogon semiberbis*, are erect and have culmed vegetative shoots growing well above ground level, whereas others, including *Leptocoryphium lanatum*, *Sporobolus cubensis* and *Elyonurus adustus*, are basal, possessing culmless vegetative shoots, which ensures them regrowth after defoliation (Sarmiento, 1992).

In most of the perennial grasses, nearly all of the above-ground parts die off during the dry season, leaving little green leaf to continue photosynthesis and transpiration. In the rainy season, the situation is reversed, with all grasses attaining their peak above-ground biomass. However, growth patterns differ,

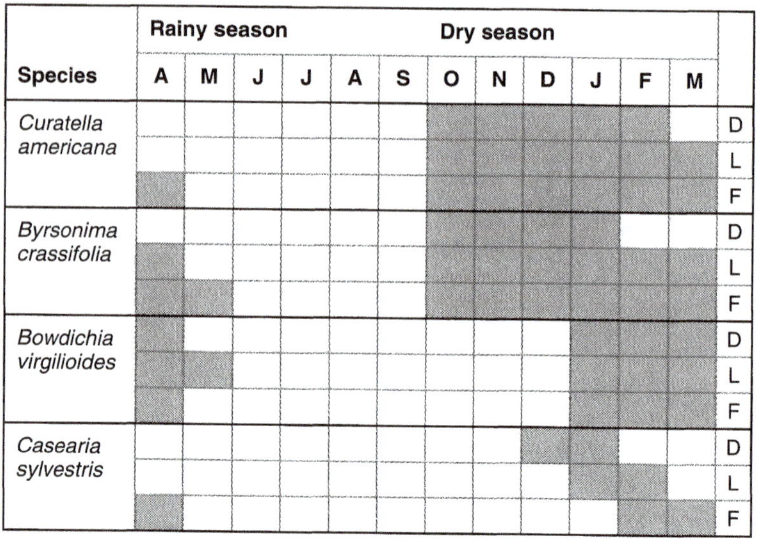

Species	Rainy season						Dry season						
	A	M	J	J	A	S	O	N	D	J	F	M	
Curatella americana							▓	▓	▓	▓	▓		D
							▓	▓	▓	▓	▓	▓	L
	▓												F
Byrsonima crassifolia	▓						▓	▓	▓		▓		D
							▓	▓	▓	▓	▓	▓	L
	▓	▓					▓	▓	▓	▓			F
Bowdichia virgilioides	▓	▓								▓	▓	▓	D
	▓	▓							▓	▓	▓	▓	L
	▓	▓	▓										F
Casearia sylvestris							▓	▓	▓	▓			D
									▓	▓	▓	▓	L
	▓										▓	▓	F

Figure 3.3 The phenological cycles of four woody species from the Venezuelan *llanos*. D = leaf drop, L = leaf flush, F = flowering (based on Sarmiento *et al.*, 1985).

with some species growing immediately after the first rains, or even earlier if they are burned, whilst others have time lags of some months (Sarmiento, 1992).

Although perennial grasses all flower during the rainy season, there are temporal variations related to the period of flowering and maximum growth. Accordingly, perennial grasses of the Venezuelan *llanos* have been classified into four phenological groups: precocious, early, intermediate and late (Sarmiento, 1983b) (Table 3.1). The precocious species are highly dependent on fire. Very few tussocks will flower when the savanna does not burn. For example, burning has been found to increase the percentage of flowering shoots of *Sporobolus cubensis* from 0.05 to 7% (Canales and Silva, 1987). Some precocious grasses flower immediately after a fire, even during the rainy season, but only once a year; whereas others, such as *Imperata brasiliensis*, will flower after any fire event (Sarmiento, 1992). There is also a close relationship between growth form and reproductive behaviour, as well as with pattern of vegetative growth and biomass allocation. In the latter, precocious species such as *Leptocoryphium lanatum* have below-ground to above-ground ratios of 2, compared to 0.4 in *Trachypogon vestitus*, an intermediate grass (Sarmiento, 1992). This is probably because in culmless species, leaf bases remain below-ground, but also because of the storage of carbohydrates in underground organs.

Silva and Ataroff (1985) found that precocious species such as *Leptocoryphium lanatum* and *Sporobolus cubensis* produced fewer seeds per plant and per dry weight than did later flowering grasses, with the exception of *Trachypogon plumosus* (a late-flowerer with a low reproductive output but a high

Table 3.1 The four phenological groups of *llanos* perennial grasses (after Sarmiento, 1983b)

Phenological group	Flowering period	Growth form	Species
Precocious	February–May. Includes species that flower before onset of rainy season, and those that flower about two months after arrival of fire. Highly dependent on fire	Culmless	*Andropogon selloanus* *Elyonurus adustus* *Imperata brasiliensis* *Leptocoryphium lanatum* *Paspalum carinatum* *Paspalum pectinatum* *Sporobolus cubensis*
Early	May–July. Flower during first half of rainy season	Culmed and medium-tall	*Axonopus affinis* *Axonopus canescens* *Axonopus pulcher* *Panicum olyroides* *Paspalum coryphaeum*
Intermediate	August–October. Flowers during second half of rainy season	Culmed and medium-tall	*Aristida riparia* *Aristida tincta* *Schizachyrium tenerum** *Trachypogon plumosus*
Late	October–January. Reproductive peak towards the end of the rainy season or during first part of dry season	Culmed and tallest grass type	*Andropogon semiberbis* *Echinolaena inflexa* *Hyparrhenia rufa** *Melinis minutiflora**

* Exotic species.

vegetative spread). Seeds produced by precocious species germinated as soon as they reached wet ground, but the seeds from the other species remained dormant in the soil until the start of the following rainy season. Thus, all seeds germinated during the first months of the rainy season.

Although there is a degree of neighbour interference in growth between different grass types (Raventós and Silva, 1988, 1995), most *llanos* communities have a mixture of the phenological forms, and 3–12 (average 7) grass species coexist (Sarmiento, 1983b). Sarmiento and Monasterio (1983) have suggested that this pattern favours a temporal differentiation in the uptake of scarce resources such as water and nutrients.

Annual grasses, commonly of the genera *Andropogon*, *Aristida*, *Diectomis*, *Eragrostis* and *Gymnopogon*, constitute a much smaller proportion of the herbaceous layer (Monasterio and Sarmiento, 1976). Most are erect, short and largely restricted to the bare ground left between clumps of perennial grasses, where they form dense patches. They germinate at the beginning of the rainy season, and are generally found in the more stressful habitats characterised by wet soil and low nutrient content (San José and Montes, 1991).

The high light intensities and optimum leaf temperatures of 30–40 °C result in high photosynthetic rates. Torres (1984) and Baruch *et al.* (1985) have reported rates in native grasses in the order of 16–28 µmol m^{-2} s^{-1}, and as high as 34 µmol m^{-2} s^{-1} in recently expanded leaves (Chacón-Moreno *et al.*, 1995). However, these rates are limited by water stress during the dry season, and net photosynthesis can be 30–50% of rates under unrestricted CO_2 exchange (Torres, 1984; Baruch *et al.*, 1985).

3.1.3 Fauna

The *llanos* support a high diversity of wildlife, the majority of which are associated with the flooded areas, including wading birds, fish, caiman and capybara, the world's largest rodent (Figure 3.4). *Llanos* have 122 mammal species, and the majority of these (93.5%) are either partially or completely dependent on forested areas; few species depend on the savanna (Medellin and Redford, 1992).

3.2 The main savanna determinants

Using multivariate analysis, San José and Montes (1991) found that *Trachypogon* savannas and their selected species were distributed along two complex environmental gradients related to the co-occurrence of soil moisture and available nutrients, the two principal determinants of *llanos* vegetation. These constituted a soil physical and chemical gradient as expressed by changes in bulk density and magnesium concentration, and a climatic gradient of decreasing annual precipitation and increasing monthly precipitation during the dry season. At the landscape scale, the relief and its interaction with the water regime is a major determinant of the vegetation (Montes and San José, 1995). At the patch scale, water combines with nutrients to affect vegetation. Montes and San José (1992) found that there was a significant decrease in the annual hydrogen content of 44–75% as rainfall leached through the canopy, whereas bulk precipitation pH in the grass layer was similar to that of rainfall. As throughfall passed throughout the soil profile of the grove, the H^+ content decreased by 70–87%.

3.2.1 Plant–available moisture (PAM)

Soil water potential in the *llanos* reaches relatively high levels during the rainy season, but falls sharply as the dry season progresses, reaching values of the order of −2.0 MPa or less (San José and Medina, 1975; Sarmiento and Vera, 1977; Goldstein *et al.*, 1986). However, the drop in water potential is not as large at soil depths of 50 to 120 cm, falling to about −0.5 or 1.0 MPa, where a good proportion of the tree roots are found (Sarmiento and Acevedo, 1991).

The leaf water potential of grasses follows similar seasonal fluctuations, with a minimum of −1.2 to −1.8 MPa during the rainy season, falling to −3.8 MPa

(a)

(b)

Figure 3.4 The two most well-known *llanos* animals: (a) the caiman; (b) capybara.
(Reproduced by kind permission of Juan Silva, Universidad de los Andes)

(e.g. *Trachypogon plumosus*) at the peak of the dry season (Sarmiento, 1992). Even at these low values, the grasses continue to transpire and photosynthesise, albeit at very low rates. For example, Baruch *et al.* (1985) found positive carbon gains in *Trachypogon plumosus* leaves at a water potential as low as −6.0 MPa. *T. plumosus*, a late-growing species, may be drought tolerant, but precocious species behave rather as drought evaders. Precocious species, such as *Sporobolus cubensis*, have a lower drop in leaf water potential, leaf conductance and daily transpiration during the dry season.

Adaptations to PAM

Adaptations in woody plants to PAM include extended roots. In Venezuela, Foldats and Rutkis (1965, 1975) reported expansive root systems in various trees such as *Curatella americana*, *Byrsonima crassifolia* and *Bowdichia virgilioides*, for both horizontal and vertical growth. They showed, for instance, how a *C. americana* tree less than 5 m in height had roots that reached more than 20 m laterally from the trunk.

In spite of the apparent xeromorphism of their leaves, *llanos* woody species scarcely control transpiration losses, even during the peak of the dry season (Vareschi, 1960; Foldats and Rutkis, 1975). For example, daily water loss per unit leaf area in *Bowdichia virgilioides* on a clear, dry-season day may be three to four times greater than the amount transpired on a cloudy day of the rainy season (Goldstein *et al.*, 1986). Similar results have been found from measurements of leaf water potential (Meinzer *et al.*, 1983; Goldstein *et al.*, 1986). It seems therefore that transpiration depends more on atmospheric or radiation conditions than on topsoil water. Meinzer *et al.* (1983) also showed that when old and new leaves are present on trees in species such as *Curatella americana* and *Byrsonima crassifolia*, the younger leaves have significantly lower conductance to water vapour than the mature ones, suggesting the maintenance of a more favourable water budget in young leaves, making their growth and expansion in the less favourable period of the year possible.

3.2.2 Plant-available nutrients (PAN)

Plant-available nutrients are particularly low in the *llanos*. Sarmiento (1990) took the amount of exchangeable bases in the topsoil (Ca, Mg, K and Na) as simple indicators of soil fertility. Areas of *llanos* were found to have soil fertility values of between 1 and 5 me 100 g^{-1} soil, considered dystrophic, as well as < 1 me 100 g^{-1} soil, considered hyper-dystrophic, and in some cases were as low as 0.12 me 100 g^{-1}. As a result, most *llanos* plants have very low nutrient contents in all plant parts (Sarmiento *et al.*, 1985). This may be compensated by re-allocation of nutrients, inferred from mature grass leaves which drop their nitrogen and potassium content by 80% and 120% respectively (Sarmiento, 1984), and from tree leaves which decrease in leaf concentrations of N, P and K with age (Montes and Medina, 1977). Although not recorded, rhizospheric and mycorrhizal nitrogen fixation may also be important.

3.2.3 Herbivory

As in the *cerrado*, the massive extinctions that occurred in South America after the last glacial period eliminated many of the large herbivores, and left the *llanos* with an impoverished ungulate fauna without wild bovids (Ojasti, 1983). The only native grazer that remains is the capybara, and this giant rodent is restricted to areas close to rivers and lagoons (Ojasti, 1991).

In spite of the high primary production of the grass layer, seasonal savannas support few herbivores, both native and domestic, probably because of the low quality of the forage. For example, in the middle of the wet season, C_4 grasses have 4–6% crude protein and 0.12–0.18% phosphorus, compared to C_3 grasses in flooded savannas which have 6–15% crude protein and 0.18–0.24% phosphorus (Gonzalez Jiménez, 1979). This decreases in the dry season, during which average crude protein content is around 5.2% and average phosphorus content 0.15% (Tejos, 1987; Tejos *et al.*, 1990).

Little is known of the effects of herbivory on native *llanos* plants. Sarmiento (1992) describes a 15-year experiment in which a west Venezuelan *llanos* was managed as a mown lawn and clipped bimonthly to simulate severe grazing. The consequence was the total replacement of native grasses with aggressive colonisers; first those of African origin such as *Hyparrhenia rufa* and *Panicum maximum*, and then other natives such as *Paspalum virgatum* and *Axonopus compressus*. Other simulated herbivory experiments, such as that by Simoes and Baruch (1991), indicate that invasive grasses, such as *H. rufa*, are much better adapted to herbivory than native species, and are resilient under frequent defoliation.

3.2.4 Fire

Frequent fires have been occurring in the *llanos* at least since the late Holocene (Behling and Hooghiemstra, 1998). People have also contributed to past fire regimes by burning the *llanos* for various activities, including hunting. Today, fire is applied mainly by cattle-raisers in order to encourage new grass growth during the dry season.

The effects of fire on vegetation

Fire consumes about 90% (500–700 g m^{-2}) of the aerial biomass in different Venezuelan savanna communities, releasing 1.3–1.9 g m^{-2} of N, 1.1–1.6 g m^{-2} of K, 0.9–1.3 g m^{-2} of Ca, and 0.4–0.6 g m^{-2} of P (Sarmiento, 1992). Late dry season burns are generally more severe than earlier burns (Blydenstein, 1963), and high fire intensities are evident in areas with a long 'time since last fire' period (Aristeguieta and Medina, 1966).

Fire seems to have different effects on different phenological phases of savanna grasses. In perennial species, well-established tussocks are generally unaffected by burning, as their buds are protected inside basal leaf sheaths or are underground. In fact, many species are fire-dependent for reproduction, e.g. precocious species such as *Sporobolus cubensis*.

Silva *et al.* (1991) concluded that to stabilise a population of *Andropogon semiberbis*, burning frequency has to be at least 0.85, i.e. 5 burns in 6 years, below which the population begins to decline. However, in many of these species fire may affect growth (Silva and Castro, 1989). For example, Canales and Silva (1987) found that burning reduced vegetative growth, root and rhizome growth, and above-ground biomass accumulation in *Sporobolus cubensis*. Even so, adult *S. cubensis* have a higher probability of surviving until the next fire than do adults of *A. semiberbis*, and so Silva *et al.* (1991) suggest that *A. semiberbis* is a better coloniser under burnt conditions; however, once established, *S. cubensis* resistance to fire is higher.

Little work has been done on annual grasses. Canales *et al.* (1994), studying the demographic responses of the annual *Andropogon brevifolius*, found that fire exclusion substantially decreases population growth only after two years, mainly through a fall in seed production and seed-to-seedling survival. As a result, they suggest a burning frequency of once every three years, to maintain viable populations of the annual grass.

The effects of fire protection on vegetation

Areas protected from fire for more than 20 years show a significant increase in tree density, both of fire-resistant common savanna species, and of fire-susceptible species (San José and Fariñas, 1983, 1991; Fariñas and San José, 1987) (Figure 3.5). Silva and Sarmiento (1997) found that woody plant populations are more sensitive to changes in plant growth when fire is excluded, but with annual fires, are more sensitive to changes in plant survival. In fact, plant sur-

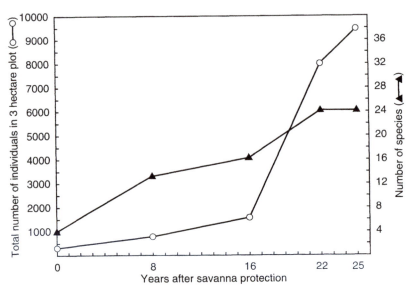

Figure 3.5 Temporal changes in the density and diversity of trees in a 3 hectare plot of *llanos* protected for 25 years (after San José and Fariñas, 1991).

vival is more important than fecundity and growth put together, and the difference is even greater in burnt areas. It is suggested that woody plant populations are maintained in burnt savannas by varying their population growth rate very close to 1 (Silva and Sarmiento, 1997). When fires are more frequent, the growth rate is below 1, and the population decreases, whereas when fire frequency lessens, the growth rate is greater than 1, and the population increases. The greater significance of plant survival therefore helps woody populations persist in burnt areas.

The exclusion of fire can have deleterious effects on the perennial grass species (Figure 3.6). Studies at the Biological Station of Calabozo, Venezuela showed that the short-term effect (after 8 years) of the lack of fire was an increase in diversity (San José and Fariñas, 1983), but that over the long term, there was higher seedling mortality, a reduction in size and vigour in adults, and a trend towards local extinction (Medina and Silva, 1991). For example, the relative density of *Trachypogon plumosus* was reduced from 57% to 14% after 20 years of fire protection, whilst *Axonopus canescens* increased from 16% to 44% over the first 16 years, and then decreased to 33% six years later (Fariñas and San José, 1987).

It is suggested that the interplay of years with fires and years without, and the patchy nature of the fires, probably determines short-term changes in the composition of the herbaceous layer, but that long-term accumulation of dead biomass may lead to the death of the whole grass layer, the process differing among species (Medina and Silva, 1991; Tilman and Wedin, 1991) (Table 3.2). At Calabozo, the tall African grass *Hyparrhenia rufa* has gradually displaced the native grasses, and after almost 30 years of fire

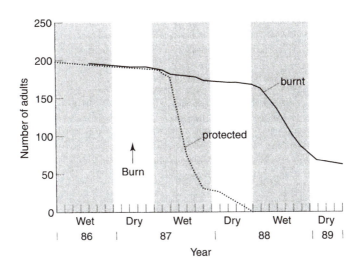

Figure 3.6 Survivorship of *Andropogon semiberbis* adults in protected and burnt *llanos*. Note that in 1988 when no fire occurred in the burnt plot, numbers of *A. semiberbis* began to fall (after Silva *et al.*, 1990).

Table 3.2 Characteristics of Venezuelan *llanos* under three different fire frequency regimes (after Sarmiento and Silva, 1997)

Characteristics	Life-form*	Annual burning	Burning every 3–5 years	Burning every 5–20 years
Dominant species	H	Native grasses	Native grasses, but change in composition	Invasive grasses
	W	Native trees	Native trees, but change in composition	Native trees
Dominant functional groups	H	Perennial grasses of diverse phenologies	Perennial grasses of diverse phenologies	Late perennial grasses
	W	Evergreen sclerophyllous trees	Evergreen sclerophyllous trees	Evergreen sclerophyllous and deciduous mesophyllous trees
Density (no. ha^{-1})	W	Low, 10–100	Increase between burns, but stable over the long term	Increases
Species diversity (no. 100 m^2)	H	High, 10–100	High, 30–40	Decreases
	W	Low, 1–10	Low, 1–10	Increases
Annual aerial biomass (g m^{-2})	H	Green: high (500) Dry: high (600)	Green: decreases Dry: increases	Green: decreases Dry: increases
Annual above-ground net primary production (g m^{-2})	H	High: > 600	Decreases	Increases
	W	Low: < 200	Increases	Increases
Seed bank	H	Transient	Decreases in size	Decreases in diversity
	W	Variable according to species	Variable according to species	Variable according to species
Forage quantity and quality	H	High, low	Increases, decreases	Increases, decreases
Number of invasive species		Low	Low	Increases
Organic matter, nitrogen, and cation exchange capacity in 1–10 cm of soil		1–2%, 0.07–0.2%, 3–8 me 100 g^{-1} soil	Increases	Increases

Table 3.2   (*Continued*)

Characteristics	Life-form*	Annual burning	Burning every 3–5 years	Burning every 5–20 years
PH, rate of saturation and sum of bases in 1–10 cm of soil		5.0–6.2, 10–50%, 1–5 me 100 g⁻¹ soil	Decreases	Increases
Microclimate	H	Dry and bright to humid and shady	Less variable, but humid	Less variable, but humid
	W	Shady, reduced in dry season	Reduction in shade less important	Reduction in shade expanded
Patchiness		High (induced by fire and fauna)	Decreases	Decreases

* H = herbaceous layer; W = woody layer.

protection and exclusion of herbivores, the vegetation continued to be a seasonal savanna, albeit with a denser woody layer (Fariñas and San José, 1987; Medina and Silva, 1991; San José and Fariñas, 1991). However, where there are indurated ironstone outcrops, tree growth has been impeded, and the vegetation remains open savanna (San José and Fariñas, 1991). This illustrates the effect of determinants at smaller, patch scales in controlling vegetation dynamics.

3.3 Human determinant of the savanna

Savannas were originally home to large populations of Amerindians. Many depended on the cultivation of manioc in combination with hunting and gathering, and although many indigenous populations were decimated after colonisation, the remaining populations continue with this lifestyle today (e.g. Gragson, 1997). With the arrival of the Spanish in the mid-sixteenth century, the predominant land use changed from subsistence resource use to cattle-rearing, and has remained so up to the present day.

3.3.1 Pre-colonial land use

Ethno-historical studies in the western *llanos* indicate that tribes such as the Jirajara and Caquetío were largely concentrated along rivers and in the gallery forests (Redmond and Spencer, 1994). Most of them were agriculturalists, but fishing and hunting were significant supplementary activities. Sweet manioc, maize, plantains and a variety of other crops were cultivated. The Caquetío planted two varieties of maize with growing seasons of just 40 and 60 days (Redmond and Spencer, 1994). However, trade was also common, and items such as food, palm products, tree resins, animal skins, pottery and turtle eggs were typically exchanged. These communities were generally led by a chief, and advised by a council of elders. Warfare was rife, for reasons such as acquiring good land, looting villages and fields, and to take captives for the slave trade (Redmond and Spencer, 1994).

The absence of large herbivores and domestic herd animals in the Americas seems to have precluded the development of the parasites and vector-borne disease organisms that evolved through thousands of years of man–animal relations in the Old World. It was rather the infectious diseases of the Old World brought by the Spaniards, and later by the African slaves, that decimated the populations. The *llanos* became quite devoid of indigenous people within a few decades of discovery, leaving the door open for Europeans and their livestock.

3.3.2 Colonisation and post–colonial land use

The *llanos* were first settled by the Spaniards in 1548. Initially the area offered little attraction to the Spaniards, as the adjacent Andes and highlands were more fertile, had a better climate, and the Indians were more abundant and compliant (Parsons, 1980). However, by the 1560s, cattle-ranching was 'on its way', and by the time of the establishment of Barinas, the first town on the margins of the *llanos*, livestock were multiplying and colonists were moving down into the lowlands. From 1640 to 1790, the *llanos* played host to the *cumbes*, or outlaw slave communities, and contributed to its reputation as being wild, lawless country. Cattle-rearing continued to grow during the nineteenth and beginning of the twentieth centuries, estimated at around 4.8 million head in 1812, but fluctuated drastically, first with the wars of independence and then with the general political instability of the time. Over this period, the *llanos* remained relatively undeveloped, with mostly large cattle-ranching taking place. Malaria, probably brought by the Europeans, was the main barrier to savanna occupancy, together with other livestock-related diseases such as Chagas disease (Parsons, 1980), although not on the scale of, for example, trypanosomiasis in Africa.

Land tenure

The concentration of land property has been considered one of the major obstacles to agricultural development within the *llanos* (Orta, 1974). Land tenure is characterised by extensive estates (*latifundios*), most of which (80%) are privately owned. This system has been predominant since colonial times (Rodriguez Mirabal, 1987) and still remains today, although the aims of the 1961 Agrarian Reform Bill were to bring about a more equitable distribution of rural land, and to raise the standard of living (Silva and Moreno, 1993). The reform failed mainly because difficulties arose in breaking up the large estates, and instead, most land allocated to peasants was public land with little productive potential. The lack of financial and technical aid, and the limited area of agricultural land allocated to each peasant (lots of 5 ha or less), meant that it was difficult for most peasants to make a living and consequently debt problems arose. In response, many peasants migrated to cities, selling their land, ironically, to the large estate-owners, with the result that more land-ownership became private.

3.3.3 Present-day land use

Although during the last two centuries, the *llanos* in Venezuela contained most of the national herd and were the major providers of milk and meat, development has still been marginal, and still continues as such. During the last 30 years, the agricultural frontier has advanced, but only in areas with higher natural productive potential, such as the western piedmont *llanos*, where dry and mixed forests are dominant (Silva and Moreno, 1993). The existence of numerous oil wells in the mesas region has had little positive impact upon local economies and, in fact, has further encouraged abandonment of rural areas. As a result, the *llanos* remain largely undeveloped and uncolonised.

Agriculture

The *llanos* is mainly used for extensive cattle-raising with traditional, very simple management techniques of little technical input (Silva and Moreno, 1993). However, the low forage production (200–600 g m^{-2} year^{-1}) (San José *et al.*, 1985), as well as the unpalatability and nutrient content of the grasses, limits stocking rates to around 0.5 animals ha^{-1} year^{-1} (San José *et al.*, 1991b). Savannas are seasonally grazed, and in hyperseasonal savannas the herd is moved towards better-drained areas during the peak of the rainy season. At the beginning of the dry season, cattle are sold to intermediaries who move them to ranches in the piedmont. There, the animals are fed on cultivated pastures for a few months before sale to meat industries.

Farming, as with cattle-raising, generally has few modern agricultural technical inputs. The profit margins are low, partly because of government policies controlling the market prices of agricultural produce, and partly because most of the income generated goes to the intermediaries, with little going to the actual producers. The main traditional crops grown in the Orinoco *llanos* are corn, beans, manioc and cotton, with corn being predominant. A large proportion of this production is for the national markets and agro-industry, with only a small fraction being consumed on the farms.

Despite direct and indirect subsidies to agriculture, mostly from oil revenue, traditional crops such as corn, which are an important element of the *llanos* diet, have shown significant fluctuations in production, particularly stagnation or deterioration, during the last 30 years (Silva and Moreno, 1993). As a result, corn inputs have increased, and competition from imported wheat has been an additional problem for corn producers. Unfortunately, the introduction of new crops, such as peanut and sorghum in the 1970s, has also been unsuccessful (Silva and Moreno, 1993), and producers have only been able to persist through government subsidies.

Petroleum

Oil was discovered in the *llanos* at the beginning of the twentieth century. A vigorous petroleum industry plays a key role in the economic revitalisation of extensive sections of the *llanos*, and exploration continues to increase in the *llanos* basin of Colombia (Wiman, 1990). However, little seems to be known about the impacts of these activities. It has been estimated that the carbon released from intensive petroleum extraction and refining activities in the Orinoco *llanos*, where the petroleum belt covers 25% of the whole area, was around 0.057 Pg year^{-1} in 1990. This is equivalent to 6% of the total carbon emission in the northern Orinoco *llanos* (MARNR, 1995).

Plantations

Pine plantations have had some success in the *llanos* region. For example, the programme of pine plantations in the southern mesas of the State of Monagas, initiated in 1968, has reached 70% of its goals in terms of production. However, the delay in building a paper mill has meant that most of the pines are used for lumber and resins in small-scale operations, and not as the programme aimed, to provide national paper needs (Silva and Moreno, 1993).

3.4 Current management problems and strategies

In an attempt to diversify the local economy, the two major management endeavours in the *llanos* have been a scheme to hold back excess winter water to maintain moisture through the dry summers using modular dams, and to harvest wildlife. The latter has been tied up with conservation efforts in the region. However, a major threat to biodiversity and productivity is the spread of invasive grasses, as discussed below.

3.4.1 Floodplain management

In the early 1970s, a government programme was started in Venezuela to improve the productivity of the *llanos* by using systems of dikes, or *módulos*, to control flooding, and to keep a reservoir of water for the dry season. Studies of the ecological changes of the land affected were carried out, and showed a significant increase in the contribution of nutritious species (e.g. *Leersia hexandra, Hymenachne amplexicaulis*), higher animal production and positive changes in the capacity of the land for production (Gil Beroes, 1976; López-Hernández *et al.*, 1983; Berrade and Tejos, 1984). López-Hernández *et al.* (1994) showed that although diking did not induce a dramatic change in nutrient budgets, potassium, a critical plant element did decrease significantly, and may be related to the increase in biomass. However, no evaluation was made of the socio-economic impacts of the dikes, and as a result of political changes, the programme has been practically abandoned (Silva and Moreno, 1993). A study currently being undertaken by researchers at the University of Merida, Venezuela, is assessing the effects of these *módulos* on the local environment, and hopefully results should be disseminated in the near future.

3.4.2 Wildlife harvesting

The two main animals exploited in the *llanos* are the capybara (*Hydrochaeris hydrochaeris*) and the caiman (*Caiman crocodilus*). The capybara is the largest living rodent, and its efficient digestion of herbage (similar to ruminants) enables it to attain high population densities of up to 200 individuals per square kilometre (Ojasti, 1991). Competition with cattle is normally minimal as its preferred food plants grow in swampy areas, not frequented by cattle. The capybara is hunted by many Indian communities and by farmers principally for its meat, although the hide is an important by-product. Yearly sustained harvest levels in well-managed cattle ranches with water retention systems and control over poaching can produce a sustained harvest of 1200 kg km^{-2} year^{-1} (Hoogesteijn and Chapman, 1997).

The caiman's patterns of movement and habitat use are closely tied to the annual flooding of the *llanos*, and in the dry season, densities can reach 50–300 per hectare. They are hunted for their hides. The history of commercial caiman harvest, beginning in the 1950s, has been dynamic, involving periods of uncontrolled harvests, leading to population reductions and then subsequent recovery (Figure 3.7). In response, a caiman management programme was started in Venezuela in 1983, based on the harvest of adult males on private lands (Thorbjarnarson, 1991). Although the official programme is not believed to have had a negative effect on wild populations (Thorbjarnarson, 1991), critics have pointed out a number of problems, including permits being given to ranches with dubious ownership papers, unrealistically high harvest estimates, and the failure to control harvesting by illegal hunters (e.g. Rivero Blanco,

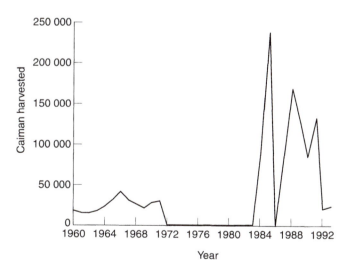

Figure 3.7 The number of caiman legally harvested in Venezuela between 1960 and 1993 (after Hoogesteijn and Chapman, 1997).

1990; Luy, 1992). The latter is a major problem for both the capybara and caiman. Poachers use four-wheel-drive vehicles or horses in the dry season and motor boats in the wet season, making the large ranches and parks difficult to patrol. There is little respect for private property and light punishment for offenders; as a result, poaching, particularly by organised bands of hunters, has steadily risen.

Wildlife and conservation

Traditional methods of conservation concentrate on protecting plants and animals through the establishment of national parks and reserves. In the Venezuelan *llanos*, there are two national parks – Aguaro-Guariquito (570 km^2) and Santos Luzardo (584 km^2) – which together cover about 3.8% of the Venezuelan *llanos*. However, these areas only protect a small proportion of the habitat and are difficult to patrol (Hoogesteijn and Mondolfi, 1992). An alternative strategy may be the application of wildlife conservation regulations by private land-owners on their properties.

Hoogesteijn and Chapman (1997) calculated that ranches could potentially increase their income by 61%, although as yet it varies among ranches from 25 to 52%. Because the economic benefits of capybara and caiman harvests can only be realised by conserving wildlife habitats, their analysis suggests that the economic gains from the harvest of these animals could motivate ranchers to improve wildlife habitats. The idea is not to replace efficiently operated and managed national parks, but rather to augment them with a vast amount of privately owned land, where wildlife can coexist with cattle. However, Hoogesteijn and Chapman (1997) do point out that their analysis is optimistic, and that if prices fall or fluctuate widely, or if the economic stability of the country changes, land-owners' incentives will decrease. The incorporation of cattle ranches into a conservation programme will require that government and conservation agencies assist land-owners.

3.4.3 The expansion of invasive grasses

The *llanos* have been invaded by a number of African grass species and, as in the *cerrado* (Section 2.4.4), there are potential detrimental effects on productivity, agriculture and conservation. *Panicum maximum* is one of the most widespread invasive grasses in the *llanos*, as well as throughout the Neotropics. It has a high productivity, reaching an annual production of 60–70 t^{-1} ha^{-1} under grazing, and a wide tolerance to a range of environmental factors (Sarmiento, 1992). Severe defoliation, for example, was found to promote its growth and palatability, independent of absolute leaf age (Chacón-Moreno and Sarmiento, 1995; Chacón-Moreno *et al.*, 1995).

Hyparrhenia rufa is another African grass that has displaced native species from the wetter and more fertile habitats (Parsons, 1972; Fariñas and San José, 1985). This displacement has been attributed to higher photosynthetic rates, larger proportions of assimilates allocated to leaves, an opportunistic use of water and higher germination capacity of the African grass (Baruch *et al.*, 1985, 1989). In addition, *H. rufa* has a greater ability than natives such as *Trachypogon plumosus*, to tolerate and compensate for herbivory. In simulated experiments, *H. rufa* responded by increasing resource allocation to the above-ground organs, increasing photosynthetic rate and number of tillers, and changing the basic architecture of the plant to a prostrate growth form, demonstrating its high plasticity (Simoes and Baruch, 1991). *T. plumosus*, on the other hand, did not compensate for defoliation but decreased growth and tiller number, and increased leaf senescence.

Like *H. rufa*, *Melinis minutiflora* also displaces native species in areas of more fertile savanna. Bilbao and Medina (1991) propose that nutrients may in fact be the key factor in determining invasion (Table 3.3). *M. minutiflora* is able to produce more biomass under a similar nutrient supply than native grass species, but is not as efficient in 'foraging' for nutrients because of its high shoot/root ratio, i.e. *M. minutiflora* has fewer roots available for nutrient absorption per unit leaf area than native species. Therefore, *M. minutiflora* is only able to out-compete native species where nutrients are in relatively high concentrations. An increase in soil mineralisation following disturbance (e.g. fire), or an accumulation of organic matter at the soil surface in fire-protected sites, would lead to a higher concentration of nutrients within the upper soil layers. At these sites, *M. minutiflora* would be able to have a higher growth rate than native species and ultimately displace them.

Table 3.3 Nutrient and photosynthetic characteristics of *Hyparrhenia rufa*, *Melinis minutiflora* and *Trachypogon plumosus* grown under optimum conditions of water and nutrient supply (after Baruch *et al.*, 1985). Means ± standard error are presented.

Nutrient and photosynthetic characteristics	Hyparrhenia rufa	Melinis minutiflora	Trachypogon plumosus
Nitrogen concentration (% dry weight)	2.74 ±0.04	3.09 ±0.08	1.62 ±0.09
Phosphorus concentration (% dry weight)	0.18 ±0.01	0.20 ±0.01	0.11 ±0.04
Potassium concentration (% dry weight)	1.96 ±0.09	2.13 ±0.13	1.21 ±0.08
Leaf net photosynthetic rate ($\mu mol\ m^{-2}\ s^{-1}$)	31.9 ±0.6	30.5 ±0.7	26.2 ±1.6

3.5 Concluding remarks

The *llanos* savanna ecosystem is driven by PAM. This is demonstrated by the attempts of governments to trap water during the wet season using dams, and the dependence of both domestic cattle and wildlife on the seasonal moisture availability. The dominant herbaceous layer has undoubtedly had the most attention, and much research has focused on grass population dynamics and phenology. Although relatively little exploited, the *llanos* will face increasing pressure from development in the future, and strategies for management, such as outlined for wildlife harvesting, will need to be implemented. Conservation may be further aided by encouraging sustainable land-use practices such as eco-tourism and sport hunting, which are as yet limited, but potentially important, revenue-earning activities.

Key reading

Hoogesteijn, R. and Chapman, C.A. (1997). Large ranches as conservation tools in the Venezuelan *llanos*. *Oryx*, **31**(4): 274–284.

Medina, E. and Silva, J.F. (1991). Savannas of northern South America: a steady state regulated by water–fire interactions on a background of low nutrient availability. In: Werner, P.A. (ed.), *Savanna Ecology and Management. Australian Perspectives and Intercontinental Comparisons*. Blackwell, Oxford, pp. 59–69.

Sarmiento, G. (1984). *The Ecology of Neotropical Savannas*. Harvard University Press, Cambridge, MA.

Silva, J.F. and Moreno, A. (1993). Land use in Venezuela. In: Young, M.D. and Solbrig, O.T. (eds), *The World's Savannas: Economic Driving Forces, Ecological Constraints and Policy Options for Sustainable Land Use* (MAB 12), UNESCO and Parthenon Press, Paris, pp. 239–257.

Chapter 4

The *miombo* of central Africa

4.1 Introduction

4.1.1 Distribution, climate and soils

Miombo savanna woodland is one of the most uniform and extensive vegetation types in Africa, and is found in seven central and southern African countries: Tanzania, Democratic Republic of Congo, Angola, Zambia, Malawi, Zimbabwe and Mozambique (White, 1983) (Figure 4.1). Covering an area of approximately 2.7 million km^2 (Millington *et al.*, 1994), its distribution largely coincides with the flat to gently undulating African (early Tertiary) and post-African I (Miocene) planation surfaces that form the central African plateau (Cole, 1986). The soils are predominantly infertile, comprising oxisols, ultisols and alfisols, which have a low cation exchange capacity, low total exchangeable bases, high acidity, low available phosphorus and a laterite or gley horizon near the surface (Nyamapfene *et al.*, 1988). The climate is strongly seasonal, with more than 95% of the 700–1400 mm mean annual precipitation falling during the 5–7 month summer season (Frost, 1996).

The *miombo* woodlands are interspersed with broad, grassy depressions called *dambos* or *mbunga*. These seasonally waterlogged bottomlands are principally poorly drained but comprise fertile vertisols, and are formed through differential weathering and subsurface removal of material by the lateral flow of groundwater (Boast, 1990). They can cover up to 40% of the landscape in some areas, and are important areas for cultivation and livestock grazing (Bell *et al.*, 1987).

In the more fertile lower catenary positions of slopes, open woodland dominated by *Combretum* species with a tall grass understorey occurs, and where there are poorly drained depressions, *Acacia*-dominated woodlands are present (Campbell *et al.*, 1996). Fire-induced tall grass savanna (*chipya*) with scattered fire-tolerant trees (e.g. *Pterocarpus angolensis*, *Burkea africana* and *Erythrophleum africanum*), and mixed woodlands and thickets dominated by *Combretum*, *Acacia*, *Afzelia quanzensis* and *Pericopsis*, occur on nutrient-rich soils derived from limestone and mica schists (Campbell *et al.*, 1996). Spinescent, evergreen thickets associated with high (up to 15 m) termitaria are found throughout the *miombo* region.

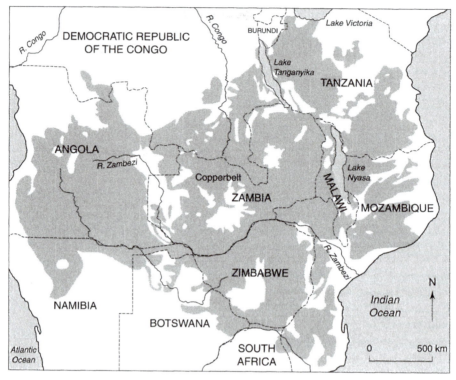

Figure 4.1 The distribution of *miombo* in central Africa.
(Reproduced by kind permission of Justin Jacyno)

4.1.2 Vegetation structure and composition

Structurally, *miombo* is quite uniform over large areas (Figure 4.2). The woody layer is typically composed of a continuous upper canopy of umbrella-shaped trees, scattered sub-canopy trees, and irregularly distributed understorey shrubs and saplings. Canopy height is normally 10–20 m, but where scrub *miombo* occurs, it can be as low as 3 m (Malaisse, 1978a). The herbaceous layer is generally 1–2 m tall, and dominated by annuals, with C_4 grasses and sedges making up 90% of the biomass (Chidumayo, 1993a).

The *miombo* vegetation is characterised by the presence of broad-leafed, thornless trees of the genera *Brachystegia, Isoberlinia* and *Julbernardia* (family Fabaceae, subfamily Caesalpinioideae), and a herbaceous layer comprising mainly Andropogoneae species (Frost, 1996). The diversity of the canopy tree species is low, although the overall species richness of the flora is high. It is estimated that the *miombo* region has around 8500 species of higher plants, over 54% of which are endemic. Of these, 334 are trees, and Zambia has not only the highest diversity of trees, but is also the centre of endemism for *Brachystegia,*

Figure 4.2 The *miombo* vegetation.
(Reproduced by kind permission of Paul Desanker, University of Virginia)

with 17 species (1 in Kenya, 6 in south-eastern Tanzania, and 11 in western Tanzania) (Rodgers *et al.*, 1996). Species diversity and localised endemism is high in many herbaceous genera, such as *Crotalaria* (over 200 species).

The present-day distribution of *miombo* reflects its history, principally past climatic changes and the history of anthropogenic activities in the region. Although there is a paucity of well-preserved and accurately dated sediment cores, *Brachystegia* pollen has been found in various sites near the centre of the *miombo* distribution as well as beyond the southern limit of *miombo* woodland, with dates of up to 38 000 years BP (Vincens, 1991). Studies suggest that prior to the last glacial maximum, *Brachystegia* was widespread across central Africa (Scott, 1984). However, the onset and duration of the glaciation, around 18 000 years BP, probably caused air temperatures to fall by about 3–5 °C, and drier conditions to prevail, with rainfall 250 mm lower than today (Vincens *et al.*, 1993). This may have been accompanied by a shift in the *miombo* vegetation to a more open physiognomy, dominated by Compositae, *Cliffortia* and *Cyathea*, all common in the present-day montane flora of central Africa (Scott, 1984). It seems, therefore, that the *miombo* woodland of the central plateau was probably displaced to lower altitudes as the climate became cooler and drier.

With the amelioration of climate during the Holocene, *Brachystegia* once more became a prominent component of the plateau vegetation. The late Holocene saw subsequent expansions and contractions of *Brachystegia*, with a marked northward contraction in the past 1000 years from sites more than 400 km south of its present distribution in Zimbabwe (Scott, 1984). This may have been the result of a sudden change in climate during that period, of which there is no evidence, or relatively small shifts in temperature and moisture may have adversely affected *Brachystegia* populations at its geographical limits (Scott, 1984).

Plant life strategies

Tree crown cover ranges from 60 to 90%, and 12–20% of this is made up of two or three crown layers of overlapping trees of similar height or of trees at different heights (Araki, 1992). Leaf biomass ranges from 200 to 300 g m^{-2}, implying a Leaf Area Index (LAI) of 2–3 m^2 m^{-2}. *Miombo* trees are briefly deciduous, with leaf fall peaks during August–October while leaf flush occurs during September–November. In drier areas, trees may be completely deciduous while in moister areas they may be virtually evergreen (Campbell *et al.*, 1996).

Flowering in the *miombo* takes place throughout the year, but peaks towards the end of the dry season, after which seeds are dispersed (Malaisse, 1974; Chidumayo, 1993a) (Figure 4.3). There is a great diversity in fruit types and dispersal mechanisms among *miombo* trees, with the dominant canopy trees, including *Brachystegia, Isoberlinia* and *Julbernardia*, dispersing seeds by explosive dehiscence of the pod (Chidumayo and Frost, 1996). In these species, dispersal distance is limited to around 20 m, although seeds of wind-dispersed species such as *Albizia* and *Pterocarpus* have been found up to 103 m away (Malaisse, 1978a).

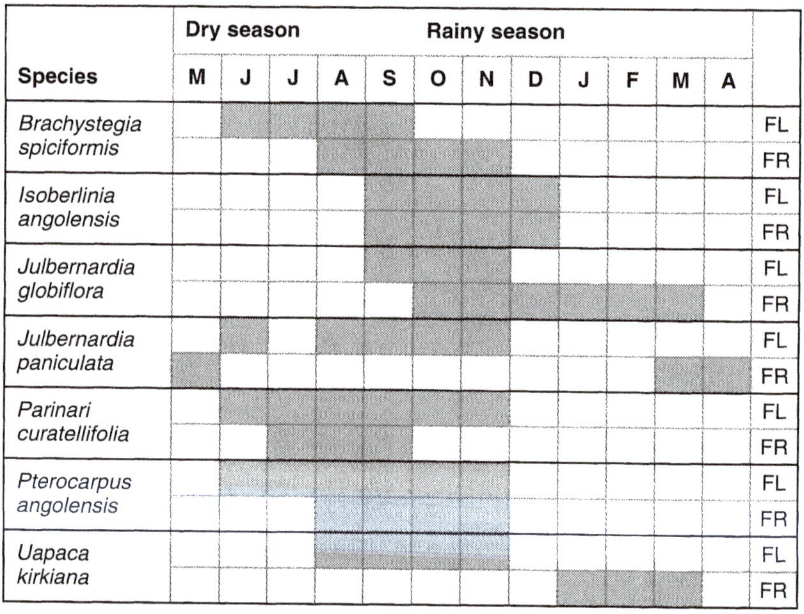

Species	Dry season					Rainy season							
	M	J	J	A	S	O	N	D	J	F	M	A	
Brachystegia spiciformis													FL
													FR
Isoberlinia angolensis													FL
													FR
Julbernardia globiflora													FL
													FR
Julbernardia paniculata													FL
													FR
Parinari curatellifolia													FL
													FR
Pterocarpus angolensis													FL
													FR
Uapaca kirkiana													FL
													FR

Figure 4.3 Reproductive phenology of some *miombo* trees. FL = flowering, FR = ripe fruits (after Chidumayo, 1997).

Seed germination rate among *miombo* trees, as determined by seedling emergence, ranges from 10 to 99%, and the germination period for most seeds is the first 2–6 weeks, after which most die and decompose (Chidumayo, 1993a). Studies have suggested that seed dormancy may be rare in many woody species of *miombo* woodland (Ernst, 1988). This led to the hypothesis that the banks of stunted seedlings in the grass layer may be the major source of regeneration in *miombo*, and a critical stage in the maintenance of woody plants (Boaler, 1966; Chidumayo, 1989a). These tree seedlings experience a prolonged period of successive shoot die-back, attributed to drought and fire (Trapnell, 1959; Boaler, 1966; Chidumayo, 1989a, 1990, 1991, 1992), adding to their normal slow growth rates (Chidumayo, 1992).

However, stumps and roots of almost all *miombo* trees produce sucker shoots once the above-ground parts have been removed or killed (Boaler and Sciwale, 1966; Strang, 1974; Banda, 1988), and these grow faster than the shoots of old stunted seedlings (Chidumayo, 1993a; Grundy, 1995a). Generally, removal of the canopy allows both stump regeneration and seedling growth (through higher light levels), and as a result, regrowth *miombo* has a higher tree density (Fanshawe, 1971). As the age of *miombo* regrowth increases, the number of shoots decreases as a result of inter-shoot competition, and therefore stem density per plant declines slowly with time (Chidumayo and Frost, 1996).

4.1.3 Fauna

The *miombo* and associated vegetation types support a large range of animal species, including large mammals such as elephant (*Loxodonta africana*), Liechtenstein's hartebeest (*Alcelaphus lichtensteini*), sable antelope (*Hippotragus niger*), buffalo (*Syncerus caffer*), impala (*Aepyceros melampus*), black rhinoceros (*Diceros bicornis*) and wildebeest (*Connochaetes taurinus*). Carnivore species include the leopard (*Panthera pardus*), and the *miombo* is rich in bird-life. Invertebrate species are also diverse, most conspicuous of which are the termites, whose mounds are found dotted over the landscape.

4.2 The main savanna determinants

The dynamics of *miombo* woodlands have been interpreted largely in terms of a single-state equilibrium model of succession to a regional climax vegetation, with fire, humans or wildlife being the main agents of disturbance (Strang, 1974; Lawton, 1978). Of late, multi-state models have been used to describe *miombo* dynamics, in which, following either cultivation field abandonment or the combined impact of elephants and fire, there is a transition to open woodland dominated by *Combretum* species, or to grassland due to fire (Stromgaard, 1986; Starfield et al., 1993). A longer than normal fire-free period is needed to return the vegetation to *Brachystegia*-dominated woodland. Unfortunately, many of these studies have taken place in a number of sites presumed to differ in only one factor, e.g. fire regime, or have been short-term studies (Trapnell, 1959; Lawton, 1978; Chidumayo, 1988a). The latter are unlikely to produce many insights given that most *miombo* tree species are relatively long-lived with life-spans of between 60 and 120 years (Frost, 1996). Vegetation change in response to various factors can only be seen through long-term monitoring at a range of sites, and through the use of simulation models of plant community dynamics, which are currently being developed by researchers (e.g. Desanker and Prentice, 1994).

4.2.1 Plant–available moisture

The *miombo* region can be delimited by an average annual rainfall of between 600 and 1500 mm, and a dry season lasting six to seven months. However,

Table 4.1 Some characteristics of wet and dry *miombo* (based on Frost, 1996)

Characteristics	Wet *miombo*	Dry *miombo*
Annual rainfall (mm)	> 1000	< 1000
Occurrence	Over much of eastern Angola, northern Zambia, south-western Tanzania and central Malawi	Over Zimbabwe, central Tanzania, and the southern areas of Mozambique, Malawi and Zambia
Canopy height (m)	> 15, reflecting the generally deeper and moister soils	< 15
Floristic diversity	High	Low
Dominant woody layer species	*Brachystegia boehmii, B. floribunda, B. glaberrima, B. longifolia, B. spiciformis, B. wangermeeana, Julbernardia globiflora, J. paniculata, Isoberlinia angolensis, Marquesia macroura*	Dominant species of wet *miombo* absent or local in occurrence. *Brachystegia spiciformis, B. boehmii, Julbernardia globiflora* dominate
Dominant herbaceous layer species	Varies. Includes grasses of genera *Hyparrhenia, Andropogon, Loudetia, Digitaria* and *Eragrostis*, sedges, leguminous shrubs such as *Eriosema, Sphenostylis, Kotschya, Dolichos* and *Indigofera*, and suppressed canopy tree saplings	Mixture of mainly C_4 grasses such as *Hyparrhenia, Andropogon* and *Loudetia* species, bracken (*Pteridium aquilinum*) and shrubs, e.g. *Aframomum biauriculatum*
Mean woody biomass (Mg DM ha^{-1})*	90	55
Rate of above-ground biomass production (Mg ha^{-1} year^{-1})	2.2–3.4	1.2–2.0

* DM = dry matter.

temperature may be more important than precipitation in determining the southern limit of its distribution. *Miombo* trees are frost-sensitive, unable to tolerate absolute minimum temperatures of less than −4 °C. This may be the significant factor determining the southern limit of *miombo* distribution in the Zambezian phytoregion (Werger and Coetzee, 1978).

Structural differences and species compositional changes occur along rainfall gradients, from the drier fringes of the *miombo* region, to the wetter core areas. White (1983) divided *miombo* into wet and dry types. Some characteristics of these two forms are shown in Table 4.1.

Soil type is an important factor determining PAM. Soils in the *miombo* are of eluvial origin, found on basement quartzites, schists and granite rocks (Fanshawe, 1971). The soil texture is sandy loam, sandy clay loam and sandy clay (Chidumayo, 1993a), indicating a gradient of soil moisture characteristics, particularly of water drainage. Indeed, Savory (1962) found that this edaphic and soil moisture gradient determines the distribution of *miombo* species, with certain species having specific preferences (Table 4.2).

Table 4.2 Edaphic preferences by *Brachystegia* spp. in the Zambian Copperbelt (based on Savory, 1962)

Species	Soil preference
Brachystegia boehmii	Clay loam
Brachystegia floribunda	Heavy-textured soil
Brachystegia longifolia	Deep sandy soil
Brachystegia spiciformis	Deep well-drained soil
Brachystegia utilis	Deep loams

Particular *miombo* species also have preferences in respect of topography and soil moisture. *Brachystegia boehmii* and *Parinari curatellifolia*, for example, are indicators of shallow soils with partial waterlogging or a high water-table, whereas other species, such as *Brachystegia microphylla* in wet *miombo* and *Brachystegia glaucescens* in dry *miombo*, prefer well-drained rocky habitats of the higher altitudes. Since edaphic changes can occur over short distances, the result is a complex pattern of species dominance, with hills and escarpments having a greater small-scale substrate heterogeneity and a more complex sequence of species dominance than plateau landscapes (Chidumayo, 1997).

Miombo species have both horizontally and vertically extensive root systems (Malimbwi *et al.*, 1994). The maximum lateral root has been recorded for *Julbernardia globiflora*, at 27 m (Strang, 1965), and most of the dominant trees have tap roots exceeding 5 m in depth.

4.2.2 Plant-available nutrients

Miombo soils have low concentrations of organic matter, macro-nutrients and exchangeable bases, with levels generally higher in dry than in wet *miombo* (Stromgaard, 1984; Chidumayo, 1993a). The low levels of organic matter are probably a result of the preponderance of low-activity clays, high soil tempera-tures, frequent fires, and the abundance of termites (Trapnell *et al.*, 1976; Jones, 1989). The last of these is particularly important. Tall mounds of *Macrotermes* termites occur at densities of up to 5 ha^{-1}, covering approximately 8% of the surface area (Malaisse, 1973). Estimated to have a biomass of as much as 46 kg DM ha^{-1}, *Macrotermes* species, together with other termites, contribute to the mosaic distribution of nutrient-rich patches, within the nutrient-poor *miombo* landscape (Malaisse *et al.*, 1975).

Organic matter is collected by the foraging termites and deposited in their mounds, where fungi complete decomposition. As a result, termite mounds have higher levels of total N, acid-extractable P and cations, than surrounding soils (Trapnell *et al.*, 1976; Watson, 1977; Jones, 1989). Hence, they also support a distinct vegetation type (Fanshawe, 1968; Malaisse, 1978b), and are often the focus for animal activity. Termitarium soil is widely used by farmers as an amendment to their fields (Watson, 1977).

Adaptations to PAN

Adaptations to poor nutrient levels, and the probable higher nutrient leaching occurring in the wet *miombo*, results in a strong internal recycling of limiting nutrients within *miombo* trees. Chidumayo (1994a) measured a 30% with-drawal of foliar nitrogen (N) and 80% of foliar phosphorus (P) at the end of the growing season (February to April), followed by a further 30% removal just before leaf fall, ensuring conservation of valuable nutrients. *Miombo* trees have deep tap roots and extensive lateral roots, which may also capture nutrients lost from the upper soil levels through leaching (Chidumayo, 1997).

Miombo is notable among other savanna woodlands for the number of tree species having ectomycorrhizae (ECM) rather than vesicular–arbuscular myc-orrhizae (VAM) associations (Högberg, 1982, 1992; Högberg and Piearce, 1986) (Figure 4.4). These include species of the dominant *Brachystegia*, *Julbernardia* and *Isoberlinia* genera, species of *Marquesia* and *Monotes* (Dipterocarpaceae) and *Uapaca* (Euphorbiaceae). ECM have an advantage over VAM in infertile soils as they enable plants to take up nutrients direct from the litter. The mycorrhizae depend on carbohydrates supplied by the host plant, but the cost of maintenance to the plants has not been measured in *miombo*, and could be substantial (Högberg, 1982).

4.2.3 Herbivory

The *miombo* supports a low biomass of wild herbivores, made up of large-bodied species such as elephant (*Loxodonta africana*), buffalo (*Syncerus cafer*), and selective grazers such as Lichtenstein's hartebeest (*Alcelaphus lichtensteini*), sable (*Hippotragus niger*), and roan antelope (*H. equinus*). Specialist ungulate browsers are rare. This reflects the poor nutritional quality of the foliage. The nitrogen content of mature leaves of non-nodulated canopy trees is about 1.9%, which is significantly lower than that of the few nodulated, N-fixing tree species at 2.7%. The average phosphorus level is 0.17%. Grass nutritional quality is

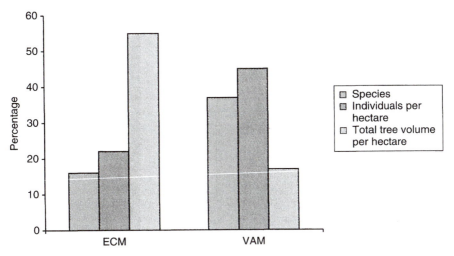

Figure 4.4 Root mycorrhizal associations of woody species (diameter at breast height greater than 4.0 cm) in *miombo* woodlands, Tanzania. The contribution of mycorrhizae is expressed in terms of species, individuals per hectare and total tree volume per hectare. ECM = ectomycorrhizae, VAM = vesticular arbuscular mycorrhizae (after Högberg, 1992).

lower than that of woody leaves for much of the year: 1.3–2.2% N during the early growing season, dropping to 0.5–0.8% N in the early dry season (Campbell *et al.*, 1996).

Foliage quality is also affected by the amount of structural carbohydrates and concentrations of secondary chemical compounds. Lignin levels are low, less than 8% (Jachmann, 1989; Mtambanengwe and Kirchmann, 1995), and total polyphenol content varies between 1 and 19%, non-N-fixing species having higher levels on average (10.2%) than N-fixing species (6.8%) (Palo *et al.*, 1993). As a consequence of foliage quality, the level of consumption by large herbivores is low, estimated at about 1% year^{-1} of available browse (Martin, 1974).

Invertebrates, mainly lepidopteran larvae, deplete greater quantities of foliage, around 2–4% year^{-1} of annual tree leaf production in normal years (Frost, 1996). However, patterns of consumption by invertebrates are different to mammals in relation to time and space. For example, invertebrates are more active in the wet season, and feeding tends to be distributed more uniformly among preferred food plants at a site, usually those of the canopy. Year to year variations in consumption are associated with periodic outbreaks of lepidopteran larvae (Malaisse-Mousset *et al.*, 1970). It is hypothesised that secondary chemical compounds may be responsible for controlling invertebrate herbivory.

Mammals, on the other hand, are normally active throughout the year. Except for elephant, most mammals feed on herbaceous and understorey plants, and feeding is patchy, dependent on plant species and parts selectivity/ specificity, and the daily or seasonal movements of the animals (Martin, 1974).

Herbivory by elephants

Elephants are well known for their habit of ring-barking, breaking, pushing over and uprooting *miombo* trees and shrubs (Thomson, 1975; Guy, 1989). Although previously thought to be more of a social display (e.g. Guy, 1976), recent evidence indicates that these actions are actually related to food specificity. Grasses are the preferred food source for elephants, but the period during which they can extensively utilise this food is limited by the long dry season. Therefore, elephants switch from grass grazing in the wet season, to foliage browsing in the dry season (Anderson and Walker, 1974; Barnes, 1982). During this time, mature individuals of particular *miombo* woody species, e.g. *Brachystegia boehmii*, are selectively pushed over. Selection for these species by elephants in Malawi was significantly positively correlated with the sugar and mineral content of tree leaves, and negatively correlated with indigestible lignin and secondary chemical compounds (Jachmann and Bell, 1985; Jachmann, 1989).

Tree and shrub damage by elephants can affect vegetation structure, transforming typical *miombo* woodland into a more open shrub-land with scattered tall trees and a higher grass biomass (Thomson, 1975; Guy, 1981, 1989) (see Section 6.2.3). This can lead to increased dry season fuel loads, more frequent and intense fires, and further woody plant suppression. However, once areas have been burnt, elephants tend to avoid them, probably because of the unpalatable scorched leaves and twigs, so less damage is done to trees there (Bell and Jachmann, 1984). This effect is greatest in early burnt areas as trees are protected for a longer proportion of the dry season. The influence of early burning on elephant movements and browsing pressure may therefore be significant for management. However, it has to be noted that burning may not always be a viable option, as burning could force elephants to browse in other undisturbed protected areas.

4.2.4 Fire

Fires occur regularly and frequently in *miombo*, with evidence of the use of fire dating back to the Early Stone Age, some 60 000 years ago (Phillipson, 1971; Clark, 1975). Fires in the *miombo* probably constitute the single largest area burned in the world – around 1 million km^2 $year^{-1}$ (Scholes *et al.*, 1996a). Dry season fires occur at most sites at 1–3 year intervals, although this depends on fuel accumulation rates, and on proximity to potential sources of ignition.

The causes of fire

A study monitoring the occurrence of fire at four dry *miombo* sites in central Zambia from 1990 to 1993 found that 15% occurred in August, 39% in September and 46% in October, with a mean frequency of 1.6 years (Chidumayo, 1993a). This has been corroborated by long-term monitoring of *miombo* forest plantation fires in the Zambian Copperbelt, which showed that the highest frequencies of fire occurred during August–October. Of these fires, few are caused by lightning. The majority are anthropogenic.

The uses of fire today have changed little from traditional practices. Most fires in the *miombo* region occur to stimulate pasture regrowth for livestock grazing, to clear land prior to cultivation, and for the collection of honey. The patterns of fire occurrence are not just related to the environment, but also to the cultural and socio-economic activities of the *miombo* inhabitants. For example, in the *chitemene* shifting cultivation region of Zambia (see Box 4.2), the annual cycle of *chitemene* activities restricts burning of wood piles close to the onset of the rainy season. Bush burning is discouraged in the early dry season so as to protect crops, and later in the dry season to protect *chitemene* wood piles until the kindling season in September–October.

Factors affecting fire

Climate
Fires generally occur during the *miombo* dry season of April to November. This period is characterised by high daily temperatures of between 27 and 34 °C, a relative humidity normally below 20% (sometimes falling under 10% towards the end of the dry season), and high average wind speeds of $13–16 \text{ km h}^{-1}$. These hot, dry and windy conditions are ideal for the propagation of fire, and so the risk of fire is greatest at this time (Chidumayo, 1997).

Fuels
The combustible fuel in the *miombo* consists of litter (small twigs and leaves), standing dead grass, and other herbs. In young *miombo* regrowth (under 6 years old), the biomass fuel is $7–12 \text{ t ha}^{-1}$, of which 55–70% is leaf litter and grass, and a significant proportion made up of wood debris, persisting from the time of felling (Chidumayo, 1997). With longer regrowth, the total fuel biomass decreases, with the proportion of herbaceous fuel increasing. For example, in old-growth *miombo*, flammable biomass ranges from 5 to 8 t ha^{-1}, 70–90% of which is leaf litter and standing grass.

The annual grass production is estimated by the mean peak standing biomass at the end of the growing season, and was found to be 1150 kg ha^{-1} in dry *miombo* and 2006 kg ha^{-1} in wet *miombo* (Chidumayo, 1993a). Maximum leaf litter fall occurs in August and September, with cumulative litter biomass reaching $2.5–3.0 \text{ t ha}^{-1}$ by the end of each year (Chidumayo, 1997). This suggests that more litter fuel is available at the end than at the beginning of the dry season, a factor which will account for differential burning regimes during the dry season. In comparison to leaf litter and grass, the annual fall of wood and pod litter in *miombo* is small, and will therefore have a less significant effect on fire. For example, in dry *miombo*, average wood litter fall is about 31 kg ha^{-1} in old growth, and only 8 kg ha^{-1} in young regrowth (Chidumayo, 1997).

The nature of the fire will depend on the timing of the burn, which is related to the fuel moisture content, the amount of fuel, and the prevailing climatic conditions. As the dry season progresses, fuel moisture content becomes progressively lower (lowest values of under 7% of dry weight occur from August to October), fuel biomass increases, and together with hot, dry and windy weather becoming dominant, fires are generally faster, of a higher combustion factor, and of a higher intensity in the late dry season compared to the early dry season. Fuel loads, and thus fire regimes, are also a function of grazing pressure. Fires tend to be more frequent and intense in areas of low woodland cover and high mean annual rainfall, where grass production is high but grass quality is low, therefore reducing grazing pressure.

The extent of fires and the homogeneity of burns in *miombo* have not been adequately studied. Mapping the extent of 13 fires at four dry *miombo* sites, Chidumayo (1997) found that, on average, each fire burned 75% of the area, regardless of the month in which the fire occurred during August–October.

Plant adaptations to fire

Miombo plants of both the herbaceous and woody strata have a number of attributes to ensure their survival from fire. In the herbaceous layer, annual plants usually survive fires through seeds buried in the soil. Seeds are also protected inside indehiscent fruits, such as in *Gardenia subacaulis*, where the seeds inside a woody fruit are only released during the rainy season once the fruit has decomposed (Medwecka-Kornas, 1980). Dormancy during the dry season is a common feature among many herbaceous species. This is normally broken by fire (Medwecka-Kornas and Kornas, 1985).

In some herbaceous *miombo* plants, the above-ground parts die back during the dry season, with the perennating buds, present either below-ground or at ground level, protected from fire by old leaf or stipe bases. Examples of this protection method can be observed in perennial bunch grasses (Graminae), sedges (Cyperaceae) and ferns (Pteridophytes). For many *miombo* under-shrubs, fire stimulates the production of vigorous new shoots and flowers, after killing the slightly lignified or herbaceous above-ground shoots. Species such as *Annona stenophylla*, *Gardenia subacaulis* and *Lannea edulis* have below-ground woody rhizomes which serve as perennating organs just below the soil (White, 1976; Medwecka-Kornas, 1980).

In the woody layer, many *miombo* trees have thick bark to insulate inner living tissues and/or a high wood moisture content, to resist fire injury. Table 4.3 shows that large stems have a bark that is three to five times thicker than that of small stems, and may explain the susceptibility of saplings and small poles to fire damage compared to larger individuals. *Miombo* trees can be divided into three groups based on wood moisture content (WMC) and tolerance to fire: those with a high WMC (> 90%) have a higher tolerance to fire; those with a moderate WMC (60–90%) have an intermediate tolerance to fire; and those with a low WMC (< 60%) are sensitive to fire. The majority of *miombo* trees have a moderate WMC.

Leaf flush is very important for the deciduous *miombo* trees, and its timing affects plant growth and productivity. The leaf flush of the majority of dry *miombo* trees occurs in October and November, avoiding fire damage, but in a few species such as *Brachystegia spiciformis*, *Isoberlinia angolensis*, *Julbernardia globiflora* and *Protea* spp., leaf flush begins in August during the late dry season. This means that these species are more susceptible to fire throughout the dry

Table 4.3 Average bark thickness and wood moisture content (WMC) of some dry *miombo* trees in central Zambia (after Chidumayo, 1997)

| Species | Average bark thickness (mm) | | WCM (%) |
	Small stems (< 15 cm GBH)*	Large stems (> 15 cm GBH)	
Albizia antunesiana	3.2	11.4	77
Brachystegia boehmii	5.2	17.5	81
Brachystegia spiciformis	4.2	14.0	70
Burkea africana	2.8	13.5	82
Dichrostachys cinerea	–	–	39
Isoberlinia angolensis	4.2	12.6	82
Julbernardia globiflora	2.6	12.6	65
Monotes spp.	2.6	10.0	68[a]
Parinari curatellifolia	4.8	14.6	–
Pseudolachnostylis maprounefolia	1.3	8.1	78
Swartzia madagascariensis	–	–	58
Uapaca spp.	3.0	10.5	95[b]

[a] Based on data from *Monotes africanus* (Chidumayo, 1997).
[b] Based on mean values from *Uapaca kirkiana* and *U. nitida* (Chidumayo, 1997).
* GBH = girth at breast height.

season than those with a late flush. For example, August fires have little effect on *I. angolensis*, but September fires reduce its productivity.

The ability to vegetatively recover from fire injury is found in the majority of *miombo* trees. Partial or complete death of above-ground parts is normally followed by resprouting from dormant buds located in the branches, stem, root collar or roots, although this will depend on the degree of damage to the apical meristems and the conditions of the dormant buds.

The effects of fire on succession

Miombo plants have been classified into three main functional groups in relation to their fire tolerance: fire-intolerant, such as forest species and including fire-tender plants, i.e. saplings of the dominant species; semi-tolerant, i.e. those tolerant to early dry season, less intense fires; and fire-tolerant, i.e. species abundant in late burnt areas (Trapnell, 1959; Lawton, 1978). These could be used as indicators of different fire regimes and succession as governed by their absence/presence and abundance, though Chidumayo (1989a) warns of caution, as the lack of statistical analyses of the original data does not render the classification totally reliable.

Changes in fire frequency and intensity therefore change *miombo* structure and composition. Frequent late dry season fires gradually lead to an open, tall grass savanna with scattered fire-tolerant canopy and understorey trees and shrubs. In contrast, woody plants are favoured by both early burning and complete protection (Chidumayo, 1988a). Grass biomass declines as the canopy closes, and so a period without fire would lead to gradual grass suppression and lower fuel loads. This, in turn, means less frequent and intense fires, less woody plant damage and therefore continued canopy closure.

The replacement of *miombo* canopy by dry evergreen forest would be expected in the absence of fire when complete canopy closure had taken place, demonstrating the status of *miombo* as a fire sub-climax. However, this has not been found to occur (Trapnell, 1959; Chidumayo, 1988a). Furthermore, such a succession might only be possible in *miombo* where dry evergreen forest species are capable of initiating succession, a situation that is highly improbable in dry *miombo* which contains no evergreen elements. Because of these observations, Trapnell (1959) concluded that 'the *Brachystegia–Julbernardia* canopy in *miombo* exists not because of fires but in spite of them'.

4.3 Human determinant of the savanna

Much of the *miombo* region has been subject to intensive use, and little unmodified woodland remains (Dewees, 1994). Humans began to settle and utilise the *miombo* region on a large scale some 10 000 years ago. Activities intensified with the emergence of the Early Iron Age culture in the late Holocene, when agriculture, and permanent and semi-permanent settlements evolved. This was also a time of large-scale population movements, and AD 2–4 saw the arrival of the Negroid people into Zambia. Further large-scale migrations into the *miombo* region took place between 1500 and 1700, with the arrival of the Luba and Lunda people of Democratic Republic of the Congo, and in the nineteenth century the Ngoni from South Africa (Langworthy, 1971).

The invasion by the Ngoni people is among a number of historic events that have driven the changing patterns of *miombo* utilisation. The others are the pre-colonial long-distance caravan trade, the great rinderpest epidemic, colonisation, the introduction of the plough, the introduction of the market economy, growing rural populations, and post-independence restructuring (Misana *et al.*, 1996).

4.3.1 Pre-colonial land use

A generalised picture of pre-colonial *miombo* land use is of a region in which agricultural production was paramount, supplemented by other forms of natural resource exploitation (Box 4.1). Traditional forms of shifting (*chitemene*) and sedentary cultivation of staple crops such as millet and sorghum were probably practised, as they still are today (Box 4.2). However, subject to environmental adversities such as droughts and pests, and constrained by inadequate grain storage, hunting and gathering were also important subsidiary activities. Apart from the drier areas of Zimbabwe and Tanzania, livestock numbers were low (as they are today) because of the constraints of poor forage quality, and the presence of animal diseases, especially trypanosomiasis or 'nagana' among cattle, spread by the tsetse fly, *Glossina* species. Trypanosomiasis occurs in both chronic and acute forms in domestic animals and people, but is benign in trypano-tolerant wildlife, which thereby act as reservoirs of the disease (Frost, 1996).

Management was through communal property regimes, where arable land was allocated by traditional authorities to individual household heads, such land acquiring a quasi-privatised status, providing permanency of tenure (Cheater, 1990). Use of areas outside these lands for grazing, or other resources, were regulated, and demarcated to distinguish between rights of adjacent communities (Scoones, 1989). This was brought about through an ethno-ecological knowledge system, where the founding ancestral spirits were regarded as the 'owners of the land', controlling the environment and resource production. Rules were put in place governing the use of natural resources (e.g. the cutting of trees and use of fire), and any breaches in these regulations were indicated by ecological irregularities such as droughts, brought about by the spirits. Given the circumstances of the time, these regimes were economically viable, ecologically sustainable and organisationally efficient, compliance being internally generated rather than externally imposed.

Box 4.1 Traditional *miombo* resource use

The *miombo* has been a rich source of resources for its inhabitants since the upper Pleistocene period, and still is today (Musonda, 1986). In many parts of the *miombo* region, the traditional use of timber was for building construction, whether it be houses, huts or other structures (Lowore *et al.*, 1995). Poles of variable sizes were used, and as shown in Table 4.4, there are clear species preferences. Wood also provided an important source of energy, especially for heat generation and cooking. Food resources include plant leaves, roots, fruits and honey, as well as mushrooms and edible animals (Table 4.5). Some, e.g. the root of the orchid *Satyria siva*, are used to prepare a thick jelly relish known as '*chikanda*' (Chidumayo, 1997), while *Rhynchosia insignis* roots are used to produce a sweet drink locally called '*munkoyo*' (Njovu, 1993). Both are widely sold throughout Zambia.

Many *miombo* trees and herbaceous plants are also a source of medicinal remedies (Chihongo, 1993) (Table 4.6). Storrs (1979) documented 106 medicinal *miombo* trees, and found that the most common parts utilised are leaves (50%), bark (66%) and roots (74%), with wood (4%) and fruits and seeds (9%) less frequently employed. In Zimbabwe, leaf litter from *miombo* woodlands is used to supplement nutrients in arable lands belonging to small-scale farmers (Nyathi and Campbell, 1993). It is particularly valuable to farmers who have no access to other fertilising inputs or additional lands, and has, as yet, only been documented in Zimbabwe (Clarke *et al.*, 1996). Many tannins, dyes, oils, resins and gums are also extracted from *miombo* trees, but their use is local, and a large proportion of these products are unknown outside their area of use, although they could have commercial potential (Clarke *et al.*, 1996).

Box 4.1 *(Continued)*

Table 4.4 Tree species preferred for building purposes in *miombo* woodland by the Bemba of northern Zambia (after Chidumayo, 1997)

Purpose	Tree species
Poles	*Erythrophleum africanum, Monotes* spp., *Pericopsis angolensis, Swartzia madagascariensis*
Walling	*Brachystegia longifolia, Julbernardia globiflora, J. paniculata, Monotes* spp., *Uapaca kirkiana, U. nitida, Diplorynchus condylocarpon*
Roofing	*Marquesia macroura, Monotes* spp., *Pericopsis angolensis, Syzygium owariense, Uapaca kirkiana, U. nitida*
Mortar	*Albizia antunesiana, Isoberlinia angolensis*
Fibre	*Brachystegia boehmii, B. longifolia, B. taxifolia, Julbernardia globiflora*

Table 4.5 Wild foods of the *miombo* (based on Chidumayo, 1997)

Food type	Species
Herbaceous plant leaves	*Amaranthus hybridus, A. spinosa, A. thunbergii, Bidens pilosa, Celosia trigyna, Cleome gynandra, C. hirta, C. monophylla, Corchorus olitorius, Portulaca oleracea, Sesamum angustifolium, S. angolense*
Tree leaves	*Afzelia quanzensis, Fagara chalybea*
Tree fruits	*Anisophyllea pomifera, Annona senegalensis, Canthium crassium, Diospyros kirkii, Fiacourtia indica, Landolphia kirkii, Parinari curatellifolia, Strychnos cocculoides, S. pungens, Syzygium g. macrocarpum, Uapaca kirkiana, U. nitida, Vitex madiensis*
Mushrooms	*Termitomyces letestui* (termite mushroom), *Amanita zambiana* (Christmas mushroom), *Cantharellus* species (chanterelles)
Herbaceous plant roots	*Satyria siva* (orchid), *Discorea hirtiflora* (wild yam), *Rhynchosia insignis*

Hunting is a traditional activity in the *miombo,* and small game such as duiker, rodents, hares and birds, including francolins and guineafowl, are a common source of meat protein for subsistence needs (Clarke *et al.,* 1996). Protein is also derived from edible insects. Termites, particularly of *Macrotermes* species, dispersed during the rainy season, are captured and eaten (Defoliart, 1995). A variety of highly valued caterpillar species are

Box 4.1 *(Continued)*

also consumed (Malaisse, 1978a; Holden, 1991). For example, the common emperor moth, *Elephrodes lactea*, feeds on the leaves of *Brachystegia*, *Julbernardia* and *Isoberlinia*. Eggs are laid in September, and hatch in October, after which fast growth to maturity means that they can be harvested during November and December (Hobane, 1994). The *miombo* has been an important zone for honey production by bees, especially the species *Apis mellifera* (Banda and De Boerr, 1993). Honey collection often involves the cutting of host trees, and traditional bee-keeping is based on bark-hives which are produced by totally ring-barking a portion of the bole, which can cause tree mortality (Clauss, 1991; Fischer, 1993).

Table 4.6 Some *miombo* woodland medicinal trees (based on Storrs, 1982)

Species	Part used	Medicinal use
Afzelia quanzensis	Bark	Relieves toothache
Albizia antunesiana	Root	Prophylactic against colds and coughs
Cassia abbreviata	Bark	Antibiotic
	Root extract	Relieves toothache
Combretum molle	Leaf paste	Treatment of wounds and sores
Dichrostachys cinerea	Bark powder	Treatment of skin ailments
	Fresh leaves	Treatment of wounds and sores
Diospyros mespiliformis	Crushed root	Treatment of ringworm
	Crushed shoot	Treatment of wounds and sores
Diplorrhynchus condylocarpon	Pounded bark	Wound dressing
	Chewed leaves on forehead	Relieves headache
	Root extract	Cough remedy
Garcinia huillensis	Bark infusion	Aphrodisiac
Hymenocardia acida	Vapour from boiling leaves	Relieves headache
Kigelia africana	Ripe fruit	Purgative
Piliostigma thonningi	Chewed fresh leaves	Cough relief
Pterocarpus angolensis	Bark paste/ash	Treatment of skin ailments

---

Box 4.2 *Chitemene* agriculture

The traditional form of agriculture in the *miombo* region is '*chitemene*' (which means 'to cut'), a form of shifting cultivation (Richards, 1939; Peters, 1950; Trapnell, 1953). Of the different types of *chitemene*, the 'large-circle' method was prominent in pre-colonial times, and is practiced among the Bemba people of northern Zambia and south-eastern Democratic Republic of Congo today, where it has been recorded in detail (Richards, 1939; Allan, 1965; Puzo, 1978).

It involves first lopping and chopping trees and branches from an area (outfield) 8–10 times larger than the growing area (infield) (Chidumayo, 1987a; Stromgaard, 1989). Only about 30% of the above-ground woody biomass is extracted, thereby allowing rapid regeneration of the stumps and trunks to woodland. The wood is then piled together and burned just before the rainy season in October, so as to allow nutrients from the ash to be washed into the soil, and has led to the infield being popularly known as the 'ash garden' (Araki, 1992).

The ash provides a valuable source of potassium and phosphorus to the largely infertile *miombo* soils (Stromgaard, 1984), as well as increasing the soil pH and thereby lowering aluminium levels, a major constraint on crop production (Araki, 1993). The heat from the burn also has a sterilising effect on the soil, reducing soil microbes that would normally compete with crops for nutrients (Chidumayo, 1987a), and any pests or diseases present in the vegetation. Nutrient supplements, such as leaf litter, manure and ter-mitarium soil, may also be applied to combat declining soil fertility over time. Common staple crops grown in the ash garden include millet, maize and cassava, but vegetables and other plants are also grown. The abandon-ment of an ash garden is normally triggered by soil reacidification to the pre-burn level (Lungu and Chinene, 1993), and fields are left fallow for periods of about 25 years.

In contrast to the 10% in large-circle *chitemene*, a small-circle *chitemene* ash garden is only 4% of the cleared land. In the latter, branches are cut at breast height and piled in small circles or narrow long strips for burning, and cultivation is by hoe. Cultivation lasts for two years before abandon-ment (Stromgaard, 1989), and consequently is practised where there is increased population pressure and/or particularly infertile soils. Although *chitemene* is the dominant form of traditional agriculture, many other types of cultivation occur (Stromgaard, 1985). Table 4.7 illustrates the rich variety of cultivation techniques of the Bemba, indicating dynamism and adaptability.

Sustainable large-circle *chitemene* has been estimated at a population density of 2–4 per km^2 (Chidumayo, 1987a). This figure has obviously been exceeded, and has led to both further clearance of old-growth *miombo* (once every two years, and breast- or knee-height cutting instead of over-head lopping) and a decrease in the fallow period from about 25 to 12 years

Box 4.2 *(Continued)*

Table 4.7 Garden types and crops grown by the Bemba of northern Zambia (after Stromgaard, 1985)

Crops grown	Ubukula	Ichikumba bukula	Chikumba	Akakumba	Chikuka	Umunkumba	Imputa/Ibala	Imputa/Chifwani	Chibela	Ibala lya musalu	Mukanda	Fiputu	Chifwani
Millet (*Eleusine coracane*)	*	*	*	*	*	*			*				
Fish poison (*Tephrosia vogelii*)	*	*	*	*	*								*
Cassava (*Manihot esculenta*)	*	*	*	*	*	*	*	*	*				*
Groundnut (*Arachis hypogea*)							*	*	*				*
Pumpkin (*Cucurbita pepo*)								*			*		
African sugar cane (*Saccharum officinarum*)													*
Yam (*Dioscorea* spp.)						*			*				
Onion (*Allium cepa*)					*					*			
Calabash (*Crescentia cujote*)	*	*	*	*	*	*					*		
Tomato (*Solanum* spp.)								*		*	*		
Cucumber (*Cucumis sativus*)	*	*	*	*	*	*							*
Cow pea (*Vigna unguiculata*)							*		*	*			
Sorghum (*Sorghum vulgare*)													*
Sweet potato (*Ipomoea batatas*)							*	*					
Tobacco (*Nicotiana tabacum*)						*						*	
Bean (*Phaseolus vulgaris*)							*	*					
Banana (*Musa sapientum*)											*		

Garden types

Ubukula: big size, first-year ash garden
Ichikumba bukula: medium, first-year ash garden
Chikumba: small, first-year ash garden
Akakumba: smaller, first-year ash garden
Chikuka: very small, first-year ash garden
Umunkumba: big trunk burned
Ibala: village garden

Imputa: mounds in ibala or chifwani
Chifwani: any old ash garden
Chibela: grass turf cultivation
Ibala lya musalu: vegetable garden
Mukanda: refuse heap garden
Fiputu: flat quadrangular mound especially for tobacco

Box 4.2   *(Continued)*

Table 4.7   *(Continued)*

Crops grown	Ubukula	Ichikumba bukula	Chikumba	Akakumba	Chikuka	Umunkumba	Imputa/Ibala	Imputa/Chifwani	Chibela	Ibala lya musalu	Mukanda	Fiputu	Chifwani
							Garden type						
Maize (*Zea mays*)	*	*	*	*	*								
Cannabis (*Cannabis sativa*)												*	
Water melon (*Citrullus vulgaris*)							*						
Mango (*Mangifera indica*)											*		

(Stromgaard, 1985; Chidumayo, 1987a). In some areas, such as north-east Zambia, fallow periods are so short that grass and young coppice has replaced woodland as a natural fallow (Stromgaard, 1989). Here, farmers cut and compost grass in mounds, with excess grass heaped around tree stumps for later burning. Soil fertility is maintained by rotation planting: nitrogen-fixing legume crops such as beans or groundnuts are grown on the mounds; in the following season, the mounds are broken open, and the compost spread flat, upon which cereal crops, mainly maize and millet, are sown. This 'fundikila system' can normally sustain soil nutrients for 4–6 years, and has been estimated to be able to support populations with densities of 12–20 km^2.

The effect of trade routes

Until the end of the eighteenth century, trade in products was limited to iron tools and some exotic luxury goods assimilated into categories of bride wealth (Gray and Birmingham, 1970). But by 1825, active trade routes, such as the central route in Tanzania, were firmly established, dominated by Arabs and *wangwanas* (Arab half-castes and Swahili people from the coast who had converted to Islam), exchanging wares including cloth from India, firearms from Europe, ivory and slaves (Iliffe, 1979). Large numbers of people left their farms to work the routes, which led to demand for foodstuffs and goods such as iron hoes and ivory, and to massive clearing of *miombo* woodland both for expanding agricultural production and for woodfuel supplies to the iron-smelting kilns (Misana *et al.*, 1996). Traditional shifting cultivation systems, where woodland fallows were left for at least eight years, were shortened to bush and grass fallows (Lyaruu, 1995; Yanda and Mung'ong'o, 1995). This resulted in both a decrease in soil productivity, and an expansion in *miombo* clearing. Ivory exports rose during the nineteenth century (Koponen, 1994), and considering the destructive nature of elephants, probably led to regeneration of *miombo* in many areas.

Migrations and disease

The mid-nineteenth century saw the Ngoni people invade the *miombo* region, their main impact being a change to more permanent settlement patterns, resulting in localised extensive clearing of woodland. However, these territorial expansions were halted at the end of the century by the outbreak of the great rinderpest epidemic (Misana *et al.*, 1996). More than 90% of the region's cattle herd was wiped out, famines ensued, and many pastoralist populations perished. This eased the pressures on the land, allowing woodland regeneration, an ideal habitat for tsetse fly, and the proliferation of trypanosomiasis. Consequently, populations became more and more concentrated in smaller, safer areas, and probably increased local clearing of *miombo*.

4.3.2 The impact of colonisation on land use

With the colonisation of various *miombo* countries from the late nineteenth century, came the introduction of a wide range of policies and legal acts affecting land use (Misana *et al.*, 1996). Mass populations at risk of tsetse fly, e.g. in the Luangwa Valley in Zambia (Vail, 1977), were evacuated and resettled, leading to regenerating woodland in some areas and clearance in others. Large-scale clearing of *miombo* to eradicate tsetse fly was also initiated, and together with the introduction of the ox-plough, further encouraged the opening up of *miombo*.

Land tenure

At the same time, most colonial governments adopted land allocation and division policies which alienated the indigenous population and strongly influenced land use (Misana *et al.*, 1996) (Figure 4.5). The primary change was the appropriation of the best agricultural (i.e. most fertile with reliable rainfall) and advantageously located land (in terms of urban markets and service centres) to the white settler populations for both production and speculation (Moyo *et al.*, 1993). Concomitantly, communal lands were demarcated for the indigenous people, under titles such as 'native reserves', 'tribal trust lands' and 'communal lands'. For example, the Zimbabwean Land Apportionment Act of 1930, set aside $198\,539\,km^2$ to 50 000 white settlers and $117\,602\,km^2$ to 1 080 000 natives (Murphree and Cumming, 1993).

African wildlife reserves were also established in Zimbabwe, Zambia and Malawi, which further marginalised a large number of people, and increased landlessness (Moyo *et al.*, 1993). For example, by the late 1950s, 30% of African Zimbabweans were already landless. The overall result was overpopulation in communal lands, leading to severely deforested and degraded woodlands, and the reversion to woodland in settler areas where the land use was limited (Vail, 1977).

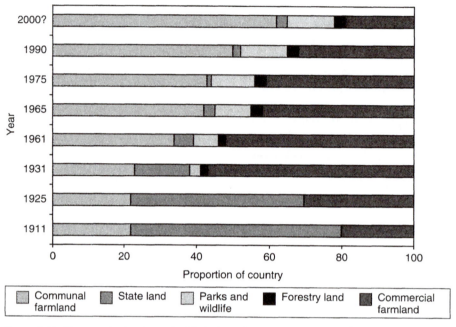

Figure 4.5 Changes in land use in Zimbabwe from 1911 to 2000. Resettlement land is included within communal land and small-scale commercial farm land within commercial farm land (after Murphree and Cumming, 1993).

Changes in agriculture and wood use

Colonisation also brought about an increase in semi-permanent and permanent forms of cultivation, in order to feed the growing urban populations (Misana *et al.*, 1996). Policies encouraged the intensified cultivation of export crops such as tea, tobacco, coffee and cotton, and led to accelerated conversion of wood-land to arable land (Moyo *et al.*, 1993). Activities such as mining, particularly in the Copperbelt area of Zambia, and the consequent rapid urbanisation, brought about a huge requirement for woodfuel from the surrounding wood-lands (Chidumayo, 1989b). Further deforestation took place to provide wood for the growing construction of roads and railways, and adjacent urban popula-tions, and to power locomotives (Misana *et al.*, 1996).

4.3.3 Present-day land use

What emerges in the above discussion is that the *miombo* has undergone periods of expansion and contraction in its history of human influence. Although Hannah *et al.* (1994) describe 61.7% as undisturbed (21.1% partially dis-turbed; 17.2% human dominated), with an annual population growth rate of 3.2% during 1980–1990, and ever growing commercial activities, human pres-sure is intensifying.

Agriculture

Most *miombo* countries have reinforced colonial agrarian systems based on cash-crops such as tea, coffee, tobacco and cotton, and large-scale farming, inherited with independence. Policy reforms in the 1980s were aimed at increasing agricultural production of both export and subsistence crops, and today, the agricultural sector is the dominant economy of the region. For example, in Zambia, 28% is *chitemene* cultivation, 12% intensive shifting cultivation, and 40% semi-permanent farming (Chidumayo, 1997). The growth in tobacco production has been particularly notable, for example accounting for 63% of total exports in Malawi in 1990. In central and western Tanzania, tobacco cultivation has increased from 228 000 ha in 1985/86 to an incredible 1 374 000 ha in 1991/92.

Higher levels of crop production are achieved in the semi-permanent system by higher inputs of inorganic fertilisers or animal manure, and ox-drawn and/or tractor soil preparation. Cultivation is for 5–10 years, and fallow periods are normally less than 10 years. This, together with tree uprooting during clearance, has led to little woodland recovery, soil erosion and water degradation. Although conversion of *miombo* to cultivation is a significant cause of deforestation in the region (e.g. 72% of cultivated land in Zambia is in *miombo* woodland), greater population densities (10–20 km^2) can be sustained by intensifying agriculture.

The introduction of high-yielding varieties of crops has had some negative impacts on the *miombo* ecosystem. For example, in the 1980s, hybrid maize production in the Northern Province of Zambia encouraged tree clearing (both cutting and uprooting), and soil tillage by either hand hoes, ox-drawn ploughs or tractors (Misana *et al.*, 1996). This working of the land brings more subsurface acidic soil to ground level, and with the encouraged use of cheap subsidised inorganic acid fertiliser, increases risks of land acidification. Furthermore, areas of maize production become uneconomic within four years, thereby requiring the same fallow period as in traditional *chitemene* cultivation. However, because of tree removal and lower cutting, recovery of woodland is much slower.

Pastoralism

Livestock husbandry in the *miombo* is low compared to other savanna regions, mainly because of the low forage quality, and the presence of trypanosomiasis. Nevertheless, some *miombo* areas have seen important increases in the bovine population. Cattle numbers in the Western Province of Zambia, for example, have increased from around 275 000 in 1964 to 500 000 in 1987 (Misana *et al.*, 1996).

As with indigenous herbivores, selective browsing by livestock can result in changes to vegetation structure and composition. Livestock readily browse woody regrowth of species such as *Brachystegia spiciformis* and *Julbernardia* spp., especially during the dry season when grass is of poor nutritional quality (Lawton, 1980). The effects depend on the type of browser and its foliage preference. For example, goats can suppress the regrowth of *B. spiciformis*, whereas cattle have a similar effect on *Julbernardia globiflora* and *Burkea africana* (Ward and Cleghorn, 1970). Repeated defoliation may make the plants less resistant to subsequent defoliation, and could result in death (Bryant *et al.*, 1991). The poor grazing potential of the *miombo* means that future growth in pastoralism may be unsustainable, and so the development of browse management techniques will be vital.

Wood consumption

Timber and crafts

Although timber still has traditional uses, today, commercial activities are increasing in the region (Clarke *et al.*, 1996) (Table 4.8). Trees such as *Pterocarpus angolensis*, and species of *Albizia*, *Faurea*, *Brachystegia* and *Julbernardia*, are selectively exploited in both the large-scale and small-scale manufacturing sectors (Brigham *et al.*, 1996). This selectivity tends to alter woodland structure, and reduce the genetic diversity of timber species because the best individuals with straight poles are selected for (Chidumayo, 1987b). Consequently, selectively felled areas are characterised by a high frequency of small stems.

Wood carvings and other hand crafts are another growing commercial activity in the *miombo* (Matose *et al.*, 1996). The most important species in this industry are *Pterocarpus angolensis* for carvings and musical instruments, and *Hyphaene petersiana* palm leaves and *Combretum zehyeri* roots for basket making. Little is known about the impact of these industries on the species, although Kotze (1993) has indicated that the long lateral roots (up to 20 m) of *C. zehyeri* individuals, harvested once a year, are sometimes cut too close to the root crown and the furrows left unfilled. Regrowth roots become exposed which may negatively affect the recovery of the harvested trees.

Fuelwood

Whether it be as firewood or charcoal, the *miombo* is a major source of woodfuel (Attwell *et al.*, 1989), and in countries such as Zambia, it accounts for 76% and 69% of the primary energy supply and final consumption, respectively. In rural areas, firewood is the primary energy source (Table 4.9), with each rural Zambian household consuming an estimated 5000 kg year^{-1}, compared to the 635 kg year^{-1} of urban households (Department of Energy, 1992). Rural subsistence firewood collection generally has a low and more local effect on *miombo* structure (within 300 m of villages) because only dead wood and specific species are collected (Grundy *et al.*, 1993). This wood is used mainly for domestic activities, as well as during cultural and social events. Some traditional beliefs

Table 4.8 Tanzanian hardwood exports from *miombo*, 1989–1994 (after Brigham *et al.*, 1996)

Value of exports (US$ million)			
1989	1991	1994	Main importing countries
1.6	1.4	4.1	Germany, Italy, Japan, Spain, UK, USA

Table 4.9 Woodfuel consumption (%) in different sectors in Zambia in 1990 (based on Chidumayo, 1997)

Sectors	Firewood	Charcoal
Rural households	79	11
Urban households	6	85
Industry	9	45
Agriculture	6	0

forbid the use of certain species, e.g. in Malawi *Psorospermum febrifugum* is thought to cause family conflicts (Lowore *et al.*, 1995).

With the emergence and growth of urban woodfuel markets came the cutting of live *miombo* trees for fuel (Monela *et al.*, 1993). In places such as the Copperbelt of Zambia, this was associated with the development of urban towns around large copper mines during the 1930s (Lees, 1962). The copper industry had always been dependent on *miombo* wood for electricity generation and running refineries. Between 1947 and 1956, 127 000 ha were clear-cut, and the copper industry still uses about 2000 t year^{-1} of charcoal in its refineries (Chidumayo, 1987c).

The introduction of charcoal in the late 1940s as a major urban household fuel has led to an ever-increasing demand for this energy source, which is now predominant in Tanzania and Zambia (Dewees, 1993; Brigham, 1994). Presently, urban households in Zambia use charcoal at an average rate of 1040 kg year^{-1}, compared to the rural household consumption of 100 kg year^{-1}. Primarily used for cooking, charcoal is derived mainly from *miombo* wood (Monela *et al.*, 1993), and in Zambia it is produced by the traditional earth-kiln method. A number of authors have outlined the techniques (e.g. Chidumayo and Chidumayo, 1984; Chidumayo, 1991; Hibajene, 1994), which involve tree felling, stem cross-cutting, kiln building and covering, wood carbonisation, kiln tending and kiln breaking to recover the charcoal. Production is typically small-scale (full- or part-time) in woodlands accessible to urban charcoal traders (Dewees, 1993). At present it requires little investment in the form of tools, and can subsidise the income of rural people by substantial amounts (Dewees, 1993; Monela *et al.*, 1993).

Charcoal production involves the use of about 94% of the above-ground wood biomass, and is therefore a significant cause of deforestation in *miombo* woodlands (Boberg, 1993; Monela *et al.*, 1993). For example, during the period 1937–1984, nearly 240 000 ha were cleared for woodfuel in the Zambian Copperbelt (Chidumayo, 1989b), and in 1990, at least 41 700 ha were deforested to provide urban areas with charcoal.

Forestry
Until 1984, the Forest Service in Zambia was responsible for the establishment of exotic forest plantations in forest reserves to supply industrial and construction wood products. The area under plantation increased from 1060 ha in 1964 to 22 500 ha in 1974 and 41 650 ha in 1980. It is projected that there will be

80 000 ha by the year 2000, largely consisting of *Pinus*, *Eucalyptus* and *Gmelina* species (Chidumayo, 1987b). Unfortunately little is known about the effects of plantation forestry on *miombo* or its recovery after logging.

Non-timber products

At a subsistence level, the harvesting of wild foods probably has little impact on the *miombo*; for example, fruits are generally only collected once they have fallen on the ground. However, with the emergence and expansion of wild plant food markets, particularly of wild fruits and *Rhynchosia insignis* roots, over-exploitation of these plants, involving tree cutting and total root removal, could seriously affect woodland structure and plant regeneration (Chidumayo, 1993a). Harvesting medicinal plant leaves rarely results in plant mortality, but bark and root harvesting can have damaging effects. In urban areas, the high demand for traditional medicines has led to fatal debarking of *miombo* trees such as *Pterocarpus angolensis* and *Cassia abbreviata* (Chidumayo, 1993a). In some areas, traditional sustainable extractive methods have been abandoned, especially for the more rare and/or lucrative species (Campbell *et al.*, 1993).

The increase in caterpillar harvesting has led to tree cutting, and clearings of up to 2 ha can be made where host trees are gregarious (Holden, 1991). Deforestation and hunting have reduced many game populations, but rodents, especially mice, and the granivorous bird, *Quelea quelea*, have grown in importance as a food source, particularly in arable landscapes (Campbell *et al.*, 1995; McGregor, 1995). Although there are examples of large-scale debarking for hive construction (e.g. up to 435 000 trees destroyed per year in the North-western Province of Zambia), the majority of beekeepers are concerned with woodland conservation and are receptive to management advice (Njovu, 1993).

4.4 Current management problems and strategies _____

The resources of the *miombo* are under increasing pressure as more and more areas are modified or transformed. The principal changes involve: the reduction in tree density through fuelwood harvesting, charcoal production, construction material, and the burning of wood for ash fertiliser; the decline in *miombo* cover due to frequent burning; and the conversion of *miombo* to permanently cultivated fields or, to a lesser extent, plantations of fast-growing non-*miombo* trees (Desanker *et al.*, 1997). The potential results of these changes include a decline in natural resource availability, and a range of environmental impacts affecting ecological functioning, carbon storage, trace gas emissions, hydrology and regional climate (Justice *et al.*, 1994).

The causes of *miombo* transformation are wide-ranging and complex. Expansion of the human population in the region has increased the demand for both agricultural and urban land, making it ever more difficult for sustainable land use (Misana *et al.*, 1996). Other factors include: deteriorating economic circumstances; inequalities in land distribution; insecure tenure; lack of credit; advances in agricultural technology and its commoditisation; commercialisation of natural products; expanding urban–rural linkages; declining terms of trade among less developed countries in world trade markets; and climatic adversities, e.g. droughts (Desanker *et al.* 1997). Many of these factors can only be resolved by policy-makers and government initiatives, although in many cases, local grass-roots community-based management schemes are being initiated.

4.4.1 Deforestation

The conversion of *miombo* woodlands to arable land, and the use of wood for fuel, are two major causes of deforestation in the region. Timber concessions continue to be a source of revenue for many governments in the region. Exploitation by private concessionaire companies has led to widespread degradation of *miombo*, particularly in Zambia, Angola and Zimbabwe. This has been exacerbated by the low royalty fees being charged and a variety of subsidies accorded to the concessionaires and woodfuel sellers, which do not correspond to the value of the wood, and encourage inefficient exploitation and the increased sale of licences by authorities to obtain badly needed revenue.

In some countries, such as Malawi and Zimbabwe, recent adjustments to stumpage prices do reflect the actual market value of the products and protect forests from further damage, although the tobacco industry is still heavily subsidised on indigenous wood use. Other factors contributing to deforestation include the movement of refugees. For example, in Tanzania, an influx of refugees from neighbouring Rwanda and Burundi has had a devastating impact on some *miombo* woodland areas in Kagera, Kigoma and Tabora regions. War has also contributed to the high rates of urbanisation in the region, further increasing the demand for fuelwood.

All the national forest policies in the *miombo* countries focus more on regulating woodland use and protecting against careless tree cutting and encroachment, than on sustainable use. A licensing system to control illegal forestry activities, such as commercial woodfuel collectors and charcoal-makers, does not work because policing is difficult over large areas. However, there are strategies for managing wood production at more local levels, and these involve traditional practices, natural regeneration and plantation regeneration.

Traditional woodland management practices

The most common and traditional conservation practice in *miombo* areas is to selectively retain some trees, or an area of woodland, in or near fields and villages at the time of clearing (Shepherd, 1992). This can be to provide shade, fruit and other products, but sometimes is simply because the trees are considered too hard to cut down, or have spiritual significance (Chidumayo, 1988b; Grundy *et al.*, 1993). Management of natural resources is most commonly and traditionally enforced by means of religious sanction, and there are various taboos on cutting individual trees or trees within reservations and sacred groves.

Bush fallows, i.e. natural woody regrowth on abandoned cultivated land, were common in traditional farming systems, but have become rarer because of land shortages (Campbell *et al.*, 1993). They are a useful source of many products such as wood, ash and browse, and can be enriched by inter-planting with other trees such as fruit trees (Chidumayo, 1988b). Exotic fruit trees are also raised from seed and seedlings, and in some areas, the nurturing of self-sown seeds is also carried out (Chidumayo, 1997).

Natural regeneration

The method of wood harvesting influences the characteristics of natural regeneration in *miombo* (Chidumayo *et al.*, 1996). For example, pollarding, used in woodland management in parts of southern Tanzania, and in Zambian *chitemene* shifting cultivation, hastens regeneration of biomass and provides good fallow, but the regrowth has low pole and timber value (Chidumayo, 1997). For this, stump coppicing (cutting at base) is better (Grundy, 1990). Selective cutting is predominant for specific wood products such as poles and timber, as well as the harvesting of edible caterpillars (Chidumayo *et al.*, 1996). However, this method can suppress the regeneration of stunted seedlings in the herbaceous layer, and may be the cause of the poor regeneration of valuable timber species such as *Pterocarpus angolensis* and *Baikiaea* spp. (Chidumayo, 1997).

Removal of the entire canopy with or without reserved trees, i.e. clear-cutting, is the best way of encouraging regrowth from stump coppices and seedlings (Chidumayo *et al.*, 1996). However, clear-cutting causes much more ecological damage than selective harvesting, particularly on catchment hydrology (Chidumayo *et al.*, 1996). The impact is heightened if cultivation and/or fire follow. To combat this, the Forest Department of Zambia has shown how a system of cleared strips alternating with shelterbelts orientated across the slope in forest reserves harvested for woodfuel may help to reduce damage (Serenje *et al.*, 1994). Once adequate regrowth occurs in the strips (usually after 10 years), the shelterbelts can be cleared, with the regrowth strips performing the role of shelterbelts (Serenje *et al.*, 1994).

Particular attention is now being given to the development of wood yield models for estimating potential timber production from regenerating *miombo* (Chidumayo, 1990, 1991, 1993c; Lowore *et al.*, 1994). Little is known of the growth rates and woody biomass yield of indigenous trees, which is one of the reasons for the plantation of exotic species, such as *Eucalyptus* (Chidumayo, 1988c). Therefore, investigating yield from *miombo* regrowth would be especially useful for sustainable exploitation and long-term management. Grundy (1995b), for example, has demonstrated a good relationship between some stem characteristics and total woody biomass in Zimbabwean *miombo*, from which reliable biomass tables have been constructed.

Plantation regeneration

Another way of sustaining the resources of the *miombo* is to plant *miombo* trees, especially where complete uprooting of trees has destroyed any natural regeneration. *Miombo* species show differential performance under plantation conditions. For example, *Brachystegia* and *Julbernardia* species have slow growth rates from seed, so perform better as coppice regrowth (Chidumayo *et al.*, 1996). The fruit trees *Parinari curatellifolia* and *Strychnos cocculoides*, on the other hand, grow better in plantations, whereas the timber tree *Pterocarpus angolensis* does well under both types of regeneration (Chidumayo, 1997).

The promotion of indigenous plantations has been dampened by the occurrence of various pests and diseases. Fungal outbreaks and insect attacks have seriously reduced the viability of plantations in the past (Piearce, 1993). Consequently, careful species selection, together with disease and pest control, will be vital for the success of future plantation silviculture (Chidumayo, 1993a).

With the increase in charcoal production in the *miombo*, regeneration through planting on charcoal 'spots' is currently being investigated (Chidumayo, 1988a, 1994b). These 'spots' refer to the kiln site after the charcoal has been recovered, and consist of burnt and charcoal soil (Lees, 1962). They normally cover 2–3% of the cut-over area (Chidumayo, 1993b). The altered edaphic conditions make seedling establishment difficult, and studies show that dominant *miombo* trees such as *Julbernardia*, *Brachystegia* and *Isoberlinia* have poor regenerative capacities in these areas (Chidumayo, 1994b).

Plantations may also be useful as a source of litter to improve soil fertility on already cleared land (Nyathi and Campbell, 1994). The dominant *miombo* trees contain low levels of nutrients in their leaf matter (King and Campbell, 1994), but nitrogen-fixing exotic tree legumes such as *Sesbania sesban* and *Leucaena leucocephala* var. *cunningham* produce high quality litter, and have a potential use in livestock feeding systems and as a source of soil macro-nutrients (Nyathi and Campbell, 1994).

4.4.2 Grazing management

Government policies on livestock in many *miombo* countries have emphasised colonial policies of destocking, particularly in areas of high livestock concentration, and bush clearing, to control tsetse flies. However, many pastoralists have refused to destock livestock, which is their chief source of livelihood, and government encouragement to migrate to less populated areas has caused conflicts with local farmers. Additionally, bush clearance for tsetse fly has hindered regeneration of *miombo*, and led to further woodland clearance.

Since rainfall variability is a major factor influencing herbage production, opportunistic rangeland strategies involve variable stocking rates and livestock mobility across extensive areas (Behnke *et al.*, 1993). However, this practice is becoming increasingly difficult because of the development of sedentary agriculture and strict veterinary regulations (Chidumayo *et al.*, 1996). Common methods of improving plant quality include the cultivation of forage legumes and multipurpose trees on rangelands, the collection of supplements such as wild fruits and nuts during the dry season, and the retention of important browse trees during woodland clearance (Chidumayo *et al.*, 1996).

Rotational grazing may also be necessary to ensure species diversity, conservation and sustainable production of browse for herbivores in *miombo* woodland. Where bush encroachment occurs due to prolonged grazing pressure, late dry season fires are used (Chidumayo, 1997). Since tsetse fly severely limits livestock rearing in some *miombo* areas, Chidumayo *et al.* (1996) suggest further research into the use of trypanosome-tolerant cattle.

4.4.3 Fire management

Fire is probably the most important problem in the *miombo*. There are no policies regarding fire management in the *miombo*, although in Zimbabwe, the use of fire in small-scale farming areas is prohibited by law (Clarke *et al.*, 1996). It is generally agreed that early dry season burning is one of the best ways of protecting and managing *miombo*. However, farmers burn pastures at the end of the dry season to promote new grass shoots for livestock, and in many *miombo* cultures, late season burning has symbolic significance (Schoffeleers, 1971). For example, in Malawian cosmology, the appearance and movement of the rains and bush fires appear to replicate a cyclical movement between sky and earth which is associated with a similar movement of the spirits. Prescribed burning may be a way of reducing damaging fires, but further work needs to address this issue.

4.4.4 Community–based management

The majority of people in the *miombo* region live on customary or communal land, and depend on the resources in these areas. However, the state controls all these lands and any decisions on its use. Increasing population in communal areas has resulted in widespread deforestation, while private land remains idle or fallow, particularly in Zambia and Zimbabwe. Privatisation of land has also reduced access rights to trees, and together with government protection of forest reserve resources, has put enormous pressure on communal lands. The latter has led to widespread conflicts throughout the region. Many local people believe they have the right to use forest reserves by virtue of their proximity or historical connection to the land.

Further conflicts have arisen as a result of areas being demarcated for conservation. Wildlife policies in Tanzania, Zambia and Zimbabwe have, until the 1970s, mainly focused on animal and habitat preservation through the establishment of conservation areas such as national parks. Such protectionist policies have excluded local people from these areas, limiting their utilisation of resources, and disregarding local hunting cultures. This has led to increased poaching and subsequent animal loss, especially elephants and black rhinoceros, and the encroachment on national parks for grazing, wood cutting and farming.

Recently, there has been a trend from centralised and state-driven woodland management of the colonial period, towards community-based natural resource management programmes (Matose and Wily, 1996). These were initially directed at wildlife use. For example, the CAMPFIRE programme of Zimbabwe granted 'appropriate authority' to rural councils under the Parks and Wildlife Act of 1975, and sought to ensure that revenue derived from wildlife, such as that obtained through hunting concessions and ecotourism, reached councils and communities, and not just the Treasury. This promoted investment in wildlife schemes and support from NGOs and the Department of National Parks and Wildlife Management (Matose and Wily, 1996).

The CAMPFIRE initiative provided a useful example to other *miombo* countries, and similar schemes have evolved (Rodgers and Salehe, 1996). Community-based local-level management is also increasingly being extended to woodland resources in general (Bradley and McNamara, 1993; Dewees, 1994). However, Matose and Wily (1996) state that the way forward for *miombo* management will additionally require a range of institutional changes. These could come about through policy reform, enabling legislation, and institutional development. Changes need to address community-based woodland management authority and implementation, and investment in natural resource management and sustainable utilisation.

4.5 Concluding remarks

There is a substantial amount of information available on *miombo* ecology and management, particularly in respect of vegetation characteristics and dynamics, and resource use. It is clear that the *miombo*, as with many other African savannas, has had a long history of exploitation, first by different peoples, and then by different land-use interests. Although there has been a shift from the centralised structure of past times to more community orientated management, success has been hindered by institutional constraints. Further problems will undoubtedly arise with the increasing impact of war in the region.

Key reading

Campbell, B. (ed.) (1996). *The Miombo in Transition: Woodlands and Welfare in Africa*, Center for International Forestry Research (CIFOR), Bogor, Indonesia.

Chidumayo, E.N. (1997). *Miombo Ecology and Management. An Introduction*, Stockholm Environment Institute, Stockholm.

Murphree, M.W. and Cumming, D.H.M. (1993). Savanna land use: policy and practice in Zimbabwe. In: Young, M.D. and Solbrig, O.T. (eds), *The World's Savannas. Economic Driving Forces, Ecological Constraints and Policy Options for Sustainable Land Use. Man and the Biosphere Vol. 12.* UNESCO, Paris, pp. 139–178.

Chapter 5

The savannas of West Africa

5.1 Introduction

5.1.1 Distribution, climate and soils

The West African savannas cover an area of about 4 950 000 km², and include portions of Senegal, Mali, Mauritania, Guinea, Ivory Coast, Togo, Ghana, Niger, Benin, Burkina Faso, Central African Republic, Nigeria, Chad and Cameroon (Figure 5.1). The region can be divided into three main zones: the Sahel, the Sudan and the Guinea (White, 1983). From the border with the desert region, the Sahelian zone forms a flat or gently undulating landscape below 600 m. Rainfall is irregular, low (less than 500 mm), and there is a dry season of 7.5–10 months. Except near the coast, the mean annual temperature is between 26 and 30 °C.

As you move southwards, rainfall increases, and seasonality becomes more distinct. Generally, the Sudan zone has a variable rainfall of around 500–1000 mm and a dry season lasting 5–7.5 months. The Guinea zone receives a mean annual rainfall of 1000–1750 mm, locally up to 2000 mm, and

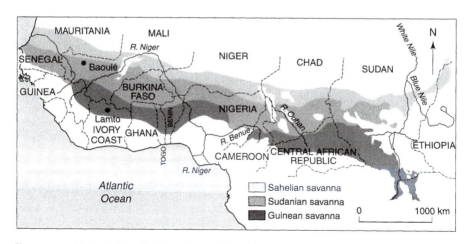

Figure 5.1 The Sahelian, the Sudanian and the Guinean savannas of West Africa.
(Reproduced by kind permission of Justin Jacyno and Kevin Dobbyn)

experiences a dry season of 2.5–5 months. Soil type varies between the zones, but in general soils are infertile, indicated by the low soil organic matter, cation exchange capacity, and nutrient content, particularly phosphorus. They are weakly structured, with a sandy surface but clayey compact subsoil, and a hardpan is common (Menaut *et al.*, 1985).

5.1.2 Vegetation structure and composition

The vegetation types of the three West African savanna zones are described below and shown diagrammatically in Figure 5.2. Note, however, that within these zones, and on their boundaries, there is a wide floristic and structural heterogeneity aligned to variations in the principal determinants. In general, trees and shrubs form an open woodland with a canopy of 40% or more, but never closed. Other vegetation types in the region include permanent swamps, floodplains and riparian forests. The latter are semi-evergreen in the south and semi-deciduous in the north, with many forests being degraded to woodlands as a result of human activity.

Sahelian vegetation

In the northern Sahel, where rainfall is less than 250 mm year^{-1}, a grassland with widely scattered bushes and small trees is prevalent, with a total crown cover of about 10% (White, 1983). The principal woody species are *Acacia tortilis, Commiphora africana, Balanites aegyptiaca, Boscia senegalensis, Leptadenia pyrotechnica, Acacia laeta* and *Acacia ehrenbergiana*. Most of these never reach heights taller than 2–5 m. Annual grasses are dominant, and include *Cenchrus biflorus, Schoenefeldia gracilis, Aristida stipoides* and *Tragus racemosus*.

In the driest areas, where rainfall is about 100 mm year^{-1}, certain desert grasses such as *Panicum turgidum* and *Stipagrostis pungens* become locally dominant, and all but a few woody plants, such as *Acacia tortilis*, are absent. The transition to desert grassland is not a gradual one related to rainfall, but is modified by local edaphic factors, particularly the relief of the sandy covering. Southwards, where rainfall is between 250 and 500 mm year^{-1} and sandy soils

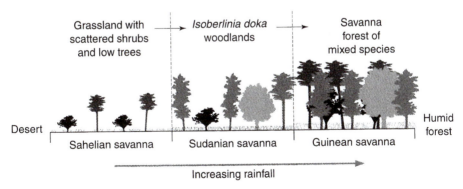

Figure 5.2 The vegetation of the different West African savannas.

prevalent, the vegetation becomes more woody, although still stunted. The same tree and grass species are present, although the perennial grass *Andropogon gayanus* can sometimes occur in almost pure stands on deeper sands.

Sudanian vegetation

The most common savanna formation in this region is the *Isoberlinia doka* woodlands. They dominate in the more southern areas of the zone, although locally, other species such as *Afzelia africana, Burkea africana, Daniellia oliveri, Erythrophleum africanum* and *Terminalia* spp. become important. Most species in this zone have a wide tolerance, rendering it difficult to recognise distinctive zones and vegetation patterns. However, some species characteristic of the drier northern parts include *Acacia albida, A. macrostachya, A. nilotica, Albizia cheva-lieri, Bauhinia rufescens, Commiphora africana, Lonchocarpus laxiflorus, Piliostigma reticulatum* and *Sclerocarya birrea*. Species occurring in the wetter southern region include *Acacia dudgeonii, A. gourmaensis, Maranthes polyandra, Monotes kerstingii, Terminalia glaucescens* and *Uapaca togoensis*. Tall grasses, both perennial species such as *Andropogon* spp. and *Hyparrhenia* spp., and annuals including *Andropogon pseudapricus, Ctenium* spp. and *Loudetia* spp., characterise the herbaceous layer.

Guinean vegetation

The northern part of the Guinea zone is characterised by communities generally of only one tree species (as in the southern Sudan zone), typically *Isoberlinia doka*, but also *I. tomentosa, Monotes kerstingii* and *Uapaca togoensis*. The south-ern Guinea zone, on the other hand, has a greater mixture of tree species, including *Afzelia africaba, Burkea africana, Butyrospermum paradoxum, Daniellia oliveri, Hymernocardia acida, Lophira lanceolata, Parinari polyandra, Pterocarpus erinaceus, Terminalia glaucescens* and *Vitex doniana*. Grasses are gen-erally taller in the southern part, growing up to 1.5–3 m. Common species include *Beckeropsis uniseta, Monocymbium ceresiiforme, Andropogon gayanus, A. pseudapricus*, with other species of this genus, and of the genera *Hyparrhenia* and *Pennisetum*.

Plant life strategies

The annual cycle of vegetation is controlled by the dry season which occurs from December to February. The growth of woody plant species occurs throughout the year, but most top growth takes place during the wet season. In the woody plants, most leaf fall occurs during the dry season, but is spread over a period that varies from year to year depending on the occurrence of fire, which stimulates more rapid leaf fall (César and Menaut, 1974). Based on the period of the year and the length of time woody plant species maintain green leaves, woody plants of the West African savanna have been divided into three main phenological groups: (1) evergreen species; (2) semi-evergreen species; (3) deciduous species (De Bie *et al*, 1998) (Figure 5.3).

Evergreen species have no fixed period for reproductive activities, and can flower at any time of the year, or continuously bear flowers, e.g. *Khaya*

Phenological groups	Wet season				Early dry season				Late dry season				Examples of species
	J	J	A	S	O	N	D	J	F	M	A	M	
Evergreen species, riparian													*Acacia albida, Daniellia olivieri, Dichrostachys cinerea, Sesbanea sesban*
Evergreen species, upland													*Boscia senegalensis, Ficus glumosa, Syzygium guineense, Vitex simplicifolia*
Semi-deciduous species													*Diospyros mespiliformis, Piliostigma thonningii, Prosopis africana, Ziziphus mucronata*
Deciduous species, sprouting in middle dry season													*Acacia seyal, Cassia sieberana, Gardenia ternifolia, Terminalia macroptera*
Deciduous species, sprouting in late dry season													*Acacia polycantha, Albizia chevalieri, Burkea africana, Pterocarpus erinaceus*
Deciduous species, sprouting in early rainy season													*Acacia nilotica, Commiphora africana, Dalbergia melanoxylon, Strychnos spinosa*

Key ▒ All species with leaves ▒ Some species with and some species without leaves ▒ All species without leaves

Figure 5.3 Woody plant phenological groups (after De Bie et al., 1998).

senegalensis (De Bie *et al.*, 1998). Semi-evergreen species generally flower in the dry season, but there is variation. Deciduous species normally flower at the end of the dry season and at the beginning of the wet season. Since there seem to be large differences between species within the same phenological group, as well as for the same species at different sites (De Bie *et al.*, 1998) (Figure 5.4), flowering and fruiting phenological patterns are less obvious. In view of this, flowering and fruiting patterns have not been classified (although see Seghieri *et al.* (1995) for phenological groups based on flowering periods in a Cameroonian savanna). De Bie *et al.* (1998), comparing phenological patterns of individual species along a rainfall gradient, have concluded that there is a relationship between the phenology of deciduous woody plants and rainfall, either directly or indirectly, e.g. through higher air humidity. Plants begin to drop their leaves earlier and sprout later with declining rainfall.

Herbaceous species generally wither and die during the dry season. However, rapid regrowth and the first wave of flowering can occur even before the onset of the rains. Depending on the time of emergence (early or delayed), the time of flowering (early or late) and the length of the vegetative state, six different phenological types have been identified (César, 1971; Menaut and César, 1982) (Figure 5.5).

Even though herbaceous plants are only a small part of the above-ground biomass, they make up nearly all of the above-ground net production. It is only in the more dense savanna that the production of woody plants reaches 10–20% of the herbaceous layer. In the woody component, the above-ground production is always greater than the below-ground production. In the herba-

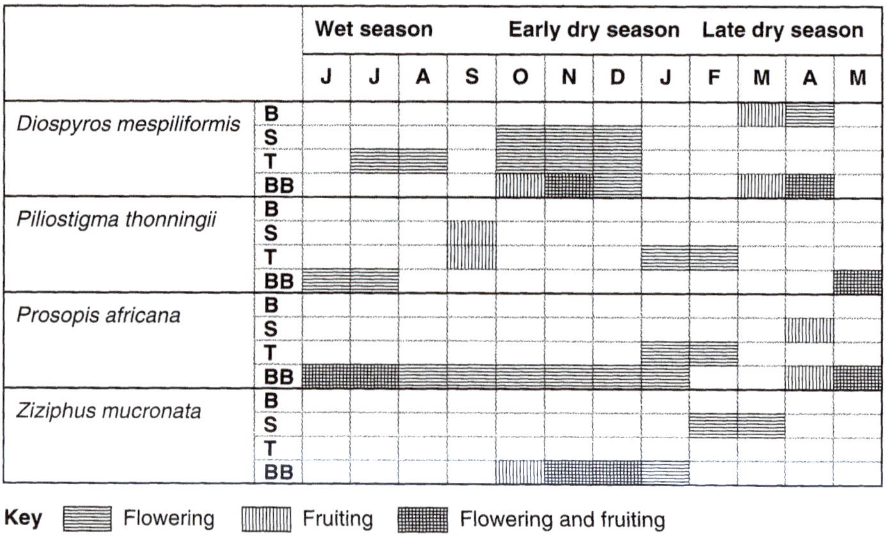

		Wet season				Early dry season				Late dry season			
		J	J	A	S	O	N	D	J	F	M	A	M
Diospyros mespiliformis	B										▨	▨	
	S												
	T		▨	▨									
	BB					▨	▨				▨	▨	
Piliostigma thonningii	B												
	S				▨								
	T				▨								
	BB	▨	▨										▨
Prosopis africana	B												
	S									▨			
	T												
	BB	▨	▨								▨	▨	
Ziziphus mucronata	B												
	S									▨	▨		
	T												
	BB					▨	▨	▨					

Key ▨ Flowering ▨ Fruiting ▨ Flowering and fruiting

Figure 5.4 Flowering and fruiting phenology of four semi-deciduous woody species from different sites in West Africa. Sites are B = Bissiga, S = Sourou, T = Tissé, BB = Boucle de Baoulé (after De Bie *et al.*, 1988).

	Short cycle: the above-ground parts wither rapidly and disappear by the time the dry season starts.	**Long cycle**: the above-ground vegetation remains until the next fire.	
Flower early: soon after emergence, either immediately after fire or with the first rains.	Short, early flowering, e.g. *Cyperus tenuiculmis*	Long, early flowering, e.g. *Imperata cylindrica*	
Flower late: during the second part of the wet season or even at the beginning of the dry season.	Short, late flowering with early emergence, e.g. *Scleria lagoensis*	Long, late flowering with early emergence, e.g. *Hyparrhenia smithiana*, most abundant group; provides fuels for fires	**Early emergence**: emerge early in spring
	Short, late flowering with delayed emergence, e.g. *Aspilia bussei*	Long, late flowering with delayed emergence, e.g. *Tephrosia elegans*	**Delayed emergence**: emerge later during the main wet season but grow during the second part

Figure 5.5 Herbaceous plant phenological groups (adapted from Menaut and César, 1982).

ceous layer, the below-ground production is greatest in the more grass-dominated formations, but this decreases, and above-ground production becomes prominent in the more woody-dominated formations (Menaut and César, 1979). Mordelet and Menaut (1995) found lower grass production under tree clumps, varying from 51 to 90% of the production in an open situation. They hypothesise that this may be due to tree canopies shading grasses and thereby limiting photosynthesis.

5.1.3 Fauna

The West African savannas are known for their high diversity of wildlife, especially birds (Keith and Plowes, 1997) and ungulates, such as elephants, giraffes, several species of antelope and warthog (De Bie, 1991). However, populations of all these mammals have become much smaller as their habitats have either disappeared or become fragmented, and in some countries they are locally extinct (see Section 5.4.1).

5.2 The main savanna determinants

Plant-available moisture is undoubtedly the most important factor differentiating the Sahelian, Sudanian and Guinean savannas. Water is an important determinant of vegetation in the drier parts of the Sahelian and Sudanian savannas, but may play a lesser role in the southern savanna zones. Recent studies investigating root distributions and water use patterns of trees and grasses in the humid savannas of the Lamto Reserve, Ivory Coast (Lamont and Bergl, 1991; Mordelet *et al.*, 1993, 1997; LeRoux *et al.*, 1995) (Figure 5.6), have found no evidence for below-ground partitioning of water, neither spatially nor temporally. This may be because soil water availability in these savannas is high during most of the year, and plants have particular strategies for facing water stress (see below). Therefore, in humid West African savannas, soil-available nutrients may be the driving environmental force (LeRoux *et al.*, 1995; Abbadie *et al.*, 1996).

5.2.1 Plant–available moisture

Rainfall and its seasonality is extremely important for plant growth in West African savannas, especially in the drier north. Droughts tend to induce the replacement of perennial plant species by annual ones, and of species with high water requirements by species which depend on less (De Bie *et al.*, 1998). Nevertheless, many species have strategies to cope with the dry conditions. Leaf

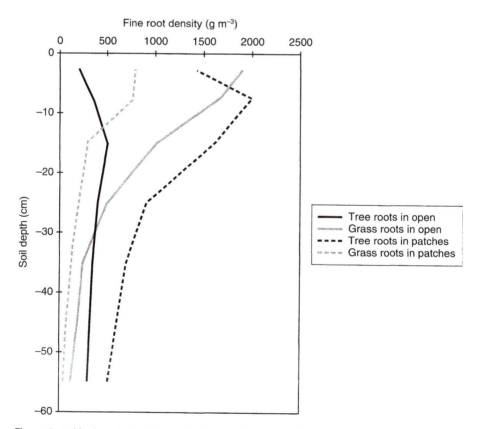

Figure 5.6 Vertical distribution of fine tree and grass roots in open savanna and beneath tree patches in a Guinean savanna (based on Mordelet *et al.*, 1997).

shedding at the start of the dry season is the strategy deciduous species use to avoid drought conditions. As rainfall declines from south to north, deciduous species shed their leaves earlier and start sprouting later.

Evergreen and semi-deciduous species maintain their leaves throughout the dry season, but avoid water loss through scleromorphic features, reduced physiological activities, and by accessing deep water reservoirs using tap roots, e.g. *Vitellaria paradoxa* (Childes, 1989). De Bie *et al.* (1998) suggest that water availability is not a limiting but a selective factor, as evergreen, semi-evergreen and deciduous strategies all aim for maximum survival and reproduction but along different lines. The adaptability of plants to the environment may also be indicated by the fact that some species follow different strategies in different sites, showing some plasticity. The reason for this difference is still unknown.

5.2.2 Plant-available nutrients

In the Lamto Reserve, nitrogen requirements of the herbaceous stratum have been estimated at 70 kg ha^{-1} year^{-1} (Abbadie, 1984). In the field, as well as during incubation under optimal conditions, net N mineralisation rates are very low (~5 kg ha^{-1} year^{-1}), and nitrification does not occur (Abbadie and Lensi, 1990). Even with increased ammonification in particular sites where the soil has been processed by fauna (Abbadie and Lepage, 1989), the mineralisation of soil organic nitrogen cannot be the only source of nitrogen assimilated by grasses.

Abbadie *et al.* (1992) confirm that nitrogen is mostly recycled internally within the vegetation–soil system. They found that only a small source of nitrogen was derived through the weak mineralisation of the soil organic matter (~5 kg ha^{-1} year^{-1}), an equivalent amount from bulk precipitation (~5 kg ha^{-1} year^{-1}), about double this through non-symbiotic fixation in the rhizosphere (~12 kg ha^{-1} year^{-1}), with most nutrients becoming available through the recycling of the nitrogen stock of the dead roots before humification (~40 kg ha^{-1} year^{-1}). Therefore, savanna grasses may be independent of soil organic matter for their nitrogen supply.

Further evidence for this hypothesis comes from a study by Mordelet *et al.* (1993) on tree clumps at Lamto. They found that the higher total root phytomass under tree clumps led to a lower bulk density, higher soil organic matter content and thereafter higher soil nutrient availability (Figure 5.7). Isichei and Muoghalu (1992) found similar results in a Nigerian Guinean savanna. The nutrient patchiness beneath the tree clumps may have led to the conclusion that grass roots would be higher under tree clumps. However, grass roots show the same density profiles under canopy and in the open (Mordelet *et al.*, 1997). In fact, roots coming into the nutrient-rich patches of trees were from palm trees outside the tree clumps (Mordelet *et al.*, 1996). These results indicate that there must be a different source of nutrients for the grass component.

Termites and nutrient cycling

Fungus-growing termites play a major role in nutrient cycling as they affect the decomposition processes. Termites process 8% of the annual litter production

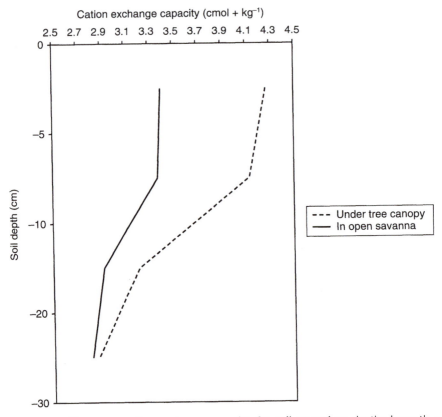

Figure 5.7 The mean cation exchange capacity for soil at various depths beneath open savanna and tree patches in a Guinean savanna (based on Mordelet *et al.*, 1993).

in the Sahelian dry savannas of Senegal (Lepage, 1972), and 28% in the humid savannas of the Ivory Coast (Josens, 1974). In Nigeria, termites remove 60% of the annual wood fall, 24% of the tree leaf annual production (Collins, 1981), and 60% of the grass litter production (Ohiagu and Wood, 1979). Lepage *et al.* (1993) found that the proportion of woody material in termite fungus combs at Lamto was much higher than the proportion of this component in the annual primary productivity. For example, in a shrub savanna, the total above-ground tree production does not exceed 17% of the total above-ground grass production (Menaut and César, 1979), while the proportion of tree material in the combs reaches 34–87%. The large use of tree material could be a response to the continuous decrease in nitrogen content of grass leaves during the growing period. In April, the protein content of grass leaves is less than 6%, and reaches 2% in November and December (Abbadie, 1984), compared to 6% all year round in tree leaves (Monnier, 1981).

At Lamto there are between 70 000 and 150 000 fungus combs per hectare, and to build them, it is estimated that Macrotermitinae collect 1.5 tons ha^{-1} year^{-1} of vegetation (Josens, 1974). The combined organic matter consumption by soil- and litter-feeding termites is estimated at about $12 \, \text{g m}^{-2} \, \text{year}^{-1}$ for a dry Sahelian savanna (Josens, 1983). Considering they selectively choose their plant material, the nutrient content of the vegetation must be high, and so the output of nutrients to the ecosystem, through predation and building activities, must be substantial. Unfortunately, the quantitative impact of this process is still unknown.

5.2.3 Herbivory

Herbivory in the West African savannas is mainly carried out by large ungulates such as elephants, buffalo and species of antelope, and according to their diet composition, they can be classified as grazers, mixed feeders and browsers (Table 5.1). The main limitation to them is the dry season, during which the overall phytomass available for consumption decreases. The quality of both perennial and annual grass species also decreases during this season, especially in crude protein content (Afolayan and Fafunsho, 1978; De Leeuw, 1979; LeHouérou, 1980). The quality of woody foliage is higher than that of grasses and more constant, and in the dry season, more woody foliage is available. However, the presence of secondary compounds, such as tannins, depresses digestibility, which positively limits consumption.

Resource separation by herbivores

The variation in resource availability, especially in the dry season, leads most herbivores to a degree of food preference, and some form of separation in resource use. For example, De Bie (1991) has recorded habitat separation for some species in the Bauolé Reserve, western Mali (Figure 5.8). Here, in the dry season, species concentrate near water sources, moving down the catena for their forage and water requirements. Warthog shows the most plasticity in

Table 5.1   Different classes of herbivores in the West African savannas (after De Bie, 1991)

Grazers
  Warthog (*Phacochoerus aethiopicus*)
  Hippopotamus (*Hippopotamus amphibius*)
  Reedbuck (*Redunca redunca*)
  Waterbuck (*Kobus ellipsiprymnus*)
  Hartebeest (*Alcelaphus buselaphus*)
  Buffalo (*Syncerus caffer*)
  Roan antelope (*Hippotragus equinus*) ⎫ shift seasonally between grazing and fixed
  Oribi (*Ourebia ourebi*) ⎭ feeding
Mixed feeders
  Elephant (*Loxodonta africana*)
Browsers
  Giraffe (*Giraffa camelopardalis*)
  Bushbuck (*Tragelaphus scriptus*)
  Red-flanked duiker (*Cephalophus rufilatus*)
  Grimm's duiker (*Sylvicapra grimmia*)

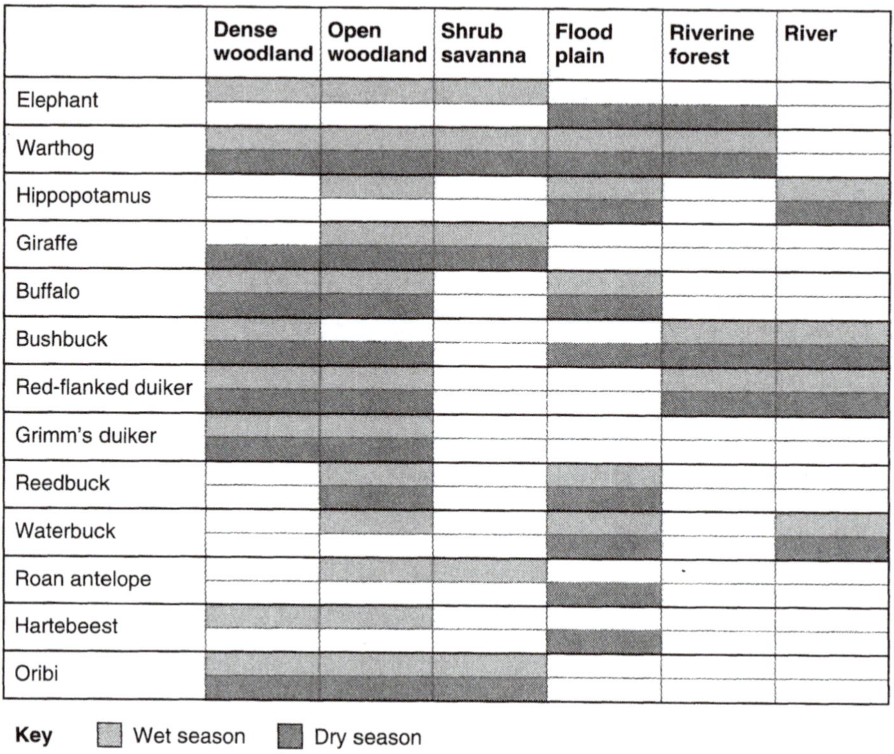

	Dense woodland	Open woodland	Shrub savanna	Flood plain	Riverine forest	River
Elephant						
Warthog						
Hippopotamus						
Giraffe						
Buffalo						
Bushbuck						
Red-flanked duiker						
Grimm's duiker						
Reedbuck						
Waterbuck						
Roan antelope						
Hartebeest						
Oribi						

Key ▨ Wet season ▰ Dry season

Figure 5.8 Habitat choice of ungulates in Sudanian savanna areas (after De Bie, 1991).

respect to spatial utilisation. However, species that are more selective feeders, such as bushbuck and waterbuck, are much less flexible (De Bie, 1991). Other mechanisms for separation between species include changes in diet to more nutritious species within the same plant community and differences in feeding level (i.e. animal size).

5.2.4 Fire

Lightning fires have been frequent in the West African savannas since about 40 000 years BP, and were most intense during periods when the global climate was changing from interglacial to glacial, resulting in drier regional conditions (Bird and Cali, 1998). Humans, however, have exercised significant control over fire regimes since at least Holocene times, and today are the source of around 90% of annual burns (Geerling, 1982, 1985). Fires are set for various reasons: to prevent bush encroachment, to promote perennial grass regrowth, to increase visibility for hunting, and to clear fields for arable use.

Effects of fire

Early dry season fires are generally of a lower intensity, and most woody and herbaceous species are relatively unaffected. Late dry season fires, on the other

hand, can reach extremely high temperatures as a result of the increased com-
bustibility of fuels and the dry climatic conditions, and have negative impacts
on both woody and fire-sensitive grass species (Hopkins, 1963, 1965;
Brookman-Amissah et al., 1980). These areas subject to intensive fire regimes
are characterised by no large trees, a low density of relatively homogeneous populations of medium-sized trees, a few small trees, a low basal area, and an abundance of tolerant grasses (Athias et al., 1975; Afolayan, 1978; Rodgers, 1979;
Geerling, 1985; Lykke and Sambou, 1998). This results in a very open savanna,
with a dense stand of fire-tolerant grasses.

In the Lamto savannas, tree distributions are highly aggregated irrespective of
soil conditions, and the spread of fire depends on vegetation patterning
(Menaut et al., 1991). Using model simulations, Hochberg et al. (1994) found
that mortality from fire acts in an inverse spatially density-dependent fashion,
enhancing tree aggregation. When a small amount of dispersal is possible, the
rate of tree population growth is greatly accelerated as compared to when no
such dispersal occurs. Granier and Cabanis (1976) also noticed a higher germination of seeds in burnt West African savanna.

Menaut (1977) has shown that both the number of woody species and the
number of individuals increase very quickly during the first three years of fire
protection, and continue to increase regularly. Later, however, the situation
changes. The increased shrub cover and its resulting shade inhibit the growth of
grass, and the habitat then becomes favourable for forest species (Brookman-
Amissah et al., 1980). This has the result of decreasing species diversity, as
shown by Lykke and Sambou (1998). Working in the savannas of south-
western Senegal, they found that fire, together with other forms of human
impact and decreasing rainfall, promote the development of a temporary
mixture of fire-sensitive and fire-tolerant species. Fire protection brought about
domination by certain species, particularly *Daniellia oliveri*, thus reducing
biodiversity.

Plant adaptations to fire

Fire burns most or all of the above-ground biomass. However, the large underground root systems of the grass species, enable them to survive even the most
intense fires, and rapidly establish new shoots before the onset of the rainy
season. In contrast, woody species which are less than 2 m in height either die
or have their growth retarded. Mature trees and shrubs beyond about 2 m are
more fire resistant and only experience die-back (Menaut and César, 1979).
Damage to semi-evergreen tree leaves is generally limited, as fires can occur at
the same time as the leaf shedding. Deciduous species are less affected because
most are bare when the fires occur. Hopkins (1963) has shown that in some
species, the heat from fire actually promotes leaf flushing.

Based on the architectural description of trees, César and Menaut (1974)
have distinguished two main strategies enabling young trees to resist fire. The
first is in species that can survive by resprouting each year from below-ground
storage structures, e.g. *Bridelia ferruginea* and *Cussonia barteri*. To recruit into
the adult population, such resprouts have to successfully establish a fire-resistant

perennial trunk which will allow further growth in height the following year. This is only achieved when the below-ground organs are strong enough to produce, between successive fires (in some cases, this can be as short as one year), a trunk reaching a height where the terminal buds are able to resist the existing fire conditions, and thick enough at its base to resist the high fire intensity in the fuel bed.

The second strategy is for young individuals to build an aerial fire-resistant thick trunk enabling it to resist all fire conditions. For example, *Crossopteryx febrifuga* has a higher intrinsic resistance to fire (bark properties) than *Piliostigma thonnongii* (Gignoux *et al.*, 1997). A 20 mm diameter stem of *C. febrifuga* survives exposure to 650 °C, whereas a 40 mm diameter stem of *P. thonnongii* would be required in order for the plant to survive the same temperature. *C. febrifuga* has a thicker stem than *P. thonnongii*. However, *P. thonnongii* grows 2.26 times faster than *C. febrifuga* between two successive fires. One relies on resistance of above-ground structures to fire, while the other relies on its ability to quickly rebuild above-ground structures. *C. febrifuga* is able to recruit in almost any fire conditions, while *P. thonnongii* needs locally or temporarily milder fire conditions (Gignoux *et al.*, 1997). Therefore, fire resistance relies on a combination of traits and cannot be measured by just one factor, i.e. bark thickness.

5.3 Human determinant of the savanna

People have been using the West African savanna for hundreds of thousands of years, and prehistoric communities relied on hunting and gathering for subsistence. It was not until the domestication of plants and animals took place, some 4000 years ago, that savanna use began to change (Clark, 1980). Subsistence agriculture and pastoralism became, and still are, the main forms of land use in the region. Historically, the major influences in the West African savannas have been the introduction of Islam into savanna societies and their functioning, and colonisation, as with much of Africa. The latter was most effective in changing population structure in the region due to the mass forced emigration of people to the Americas as slaves. This had a large effect on land-use patterns by decreasing the impact of grazing and agriculture, and probably allowing regeneration to take place. Later, colonisation produced the opposite effect, in terms of introducing commercial agriculture to the region, and therefore increasing the pressure for savanna conversion to other land uses.

5.3.1 Pre-colonial land use

Due to its geographical position, much of West Africa remained isolated and less accessible to the influences of the outside world during its early history. The main occupation of the people was agriculture, subsistence crops being millet and beans. Between the fourth and sixteenth centuries, there was a succession of kingdoms located along the southern end of the Saharan caravan route, which relied on trade in gold, ivory, slaves and cola-nuts in exchange for salt, copper, glass and linen (Connah, 1992). According to the Arab chroniclers, the oldest and most powerful of these states, the kingdom of Ghana, was founded in the fourth century by Berbers from the south-west Sahara.

One of the most influential events affecting the West African savanna region was the emergence of Islam and the Arab influence between the seventh and eleventh centuries. This saw a succession of Islamic empires in most of the region, accompanied by a growth in Arabic learning and culture. The influence of ordinary peasant farmers was replaced by a hierarchy of village and district heads, collecting taxes for the support of the central authority. Although the political and social system approximated the system prescribed by the Koran, it incorporated many traditional practices and traditions such as indigenous resource use. The trade carried across the Sahara Desert by the Berber caravans helped to introduce domesticated animals and new plants, and contributed largely to wealth (Connah, 1992).

Land tenure

As in most parts of Africa, common property regimes governed access to savanna resources such as common land, sacred groves, fishing ponds and grazing land. These systems worked through inheritance, whether it be matrilineal or patrilineal, or through a monarch distributing land to nobles, who then allocated rights to their lineage. There are many different types of indigenous tenure systems. For example, the *ton* tenure system of Guinea defines and enforces rules that seasonally regulate access to vegetation and wildlife located within village commons and on individually appropriated lands (Freudenberger *et al.*, 1997). This ensures that particular resources, such as fruits and grass, reach full maturity before they are harvested, the village institutions determining the dates and conditions of access. The system defines 'closed' periods, when all of the resource is banned by village authorities, and an 'open' season when few restrictions regulate collection of the product (Freudenberger *et al.*, 1997). Some resources, such as grass for roof thatching, are also actively protected against damage from, for example, dry season fires.

5.3.2 Colonisation

The eighteenth century saw the beginning of European interference and conquest in West Africa. Although first European contacts with West Africa took place in the mid-fifteenth century, these coastal trading posts had only an indirect effect on inland regions, which were thought too hazardous to penetrate. Coastal colonies were formed in Senegal (French), Gambia, Sierra Leone, the Gold Coast and Nigeria (British). Initially most trade was in gold and peppers, but later included products such as timber and palm oil. However, it was the discovery of the Americas that initiated the most profitable commodity – slaves (Figure 5.9). Some estimates put the total number of slaves exported from West Africa (including Angola in west central Africa) as high as 24 000 000, of which probably only about 150 000 survived the notorious Middle Passage across the Atlantic. Slavery was not abolished until 1833 in British colonies, 1848 in French colonies, 1865 in the United States, and 1888 in Brazil.

Figure 5.9 Slaves being shipped from West Africa to the Americas.

Source: Group of negro men and boys taken out of captured dhow in a state of starvation (engraving) (b & w photo) by English School (19th Century) Private Collection/Bridgeman Art Library. (Copyright © The Bridgeman Art Library, London)

After the abolition of the slave trade in West Africa, the intense competition for raw materials and markets for industrial goods in Europe, as well as concern for strategic position and prestige, led to the Berlin Conference and Treaty of 1885 and the commencement of full European sovereignity over the African continent. In the British colonies, e.g. Nigeria, traditional rulers and chiefs were maintained and allowed to run 'local government', albeit under the supervision of British officials. However, in French colonies such as the Ivory Coast, administration was directly exercised by French officials with central control. Colonisation brought about new laws of civil order, infrastructure, telecommunications and education, as well as the introduction of modern research, particularly into agriculture through new, improved crops and technology. The introduction of export crops initiated the evolution from subsistence to commercial commodity production.

Changes in public opinion after World War II led to the disengagement of European countries from their colonies. In West Africa, there was little resistance to the change, mainly because there were few settler communities, compared to East and Southern Africa. In 1960, Nigeria became self-governing, and by 1970 all of West Africa was independent.

Land tenure

The effect of colonisation in West Africa was the adoption of policies designed to replace customary tenures with state administered land *en route* to registration as private property (Elbow *et al.*, 1998). The British were less consistent than the French, and established protectorates but also vested land-ownership and management to tribal chiefs. Resource management, however, in all cases, was extremely state-centric and protectionist.

Since independence, structural adjustment programmes and associated economic and legal reforms in West Africa have served to undermine indigenous common property systems, with many governments heading for privatisation policies reminiscent of colonial times (Elbow *et al.*, 1998). Some exceptions include Nigeria and the Ivory Coast, where policies concentrate on maintaining a strong hold for the state over land-ownership and management. In Niger and Guinea, although there are aims to convert ownership to private holdings, land and resource exploitation is still to be monitored by the state (Elbow *et al.*, 1998).

5.3.3 Present-day land use

Animal husbandry and arable agriculture are the most important types of land use in the West African savannas. However, in recent times increased pressure has been put on savanna lands as a result of two main factors. The first is the severe drought period that began in the early 1970s. This has caused increased migration of pastoralists from the northern savanna lands in the Sahelian zone, southwards, particularly into the Sudanian zone. Eradication campaigns of diseases such as trypanosomiasis, and more effective medication, have also increased grazing pressure in areas not subject to grazing before.

As a result, the area used for agriculture has expanded, and conflicts over resources between pastoralists and farmers have heightened. These problems have been accentuated by the second main factor affecting savannas in West Africa: the growth in human population. Crop areas have extended, fallow periods have shortened and there has been a neglect of non-productive activities such as soil protection (LeHouérou, 1989; De Bie, 1991). Both types of agriculture, pastoralism and farming, have come into direct conflict with other land uses, namely hunting and forest resource exploitation, particularly woodfuel extraction.

Land tenure

Even though there is a move to privatisation of land tenure, common property systems are still dominant today. Interestingly, when Islam entered West Africa, it incorporated many of these tenure systems, and further enforcement of rules was by religious decree. In fact, today, Islamic Leagues are significant in contemporary resource management in West Africa, especially in managing resources shared by more than one village, and thereby averting the misfortune of the 'tragedy of the commons' (Freudenberger et al., 1997).

Seasonal activation of tenure systems is a particular feature throughout Sahelian West African savannas, with land use shifting seasonally from one ethnic group to another (e.g. Shipton, 1994). For example, pastoralists have traditional claims over resources such as trees for fodder and water during the dry season, and farmers have access during the wet season cropping period. Complementary environmental management is also evident. For instance, studies in Senegal have shown that cattle manure from pastoralist herds enhances soil fertility around the species Acacia albida, while careful management of these trees by farmers provides foliage for livestock during the harsh dry seasons (Seyler, 1993).

Agriculture

Agriculture dominates the economy and labour force of West African countries (Table 5.2), although its relative importance has been falling as each country's economy becomes more diversified. There is also more emphasis on export crops through huge capital allocations and the provision of marketing facilities, which has led to a relative neglect of food crops (Smith et al., 1994). Consequently, food which could be grown in the savanna zone is imported. Land fragmentation and land tenure problems remain chronic, and individual farmers find it impossible to cultivate large areas of land even when they have the means to do so.

Some of the fundamental problems of agricultural development relate to price fluctuations in world markets, and the weather. For example, the frequent occurrence of droughts in the 1970s led to drops in domestic food production as well as in export crops. In response, a number of West African countries have been promoting the establishment of agricultural institutions and pursuing policies to achieve self-sufficiency and autonomy in food production. Major elements in the agricultural policies introduced include improvements to rural infrastructure, and the provision of subsidised agricultural services and supplies,

Table 5.2 The total and agricultural population of West African savanna countries in 1997 (based on FAO, 1999). Figures in millions of persons

Country	Total population	Agricultural population
Benin	5.7	3.3 (58%)
Burkina Faso	11.1	10.2 (92%)
Cameroon	13.9	7.8 (56%)
Chad	6.7	5.2 (78%)
Central African Republic	3.4	2.6 (76%)
Ghana	18.3	10.4 (57%)
Guinea	7.6	6.5 (86%)
Ivory Coast	14.3	7.5 (52%)
Mali	11.5	9.5 (83%)
Mauritania	2.4	1.3 (54%)
Niger	9.8	8.7 (89%)
Nigeria	118.4	42.8 (36%)
Senegal	8.8	6.5 (74%)
Togo	4.3	2.7 (63%)

such as fertilisers, farm implements, improved seeds, extension services, farm credits, farmer training and seed multiplication.

For example, in northern Nigeria, a process of intensification of agriculture has resulted in substantial increases in productivity and farmer welfare (Smith *et al.*, 1994). This was because of an improved transport system that made it viable for farmers to market an improved variety of maize as a cash crop, and the existence of a fertiliser subsidy which favoured the improved maize. Smith *et al.* (1994) indicate the need for agricultural technologies to be targeted to West African regions where the preconditions for intensification exist. Improving crop varieties needing less fertiliser, and using legumes to increase soil organic matter, while at the same time reducing the build-up of crop pests, are some ways forward. Still, West Africa is far behind the rest of Africa in the use of agricultural inputs such as fertilisers, tractors and irrigation measures. This may be a result of the relative lack of close interdependence between the agricultural and industrial sectors of an individual country's economy.

Subsistence
Rainfall-fed agriculture is the principal form of production in West African societies. There are a diversity of staple crops: grains such as sorghum or guinea corn (*Sorghum bicolor*) and pearl or bullrush millet (*Pennisetum americanum*) are of primary importance, with maize, rice, eleusine or finger millet (*Eleusine coracana*), and *acha*, hungry rice or *fonio* (all names for *Digitaria exilis*) being more restricted in distribution. The grains all differ in terms of their environmental requirements, and so provide West African cultivators the possibility of varying their crop selection and mix in response to changing environmental and economic conditions (Burnham, 1980).

Sorghum and the millets have lower rainfall requirements and so dominate in the subhumid Sudanian and semi-arid Sahelian zones, whereas in the humid Guinean savanna, maize, sorghum and root crops, especially yams and manioc, are the main crops. Wet-rice cultivation is important in the Volta and Niger drainages, and the dry cultivation of upland rice is practised in the western part of the Guinean savanna zone.

Cash crops

Commercial plantations of tree crops are virtually absent from the West African savanna region owing to the relatively high moisture requirements of the trees. Coffee and oil palm are grown in small numbers in the wetter parts of the region but they are of limited commercial importance. The principal export crops grown in the savanna region are cotton and groundnuts (Elbow *et al.*, 1998). With regards to cotton, Nigeria produces about three-quarters of the total West African output, with other important producers including the Ivory Coast and Guinea. Groundnut is grown both as a food crop and as an export product. It is the most important commercial crop of Senegal, and one of the most important export crops in Nigeria and Niger, and production is entirely by small-scale farmers (Table 5.3). Maize, originally a subsistence crop, is now being commercially produced, and growing in importance as a major cash crop (Smith *et al.*, 1994; Freeman and Smith, 1996).

Pastoralism

Cattle are the most important livestock in the savanna lands. Pastoralism occurs largely in the Sudan and Sahel zones which are virtually free of tsetse flies. Cameroon, Chad, Mali and Nigeria have the largest cattle numbers, although populations have fluctuated over time in response to droughts (Figure 5.10). Most pastoral production is carried out by the nomadic Fulani groups, although other people such as the Tukulor, Serer, Masa, Kanembu and Chadic Arabs (Shuwa) also practise pastoralism. Although mostly cattle are raised, some groups also keep sheep and goats.

Table 5.3 Production of groundnuts and maize in some West African savanna countries in 1980, 1990 and 1998 (based on FAO, 1999). Figures in million of tonnes.

Country	Groundnuts			Maize		
	1980	1990	1998	1980	1990	1998
Benin	62 839	63 931	102 341	271 300	409 994	714 397
Burkina Faso	53 943	134 235	152 128	104 510	257 900	366 467
Chad	98 600	108 423	250 000	25 000	28 823	100 000
Ghana	142 200	113 000	135 000	382 000	552 600	1 093 000
Guinea	83 900	78 107	146 478	90 000	73 735	81 019
Ivory Coast	81 000	130 000	135 850	380 000	497 184	547 200
Mali	144 098	179 924	134 129	45 387	207 362	289 761
Nigeria	471 000	992 000	2 531 000	612 000	5 768 000	5 858 000

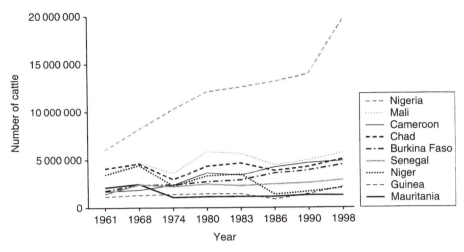

Figure 5.10 Cattle livestock numbers in some West African savanna countries from 1960 to 1998. Note the decrease in cattle populations in some countries after drought periods 1968–1973 and 1983–1985 (adapted from FAO, 1999).

Hunting

Although hunting is prohibited in many West African savanna countries, subsistence hunting still continues today, and makes up around 10–50% of the total animal protein in rural areas (De Bie *et al.*, 1987). Nowadays, commercial hunting is also important, usually undertaken by hunters from urban centres using modern equipment. This has resulted in the extermination of some mammals such as topi and oryx in the Sahel, and populations relegated to marginal lands in other western savanna regions (De Bie, 1991).

Wood extraction

The use of wood from savannas is, as in most parts of Africa, economically vital for West African countries. For example, 95% of all energy used in Burkina Faso comes from wood, and in Mali, woodfuel (firewood and charcoal) contributes between 93 and 97% of the total rural and urban energy consumption (Foley, 1987). In some savanna countries, there are afforestation programmes under way. For example, in the northern Guinean savanna zone of Nigeria, about 2000 ha of *Eucalyptus* plantations, mainly *E. camaldulensis*, *E. tereticornis* and *E. citriodora*, have been established, and more are planned (Igboanugo *et al.*, 1990).

Mining

Since around the 1960s, mineral resources have played an increasingly important role in the West African savannas. These include crude petroleum, iron ore, diamonds, gold, bauxite, tin ore, manganese and limestone. One or two of these dominate the production structure of the mining sector in some countries, e.g. iron ore in Mauritania and crude petroleum in Nigeria.

Methods of mineral exploitation vary from country to country and according to the nature and location of the resources. These can range from deep mining, alluvial workings and open-cast mines to traditional methods of hand panning. However, most of the exploration and production of minerals in West Africa is in the hands of large foreign companies and some individuals. This is probably most apparent in Nigeria, where the economy is heavily dependent on the capital-intensive oil sector, which provides 30% of GDP, 95% of foreign exchange earnings, and about 80% of government revenues.

Unfortunately, political instability, corruption between Western companies and government officials, and poor macroeconomic management have led to the government resisting greater accountability in managing the country's multibillion dollar oil earnings, which continues to limit economic growth and prevent an agreement with the IMF and bilateral creditors on debt relief. The largely subsistence agricultural sector has failed to keep up with rapid population growth, and Nigeria, once a large net exporter of food, must now import food. Agricultural production in 1996 suffered from severe shortages of fertiliser, and production of fertiliser fell even further in 1997.

5.4 Current management problems and strategies

West African savannas are under increasing pressure to be converted to other land-use systems, particularly arable and pasture land. Although the climatic fluctuations in the region are partly responsible for the increased demand on savanna lands, land tenure and reform policies in the region have not been conducive to sustainable management of resources, putting further strain on local land use and communities.

5.4.1 Conservation

Direct destruction and fragmentation of savanna lands to meet the needs of agricultural expansion and animal husbandry, is no doubt having a major impact on native plants and animals. For example, large mammals such as the black rhinoceros (*Diceros bicornis*) and the giant eland (*Tragelaphus derbianus*) are on the verge of extinction in many West African nations (e.g. Keith and Plowes, 1997); others, including elephant, topi, reedbuck and several carnivore species, are threatened over most of their distribution (LeHouérou, 1989; De Bie, 1991). Although relatively large populations are found in national parks and reserves, they are not always protected, and habitat availability for many species, especially in light of the drought occurrence in the region, is becoming increasingly constrained.

De Bie (1991) emphasises the urgent need to develop management programmes for conservation areas, including monitoring projects, and the application of fire as a management tool to reduce fuel load and the occurrence of potentially devastating burns. However, Elbow *et al.* (1998) point out that current natural resource management policies across West African savanna states are informed by the legacy of colonial administration, i.e. protecting resources through the enforced restriction of their use. This approach has not been successful in ensuring sustainable exploitation. Not only does the denial of management responsibilities over resources lead to little interest in protecting them; some populations have no choice but to continue exploiting resources through encroachment when reserves or parks are established (Elbow *et al.*, 1998). The rising conflicts between local communities and conservation areas are leading governments to undertake more community-based conservation strategies through participatory land-use planning, as has been done in the *miombo* savannas (see Section 4.4.4). However, as in most of Africa, these projects need to address the construction of new institutional arrangements to these areas involving local communities.

5.4.2 Droughts and desertification

Over the past few decades, the Sahelian zone has seen dramatic variability in climate, and averaged over 30 year intervals, annual rainfall across this region has fallen by between 20 and 30% over the decades that led to political independence for the nations and through to present time, compared to the decades before independence (Hulme, 1998) (Table 5.4). The variability even within the region is well illustrated by the 1998 FEWS Sahel Vulnerability Assessment (USAID, 1998). The good rainfall levels in 1997/98 provided favourable conditions for crop production in much of the region. For example, in Chad, Mali and Niger, cereal production was at or above the 1992/93–1996/97 average. However, the poor distribution of the rains meant that in Burkina Faso and Mauritania, production losses were large enough to reduce national cereal production levels to 7% and 14% below average, respectively.

Desertification is regarded as a major environmental problem in savanna lands (see Section 10.2.4), and the Sahelian zone of West Africa is often quoted as a seriously affected region. However, it is the confusion between desertification and drought that has led to the idea of widespread degradation of the Sahel. Although there are cases of local degradation (LeHouérou, 1989), there is no

Table 5.4 Principal droughts in the Sahel since 1740

1740	Drought, famine
1750	Drought, famine
1790	Drought
1855	Drought
1900–03	Drought
1911–14	Drought, famine
1931–34	Drought, famine
1942	Drought
1950	Drought
1968–73	Drought, famine
1983–85	Drought, famine
1987	Drought
1990	Drought

real evidence that it is occurring at regional and subcontinental scales and is irreversible, as has been stated (e.g. UNEP, 1990). This fact is important as it affects policy-making with respect to aid and economic development programmes. Using the ratio of net primary production (NPP) to precipitation (known as the rain-use efficiency, or RUE), calculated from remotely sensed vegetation indices and rain gauge data, Prince et al. (1998) have found that between 1982 and 1990, NPP in the Sahel was extremely resilient, and in step with rainfall, recovering rapidly following drought. In fact, their results indicate small but significant upward trends in RUE, probably related to the increase in rainfall in the years since the drought of 1984. Prince et al. (1998) conclude that their evidence does not support fears of extensive and progressive desertification in the Sahelian zone. Further support of system resiliency comes from the increasing number of studies indicating that people in the Sahel, as well as in other arid regions, have the capacity to adapt to climate variability (e.g. Davies, 1996).

In some areas of West African savannas, where local degradation has taken place, such as the northern Guinean savanna zone of Nigeria, herdsmen now have to rely on silvipastoralism, grazing their herds in Eucalyptus plantations (Igboanugo et al., 1990). Igboanugo et al. (1990) conclude that plants growing beneath Eucalyptus trees in plantations may play a potentially important role as a source of fodder for livestock, and since Eucalyptus trees are not edible to livestock, they have a lower chance of being damaged. However, more research needs to identify pasture species compatible with eucalypts, as it was found that some species were inhibited by plantation trees (Igboanugo et al., 1990). Kessler and Bremen (1991) also point out that agroforestry techniques are not always profitable. One has to consider carefully which properties of woody species could serve which objective, where and under what circumstances. They found that woody species influence the water balance via rainfall interception, evapotranspiration and water infiltration, and so the ultimate result for grasslands and crops would greatly depend upon local conditions.

Although the droughts of the past have had a major influence on pastoral societies in West Africa, they have also been negatively affected by socio-economic changes, many of which have originated in realms beyond the control of pastoralists. For example, pastoral resources, notably water and pasture, are shrinking through pressure from farming; herders have been experiencing decreasing security of tenure over rangeland and access to common-hold resources. The loss of control over pastoral resources has also resulted from modern infrastructure, such as boreholes and cemented wells which have weakened traditional systems of negotiation over access to water-points and other key resources (Thebaud, 1998). Unfortunately, legislation in many West African states has still not found a way of supporting traditional 'common property resource' regimes. This will be vital in promoting a viable pastoral system.

5.5 Concluding remarks

West African savannas are governed by plant-available moisture, as highlighted by the periodic occurrence of intense droughts, and their effect on plants, animals and humans. Historically, population densities were not great in the region, exacerbated by the forced removal of people as slaves bound for the American continent. This undoubtedly allowed the savannas and their wildlife to flourish. Land transformation really only began after colonisation, during which commercialisation of agriculture was encouraged. It was also during this time that pastoral communities, which had lived for centuries with climate variability, began to be seen as the causes of land degradation. Climatic fluctuations are inherent in West African savannas, and only when this is realised will land and resource management be successful.

Key reading

De Bie, S. (1991). *Wildlife Resources of the West African Savanna*, Wageningen Agricultural University Papers Nos 91–92, Wageningen Agricultural University, The Netherlands.

Menaut, J.C. and César, J. (1982). The structure and dynamics of a west African savanna. In: Huntley, B.J. and Walker, B.H. (eds), *Ecology of Tropical Savannas. Ecological Studies*, **42**, Springer-Verlag, Berlin, pp. 80–100.

Menaut, J.C., Gignoux, J., Prado, C. and Clobert, J. (1991). Tree community dynamics in a humid savanna of the Côte d'Ivoire: modelling the effects of fire and competition with grass and neighbours. In Werner, P.A. (ed.), *Savanna Ecology and Management. Australian Perspectives and Intercontinental Comparisons*. Blackwell, Oxford, p. 127–137.

Chapter 6

The savannas of East Africa

6.1 Introduction

6.1.1 Distribution, climate and soils

The East African savannas occupy a large part of the Horn of Africa, between about 16°N and 9°S, and 34°E and 51°E. They are found in south-east Sudan, eastern and southern Ethiopia, north-western and western Somalia, most of Kenya, north-east Uganda, and the dry lowlands of north and central Tanzania (Figure 6.1).

The climate of the East African savannas is semi-arid and arid, and annual rainfall is generally below 500 mm and strongly seasonal (Nicholson, 1994). In most places there are two rainy seasons related to the south-west monsoon in summer and the north-east monsoon in winter. This pattern varies over the region. In the northern part of East Africa, the rainy seasons merge and there is a single wet season. A similar event occurs in the most southern part of the savanna zone, in Tanzania, where rains beginning in November, continue until April, and thereafter there is a marked dry season of about five months. Over much of the region, however, rainfall is bi-seasonal, coming from March to May and mid-October to December, the first rains being the longest and heaviest. The dry seasons are not always well defined, occasionally being broken by showers, and the wet seasons are also unreliable, in some years failing altogether (Prins and Loth, 1988). Rainfall is therefore quite unpredictable and there are great fluctuations from year to year. Temperatures are high throughout the year, mean monthly temperatures ranging between 25 and 30 °C.

Most of the savanna region lies below 900 m, and comprises a mixture of relief types, from flat valleys and outcrops, to escarpments and low hills. The soils of the region are extremely diverse, determined primarily by the extensive area of marine sediments of Jurassic, Cretaceous and lower Tertiary, but also by Tertiary and Pleistocene lava flows, Quaternary continental deposits and Precambrian outcrops. They are generally deep, well-drained, highly weathered, and consequently nutrient-poor.

6.1.2 Vegetation structure and composition

As a result of the mosaic nature of physical factors, the vegetation of the East African savannas is also distinctly varied. A tree/shrub savanna, characterised by

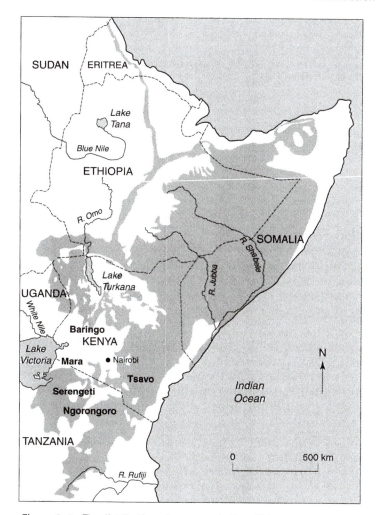

Figure 6.1 The distribution of savannas in East Africa.
(Reproduced by kind permission of Justin Jacyno and Kevin Dobbyn)

shrubs 3–5 m tall with scattered emergent trees up to 9 m tall within a grass layer, is found throughout the region (White, 1983). Although there is variation in floristic composition of this savanna type, *Acacia, Commiphora* and *Grewia* are nearly always present (White, 1983) (see Figure 1.2a). Table 6.1 lists some of the main species of the formation. The majority of species in the main canopy are multiple-stemmed bushes. For example, some *Commiphora* species have massive branches radiating from a common base.

Only a few species have well-defined trunks and these include *Acacia tortilis, Adansonia digitata* (the *baobab*), *Delonix elata, Melia volkensii, Terminalia spinosa* and *Euphorbia robecchii*. These emergent trees are normally widely scattered, and are absent in drier areas. However, in higher rainfall areas,

Table 6.1 Common plant species of the East African savannas

Type of plant	Examples
Woody species	*Adansonia digitata, Acacia elatior, A. mellifera, A. nilotica, A. reficiens, A. tortilis, Balanites orbicularis, Boswellia hildebrantii, Commiphora africana, C. boiviniana, C. campestris, C. trothea, Delonix elata, Dobra glabra, Grewia bicolor, G. tenax, G. villosa, Lannea alata, Sterculia africana, Terminalia spinosa*
Succulents	*Euphorbia cuneata, E. robecchii, E. grandicornis, Sansevieria arborescens*
Grasses	*Aristida adscensionis, A. barbicollis, Brachiara eruciformis, B. leersioides, Cenchrus ciliaris, Chloris roxburghiana*

particularly on rocky hills, these trees are denser and taller and form a woodland community. Most woody plants are deciduous, although some evergreens occur, e.g. *Boscia coriacea*. Grasses are represented by ephemeral species such as *Aristida adscensionis, A. barbicollis, Brachiara eruciformis* and *B. leersioides*, and short-lived perennials including *Cenchrus ciliaris* and *Chloris roxburghiana*. Within this landscape, there are also other vegetation types including seasonally waterlogged grasslands, scrub forest, montane forest types, and riparian forest along waterways.

Pollen and diatom analyses indicate that past changes in moisture conditions in East Africa have paralleled changes in temperature, affecting the extent and species of the savannas. Cold times have been dry, and warm times wet. Following a slightly warmer period lasting from 30 000 to 25 000 years BP temperatures decreased between 25 000 and 12 500 years BP to around 6 °C lower than today, causing the expansion of drier, savanna-type vegetation (LeHouérou 1997; Stager *et al.* 1997). Temperatures then began to increase again culminating in values similar to today around 10 500 years BP. However, from 10 000 years BP to the present day, the region saw fluctuations in climate, such as a humid period between 10 000 and 7200 years BP and increased seasonality between 7200 and 2200 years BP (Nicholson and Flohn, 1980; Stager *et al.* 1997). From around 2200 years BP human impacts on the land became dominant (e.g. Taylor, 1990), and somewhat complicate palaeoclimatic reconstructions.

The Serengeti
Of all of the East African savannas, probably the most well known and highly researched is the Serengeti ecosystem (Box 6.1) (Sinclair and Norton-Griffiths, 1979; Sinclair and Arcese, 1995). It is unique, not only for the two million wild ungulates that occupy the area and their migratory behaviour, but also for the distinctive features of the vegetation. This vegetation is composed of grasslands on the Serengeti Plain, and woodlands in the north and west. The latter are different from the tree/shrub savanna described above because of the insignificance of bushy plants, other than *Acacia* and *Commiphora*, and the relative abundance of perennial grasses. Although the Serengeti grasslands are not true savannas according to the definition outlined in Chapter 1, they will be considered in

---

Box 6.1 The Serengeti ecosystem

The Serengeti region of the East African Plateau comprises about 35 000 km^2 of grassland and savanna in northern Tanzania, extending into southern Kenya. The Serengeti plains or grasslands are organised as a continuum of vegetation with co-varying species composition and stature. For example, grasslands sampled at 105 locations throughout Serengeti National Park and the Maasai Mara National Reserve identified a core of grassland community types, each characterised by a similar species composition, that recur in different locations (McNaughton, 1983). The most widespread of these are various grasslands dominated by red oat grass (*Themeda triandra*) throughout the regions above a mean annual rainfall of about 700 mm. Short grasslands dominated by species of dropseed (*Sporobolus*) are abundant on the short-grass plains and top catenas in the western corridor, and tall grasslands of russet grass (*Loudetia*) and thatching grass (*Hyperthelia*) occur in the far north-west and the corridor (McNaughton and Banyikwa, 1995).

A fine-scale examination of the spatial distribution of short grasslands in eastern Serengeti plains identified community types even at the small scale (Banyikwa *et al.*, 1990). This indicates the spatial heterogeneity of plant communities in the grasslands, and the response of large mammals to these variations, as they forage across the landscape at certain times of the year (Senft *et al.*, 1987; McNaughton, 1989, 1991). This influence of herbivores especially affects grass species diversity (McNaughton, 1979). For example, fencing has been shown to reduce biodiversity, as taller species overtop others (McNaughton, 1984).

The woodland areas in the Serengeti ecosystem consist of an open stratum of *Acacia* and *Commiphora* thorn trees, mostly 3–7 m tall, but reaching 9–20 m in a few individuals. Understorey shrubs are poorly represented and include species of *Grewia* and *Cordia*. The grass stratum is 0.5–1.5 m tall and dominated by species such as *Digitaria macroblephara*, *Themeda triandra*, *Eustachys paspaloides* and *Pennisetum mezianum*.

---

this chapter as they are part of a greater savanna system. They are also important for the purposes of indicating the interactions among savanna determinants such as fire and particularly herbivory, and different savanna vegetation physiognomies, i.e. grassland and woodland.

Plant life strategies

The phenological cycles of plants in the East African savannas are largely controlled by the rainfall regime (Prins, 1988) (Table 6.2). Woody plants can be divided into deciduous, non-riverine evergreen, and riverine evergreen species, and grass species into sub-groups of annuals and perennials. What Table 6.2 indicates is that although rainfall in East African savannas may be strongly seasonal, plant growth is not, and only a limited number of species, mostly

Table 6.2 Phenological groups of woody and grass species from savannas in Lake Manyara National Park, Tanzania (after Prins, 1988)

Phenological group	Growth patterns	Examples of species
Woody plants		
Deciduous	Shed their leaves at end of July, two months following the end of the 'long rains', and produce new leaves when accumulated rainfall subsequent to end of long dry season is approximately 150 mm	*Acalypha fructicosa, Barleria eranthremoides, Indigofera arrecta, I. uniflora, Ocimum suave, Tephrosia villosa*
Non-riverine evergreen	New leaves emerge during the long dry season	*Acacia tortilis, Capparis farinosa, Cordia sinensis, Maerua triphylla, Ruellia megachlamys, Salvadora persica, Vernonia cinerascens*
Riverine evergreen	New leaves produced just prior to first rains that succeed long dry season	*Cardiogyne africana, Cordia ovalis, Dovyalis xanthocarpa, Gardenia jovis-tonantus, Justicia cordata, Thylachium africanum, Vangueria acutiloba*
Annual grasses		
Group I	A very short growing season and seedling emergence is during the middle of the 'long rains' in April, after total accumulated rainfall is about 400 mm	*Aristida adscensionis, Cymbosetaria* spp., *Digitaria* spp.
Group II	A medium growing season and seedling emergence is earlier, following a total accumulated rainfall of about 150 mm	*Brachiaria deflexa, Chloris virgata, Dactyloctenium aegyptium, Sporobolus* spp., *Tragus* spp.
Perennial grasses		
Group I	Sprout immediately after the first rain following the long dry season after a total accumulated rainfall of about 50 mm	*Eragrostis* spp., *Panicum maximum, Sporobolus pyramidalis*
Group II	Sprout immediately after the first rain following the long dry season, 20 days earlier than Group I	*Cenchrus* spp., *Chloris* spp., *Cynodon* spp., *Enneapogon* spp., *Enteropogon* spp., *Heteropogon* spp., *Urochloa* spp
Group III	Sprout and remain in leaf during all rainy periods. Generally present on places with a relatively high groundwater table	*Chloris gayana, Cynodon dactylon, Sporobolus* spp.

annuals and deciduous woody plants, have a well-developed seasonality. Prins (1988) suggests that this may have important implications for the herbivores in the savanna, as the perennial grasses and evergreen woody plants may act as food buffers and may 'protect' herbivores from starvation during low rainfall years.

Evidence for spatial heterogeneity, as well as temporal heterogeneity, in East African savanna plants, comes from recent studies which have documented that isolated trees may improve understorey productivity, as well as herbaceous and faunal diversity (Belsky *et al.*, 1989; Weltzin and Coughenour, 1990; Vetaas, 1992; Belsky and Canham, 1994). This increased productivity under tree canopies may be because of interactions between tree/grass competition, enhanced soil fertility, and the effects of shade (Belsky, 1992a, 1994). For example, Belsky *et al.* (1993), working in Tsavo National Park, Kenya, found that trees probably contributed to soil fertility by enriching crown-zone soils with N, most likely added in animal droppings, airborne deposits and inputs of tree litter.

6.1.3 Fauna

The East African savannas are probably best known for the huge diversity of wildlife they support. This ranges from large herbivorous mammals (23 different species occur in the Serengeti ecosystem alone), to carnivores, birds and insects. Although there is an abundance of some species such as wildebeest, other animals are under threat from various pressures, particularly illegal hunting for food and commercial purposes (see Section 6.4.1).

6.2 The main savanna determinants

Although the East African savannas conjure up images of large herds of native mammalian herbivores roaming the landscape, and apparently exerting a controlling force over them, these animals are not the dominant forces affecting vegetation pattern. They have a definite impact on their environment, significantly affecting changes in plant community composition and diversity (Lock, 1972; Edroma, 1981; Belsky, 1986a,b,c, 1992a), vegetation structure (Belsky, 1984; Smart *et al.*, 1985), biomass (Strugnell and Pigott, 1978), chemical and physical properties of soils (Lock, 1972; Hatton and Smart, 1984), soil erosion (Glover, 1963; Sinclair and Fryxell, 1985) and fire frequencies (Norton-Griffiths, 1979), but are much more likely to be influenced by existing vegetation patterns than to influence them (Belsky, 1995). The more important determinants are geomorphology, soil chemistry and soil moisture (Tinley, 1982; Cole, 1986; Belsky, 1991; Coughenour and Ellis, 1993).

6.2.1 Plant-available moisture

Plant productivity in East African savannas is limited foremost by soil moisture (Deshmukh, 1984; van Wijngaarden, 1985). There is a direct increase in productivity with increasing rainfall, accompanied by changes in species composition and vegetation structure. For example, in the Serengeti National Park, rainfall increases from less than 450 mm in the south to more than 1100 mm in the north (Belsky, 1983). Soils in southern Serengeti (the plains) are derived from volcanic ash ejected from nearby volcanoes, and are therefore high in alkalinity, sodicity and nutrient content. This decreases as one moves northwards, the increasing rainfall leaching salts out of upper soil horizons (de Wit, 1978). This rainfall/edaphic gradient determines the plant communities: short grasslands dominate the southern half of the Park, and shrubland and woodlands dominate the northern half (Belsky, 1989). The vegetation of the plains, particularly, may be determined by the shallow, extremely sodic soils, rather than low soil moisture, as trees are present in even drier savannas (Belsky, 1991).

The catenary structure of the East African savannas means that the soil sequences along the catena are vegetated by different plant communities (Jensen and Belsky, 1989). Ribbons of tall, green dense vegetation grow along water-courses in otherwise short, dry grasslands (Belsky, 1995). These can occur on the edges of permanent rivers and lakes, but also along seasonal drainage lines. Shallow sandy soils that dominate the ridge-tops of the catena are vegetated by short, shallow-rooted species, while the heavy clay soils that accumulate in the more mesic valley bottoms are vegetated by taller and more deeply rooted species. In between are medium-height grasses (Morison et al., 1948; Bell, 1970).

The mosaic nature of the East African savannas, determined primarily by PAM, is best illustrated in places like the Rift Valley. Here, the sharp relief, such as along escarpments, strongly determines the vegetation (Prins, 1989). Within Lake Manyara National Park in northern Tanzania, the escarpment rises 300–700 m from Lake Manyara on the valley bottom, up a steep slope, to the plateau on top. The patches along this gradient vary with distance above the water-table, water infiltration rate, geological substrate and soil salinity (Greenway and Vesey-Fitzgerald, 1969; Prins, 1988). Accordingly, vegetation types vary along the gradient, from waterlogged grass plains near the lake and better drained, wet ground forests near the base of the escarpment, to dry bush-lands on the excessively drained slopes, and then to arid grasslands on the upper plateau (Loth and Prins, 1986). This high diversity of patch types, and year-round availability of water from the lake, has resulted in Lake Manyara National Park having the highest density of mammalian herbivore biomass in Africa (Coe et al., 1976).

Adaptations to PAM

East African savanna plants have various adaptations to PAM including deep roots and phenological cycles (Loth and Prins, 1986). Most savanna grasses follow the C_4 photosynthetic pathway, adapted to the moisture-limited conditions of savannas (see Section 1.3.1). However, studies on East African grasses show a further step in this adaptation. Grasses occupying the most arid regions and growing during the hot dry season belong to the NAD-ME subtype of C_4 plants, whereas annual grasses showing seedling emergence in the long rainy season are of the NADP-ME subtype (Prins, 1988). NAD-ME species appear to be adapted to hot arid conditions, while NADP-ME species are adapted to hot moist conditions (Vogel *et al.*, 1986).

6.2.2 Plant–available nutrients

Data show that there is substantial local variation in plant-available nutrients (McNaughton and Banyikwa, 1995). This variation can produce considerable local patchiness in grass growth, with high growth rates in patches where mineralisation rates are high, and low rates in nearby patches. Experimental studies by Wedin and Tilman (1990) indicate a strong feedback from grassland species composition to soil processes, and suggest that local vegetation spatial patchiness can produce equally heterogeneous patterns of soil processes. For example, accumulating species may be dominant in patches with a high Ca content in the underlying soil, and the tendency of such species to accumulate Ca in their tissues will progressively enrich the spots they occupy.

PAN and grass patchiness

In the Serengeti, grasslands contain patches of variable size and species composition, or form distinct two-phase mosaics (Anderson and Talbot, 1965; Schmidt, 1975; de Wit, 1978; Belsky, 1983, 1985, 1986a). A study by Belsky (1988) found that subsurface soil concentrations of sodium had the greatest positive influence on the amount and intensity of grassland pattern (followed by presence of mound-building termites). High sodium concentrations are toxic to most plant species (Epstein, 1972), so conditions where this can be leached out of the soil improve plant growth (de Wit, 1978; Belsky, 1986a, 1988). Mound-building termites can create highly fertile patches by bringing nutrient-rich soil above ground and by adding organic matter to it (Belsky, 1983; Jones, 1990). It is estimated that the combined organic matter consumption by soil- and litter-feeding termites in a dry East African savanna is around 27 g m^{-2} year^{-1} (Josens, 1983). In many areas, these fertile patches may eventually be colonised by woody species, creating dense woodland clumps (Herlocker, 1976; Jager, 1982).

The two-phase grassland mosaic is illustrated by the *Andropogon greenwayi* community in the Serengeti plains (Figure 6.2). Patches range from < 1 m to >100 m over an area of 800 km^2 (Belsky, 1986a). One phase is found on neutral soils with rapid infiltration rates, dominated by 75–100% cover of the perennial grass *A. greenwayi*. The other has more sodic soils, slower infiltration, and only 10–70% cover by a mixture of the annual grass *Chloris pycnothrix* and other grasses. Belsky (1986a) suggested that slight topographical differences in drainage in the highly sodic soils create patches with different salt concentrations. More highly leached patches are eventually colonised by *A. greenwayi* and more poorly leached (and sodic) patches by *C. pycnothrix* and other species. As a result of continued differences in water infiltration and drainage between the two phases, differences in soil salinity and species composition between the two phases are maintained.

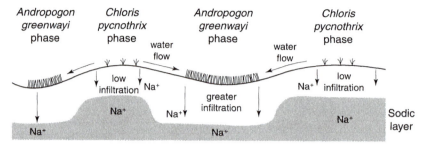

Figure 6.2 The *Andropogon greenwayi* two-phase grassland mosaic community in the Serengeti plains, Tanzania. Note that water flows over the soil surface from the *Chloris pycnothrix* phase into the *Andropogon greenwayi* phase due to slight differences in topography. Infiltration is greater in the *Andropogon greenwayi* phase, so the sodium is leached out, and together with the additional water allows increased plant growth (after Belsky, 1986a).

PAN and herbivores

Herbivores can affect plant-available nutrients both in the short and long term. Urination, for instance, can have a rapid, promotive effect on plant nitrogen content (Jarmillo and Detling, 1988), and a subsequent marked preference for grass on these patches (Day and Detling, 1990). Over decades, herbivores can create areas of nutritional sufficiency as a result of high usage (in the absence of intrinsic soil differences), promoting higher fibre and mineral values (Georgiadis and McNaughton, 1990). For example, in Tanzania, aardvarks (*Orycteropus afer*) and warthogs (*Phacochoerus aethiopicus*) excavate dens and tunnels, bringing sodic soils to the surface that are only slowly colonised by early successional species (Belsky, 1985).

6.2.3 Herbivory

Herbivory in East African savannas is best illustrated by the great wildlife migrations of the Serengeti ecosystem. Here, over 1 300 000 wildebeest (Sinclair, 1979), as well as zebra and Thomson gazelle, migrate from the northern woodland savannas to the southern shortgrass plains at the beginning of each rainy season, and return during the dry season (Maddock, 1979). The reasons for leaving the plentiful northern savannas during the wet season are not fully understood, and several hypotheses have been put forward (Table 6.3). Nevertheless, these migrations show that rather than creating them, large mammals are influenced by major vegetation patterns (Belsky, 1995).

The different types of herbivores

There are two types of herbivore population in East Africa: resident herds, occupying distinct home ranges throughout the year; and migratory herds, dominating animal biomass (Fryxell *et al.*, 1988) (Figure 6.3). The resident herds have two different types of grazing behaviour: sustained-yield grazing during the wet season; and rotational-passage grazing during the dry season. In the first,

Table 6.3 Hypotheses for migration from north to south by herbivores at beginning of the rainy season in the Serengeti ecosystem (based on Belsky, 1995; discussed in Maddock, 1979)

Hypotheses	Authors
1. The grasses in the southern plains are more nutritious, and have more nitrogen and calcium for lactating females	Kreulen (1975) McNaughton (1990)
2. The shortgrass plains allow greater avoidance from predators	Darling (1960)
3. Animals dislike the wet, muddy soils of the northern savannas in the wet season	Talbot and Talbot (1963)
4. By moving between regions, migrants more fully utilise total plant productivity in the park	Jarman and Sinclair (1979)

they occupy a limited area for an extended time, moving back and forth through the area as the rains continue (McNaughton, 1985). These grasslands are typically of short stature. When migratory herds arrive at the Serengeti plains at the onset of the rains, they show the same behaviour.

In contrast, during the dry season, resident herbivores move rapidly between medium to tall grasslands, following sporadic dry-season showers (McNaughton, 1985). The transition from a sustained-yield to a rotational-passage grazing system can be very abrupt at seasonal transitions (Inglis, 1976). Grazers track rainfall-driven pulses of primary productivity with considerable

Figure 6.3 Migratory wildebeest.
(Photograph by the author)

accuracy: grazer densities increase and decrease according to primary production on localised patches (McNaughton, 1985). In locations with above 700 mm of rainfall, resident herds are found concentrated in localised 'hot-spots' of vegetation (McNaughton and Banyikwa, 1995). These 'hot-spots' are associated with higher concentrations of the nutrients sodium and phosphorus in grass leaves, particularly important during late pregnancy and lactation for females and early growth for calves (McNaughton, 1988).

Migratory herds, on the other hand, move along a nutritional gradient that is opposite to the rainfall gradient (McNaughton, 1990). During the wet season, they concentrate in the most arid parts of the savanna, which is related to underlying higher soil nutrient levels. Calving is also most frequent during this period, suggesting that there has been natural selection for a migratory pattern synchronising reproduction with high rainfall, and the spatial distribution of nutritionally suitable forage (McNaughton and Banyikwa, 1995).

Grazing and habitat heterogeneity

Vegetation bands associated with catenas strongly influence herbivory. Bell (1970) found that large ungulates in Serengeti National Park prefer the short-grass communities on the higher parts of the catena during the wet season, but move down the catena as the vegetation senesces during the dry season. There is a 'grazing succession' (Vesey-Fitzgerald, 1960) of animal movement, with the largest species, zebra (*Equus burchelli*), moving down slope first, followed, over the following several weeks, by intermediate-sized ungulates such as topi (*Damaliscus korrigum*) and wildebeest (*Connochaetes taurinus*), and then finally by the smallest animals, the Thomson gazelle (*Gazella thomsoni*) (Bell, 1970). This grazing succession is reversed in the wet season. The sequence of the succession may depend on herbivores having differing tolerances for the lower catena grasses, which are taller and more fibrous. It may also depend on an improvement in the quality of these grasses as they are grazed and trampled by other animals (Vesey-Fitzgerald, 1960). Browsers similarly move up and down the catena in response to variations in woody species. For example, the impala (*Aepyceros melampus*) moves down the catena in the dry season, and up the catena in the wet season.

Evidence from Lake Manyara National Park indicates that the spatial relationship of patches affects herbivory (Prins and Beekman, 1987; Prins and Iason, 1988). During the day, the buffalo (*Syncerus caffer*) occupies the cool lakeshore habitats where they graze while avoiding tsetse flies, and in the evenings they enter the woodlands to graze. Passing between the grassland and woodland puts the buffalo at greater predatory risk from lions (*Panthera leo*), but the movements occur twice daily, suggesting that habitat patches are being used in response to food availability, temperature and insect pests, not for protection from predators (Sinclair, 1977; Prins and Iason, 1988). In most seasons buffalo prefer grazing in *Cynodon dactylon*-dominated patches, but in the dry season they more frequently utilise *Cyperus laevigatus* swamps, *Chloris gayana* patches, and woodlands (Prins and Beekman, 1987).

Although large mammals may not be responsible for creating many landscape patterns, they do contribute to maintaining or intensifying them. In the *Andropogon greenwayi* grassland mosaic, for example, herbivores maintain the two-phase pattern by grazing and compacting the soil (Belsky, 1986a). If this community is protected, the mosaic disappears. Similarly, woody invaders of grassland are destroyed by herbivore browsing and bush-beating, thereby maintaining distinct woodland edges (Belsky, 1984).

The role of elephant herbivory in savanna grassland–woodland dynamics

Until the middle of this century, woodlands covered much of the East African savannas. However, from the 1950s and 1960s, much of the woodlands were converted to open savanna. For example, in Tsavo National Park, large areas of *Commiphora–Acacia* woodlands and shrublands were converted to open savanna, grasslands and bare ground (Glover, 1963; Agnew, 1968), and 50% of woodlands in northern Serengeti National Park were converted to open savanna (Norton-Griffiths, 1979; Dublin, 1995). Elephants and fire were thought to be the major causes behind the conversion (e.g. Lamprey *et al.*, 1967; Laws, 1969, 1970; Laws *et al.*, 1970; Western and van Praet, 1973; Croze, 1974a; Phillips, 1974; Leuthold, 1977; Lock, 1977; Norton-Griffiths, 1979; Barnes, 1985).

The persistence of the grasslands over the recovery of woodlands has been attributed mainly to elephants (Figure 6.4). For example, there have been significant changes in the numbers and distribution of elephants in the Serengeti–Mara ecosystem over the past 20–25 years (Dublin and Douglas-Hamilton, 1987). Increased poaching within the Serengeti National Park during the late 1970s and early 1980s led to a population decline of 81% in the

Figure 6.4 Elephants in the Serengeti–Mara ecosystem.
(Photograph by the author)

park (from 2460 in 1970 to 467 in 1986) (Dublin *et al.*, 1990a). High tourist exposure and increased anti-poaching efforts provided a secure home in the Mara, and there elephant numbers have increased to about 1300 today (Dublin, 1995). The effects of these year-round residents of the Mara are pronounced.

Mara elephants eat woody species of all types, their use increasing during the dry season when grass forage is reduced (Dublin, 1986). Thickets of *Croton* provide one of the last wooded refuges for elephants, where shade and forage is available (Dublin, 1991). The constant use of the thickets has caused the formation of large pathways through the vegetation, and in subsequent rainy years, these gaps are colonised by grasses, which serve to fuel damaging fires (Norton-Griffiths, 1979). As a consequence, trees and bushes are destroyed, and over time, the thickets become increasingly fragmented.

Acacia woodlands are subject to similar pressures, especially during the dry season, when they are heavily browsed by elephant bulls (Dublin, 1995). In addition, elephants spend significant time feeding on seedlings and saplings under 1 m in height, therefore inhibiting regeneration (Dublin, 1986). There is a notable change in the elephant diet from grasses to woody species during the dry season, when wildebeest arrive in the Mara and graze the grasses down to lawn height (McNaughton, 1984; Dublin, 1986). This is probably because of a decline in grass quantity, but also quality. Woody species remain relatively high in crude protein when compared with herbaceous species as the dry season progresses (Sinclair, 1975; Pellew, 1981). Although elephant browsing increases during any period of low rainfall, the added effect of the wildebeest migration may exacerbate this pattern (Croze, 1974b).

Dublin (1986) established that 4% of all seedlings were killed annually by elephants, 4% by fire, 1% by wildebeest, and another 1% through other natural causes. The seedlings experienced the greatest impacts from elephants and other browsers in the dry season, and that more stems were removed in burned areas. The majority of seedlings removed at ground level by wildebeest or fire resprouted within six months, but those taken by elephants took much longer to recover. Multiple-burn experiments demonstrated that seedling survivorship was inversely related to the level of fire intensity. Seedling survivorship remained high after repeated cool burns (fuel loads $\leqslant 150$ g m^{-2}) but dropped significantly after repeated hot burns (fuel loads > 300 g m^{-2}) which are often characteristic of Serengeti–Mara grasslands.

Although elephants seem to have a profound effect on woody plants, Dublin *et al.* (1990b) suggest that fire alone was responsible for the decline in woodlands during the 1960s. Results from photo-points on regeneration rates in the Serengeti since the 1920s and from age distributions confirm this hypothesis (Sinclair, 1995). The unusually high rainfall had allowed substantial fuel to build up to serve fires. Elephants only exacerbated the losses further. The lack of woodland recovery in the Mara is the consequence of continued elephant pressure. The northern Serengeti, on the other hand, where elephants were virtually absent in the mid- to late 1980s, has shown a tremendous regeneration of woody species, particularly *Acacia clavigera* and *Acacia gerrardii* (Sinclair, 1995).

Although burning rates are not as severe as in the past, fires still occur in the

Mara on an annual basis. Most occur during the short dry season (January–February) when the migratory herbivores are not present. These fires inhibit the growth and survivorship of seedlings. If wildebeest numbers were ever reduced, fires would become important during the dry season, and play an important role in the inhibition of woodland recovery. Elephants also prevent woodland recovery, which may take place if elephants return to the Serengeti, as they have been doing since 1990.

Herbivore interactions: another explanation?

Recent ideas suggest that shifts in grassland–woodland may not be determined by elephants and fire only. Prins and Van der Jeugd (1993) found that a dramatic reduction in impala (*Aepyceros melampus*) numbers following outbreaks of rinderpest and anthrax enhanced seedling recruitment of *Acacia tortilis*. These authors suggest that when the browsing impact of impala and other medium-sized herbivores is high, *Acacia* seedling establishment is rare, but that this impact decreases as foliage on maturing trees grows above the reach of these herbivores. Although the influence of elephants on mature trees is indisputable (Buss, 1990), smaller herbivores may play a greater role on grassland stability, when epidemic disturbances among this group of ungulates may create narrow windows for seedling establishment (Prins and Van der Jeugd, 1993).

Van de Koppel and Prins (1998) argue that interactions between small herbivores such as impala or buffalo, and large herbivores such as elephant, may explain the dynamics of African savannas (Figure 6.5). Their model predicts that at a high plant standing crop, facilitation dominates herbivore interactions, and at a low plant standing crop, competition prevails. For example, elephants may facilitate small herbivores by opening up dense thickets (Pellew, 1983;

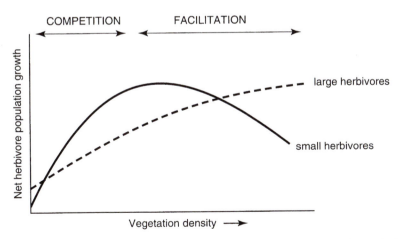

Figure 6.5 A model to show the interactions between small and large herbivores in African savannas. Note that at a low vegetation density competition favours small herbivores, but as the vegetation density increases, facilitation by large herbivores helps to increase their population (after Van de Koppel and Prins, 1998).

Jachman and Croes, 1991; Dublin, 1995; Prins, 1996; Prins and Olff, 1998), and creating gaps which promote grass production (Norton-Griffiths, 1979). Van de Koppel and Prins (1998) propose that at a low plant standing, elephant density may decrease not due to the unsuitability of the grassland, as suggested by Caughley (1976), but because of competitive exclusion by smaller herbivores. For example, De Boer and Prins (1990) report a negative influence of buffalo grazing on that of elephant, while Dublin (1995) states that the general tendency for elephants to over-utilise woodlands in the dry season could be due to direct competition with wildebeest.

The question of whether a lack of high quality food or competition causes the decrease in elephant density in grasslands has not yet been answered and needs more research. The model by Van de Koppel and Prins (1998) does not consider the influence of fire, which, as the authors state, can be regarded as a 'super-herbivore'. It is therefore probable that both elephants and fire, through complex interactions, are responsible for the woodland–grassland transitions seen in past years.

Adaptations to herbivory

Since herbivory is a dominant feature in East African savannas, many plants have various adaptations to inhibit and reduce grazing and browsing. In some cases, the solution is simple: plants such as citronella grass (*Cymbopogon*) are extremely unpalatable due to the fragrant and distasteful chemicals they contain. Woody plants also contain indigestible chemicals, such as tannins. Thorns are particularly conspicuous on many African savanna trees, including the dominant *Acacia* and *Commiphora* species.

More complex herbivore defence mechanisms include mutualism, the most well-documented being the *Acacia* and ant relationship (e.g. Cronin, 1998). *Acacia cornigera*, for example, is one of five 'swollen thorn acacias' which are colonised by the aggressive stinging ant *Pseudomyrmex ferruginea* (Young *et al.*, 1997). The *Acacia* species maintains ants using nutritious nodules on the ends of leaflets and with extrafloral nectaries, while the ants help fight against herbivore and other insect attack. Madden and Young (1992) investigated the occurrence of aggressive ants in an apparently symbiotic relationship with the savanna tree *Acacia drepanolobium* and their effects on giraffe herbivory in Kenya. Ants were found to be concentrated on shoot tips, the plant parts most preferred by giraffes, and trees with relatively more foliage had more swarming ants than did trees with less foliage. Also, the thorns of *A. drepanolobium* are significantly shorter than are the thorns of *A. seyal*, a species without symbiotic ants.

6.2.4 Fire

Although natural fires from lightning have probably occurred since the time savanna vegetation could support burning (Mworia-Maitima, 1997), it was only when people began to use fire for hunting and vegetation clearance, probably around 60 000 years ago (Cole, 1964), that fire became a major ecological

Table 6.4 Some fire behaviour characteristics under different fuel conditions in a Ugandan savanna (after Mucunguzi and Oryem-Origa, 1996)

Parameters	Fuel conditions		
	Low (0.25 kg m^{-2})	Medium (0.5 kg m^{-2})	High (1.0 kg m^{-2})
Flame height (m)	0.47	0.65	0.94
Rate of spread (m s^{-1})	0.01	0.01	0.01
Fire intensity (kJ m^{-1} s^{-1})	45	90	180

force in East African savannas. Today, people use fire to improve hunting, to clear areas for farming, to stimulate new grass growth for livestock, to make footpaths and firebreaks, to reduce weed populations, to collect honey and to make charcoal (Woube, 1998).

Unfortunately, there have been few studies on fire behaviour and its effects on savanna ecology in East Africa, compared to other African and world savannas. Most information comes from studies conducted to look at the effects of fire on grazing in savanna communities (e.g. Belsky, 1992b; Dublin, 1995). For example, Belsky (1992b) found that early dry season fire had little effect on the cover of individual grass species from the Serengeti National Park. She hypothesises that this may have resulted from fire simply duplicating the effects of intense grazing by reducing above-ground biomass.

Mucunguzi and Oryem-Origa (1996) measured fire behaviour (Table 6.4) in order to record its effects on germination of *Acacia* seeds in Uganda. They found that fire intensity and flame height increased with an increase in fuel loading. The authors also showed that *Acacia sieberiana* seeds had a higher resistance to fire than *A. gerradii*, suggesting better adaptation to fire by the former; and that survival and germination of seeds after fire increased with depth of burial.

6.3 Human determinant of the savanna

The East African savannas are considered the place where people first evolved (Clark, 1980). Stone Age people were hunters and gatherers, and although fire was first used nearly 60 000 years ago, it probably was not until the introduction of agriculture into the region that more widespread clearance and modification of savannas took place. Extensive agriculture, as well as iron-working, are thought to have been introduced into East Africa by Bantu-speaking people between AD 1 and AD 500 (Soper, 1969). This probably facilitated a more sedentary lifestyle, and allowed an increase in population numbers. The arrival of Europeans during the nineteenth century was followed by a further increase in population, mainly attributed to the introduction of western medicine and the reduction of tribal warfare. Further changes in savanna lands came about through the rinderpest epidemic, the implementation of new cash economies, the establishment of tsetse-fly control campaigns, and the large-scale hunting of native wildlife.

6.3.1 Pre-colonial land use

Prior to colonisation, the East African savannas were used for subsistence agriculture and pastoralism, supplemented by hunting and gathering activities. The transition from Early Iron Age cultures to more complex societies was a continual process, in part a result of the Bantu-speaking cultivators expanding their numbers and developing skills needed to colonise new environments (Iliffe, 1995). However, it was also because of the gradual drift southwards into East Africa of Nilotic-speaking people from southern Sudan. Although pastoralism probably began in the East African savannas in the late third century BC (Clutton-Brock, 1995), these southern Nilotic pastoralists probably arrived during the first century BC, and established pastoralism as the basic lifestyle.

The eastern Nilotic pastoralists expanded slowly behind them, although their most powerful group, the Maasai, came to dominate the Rift Valley only in the seventeenth and eighteen centuries (Iliffe, 1995). Western Nilotes were cultivators and pastoralists, and began their expansion into East Africa in the second century AD, mostly into the Great Lakes region. The heterogeneous nature of the landscape resulted in an exceptionally uneven distribution of people, with islands of intensive cultivation amidst vast tracts of pasture lands. The interior of East Africa was one of the most isolated parts of Africa until the eighteenth century, after which long-distance trade, mostly in ivory and slaves, drew the region into the world economy (Iliffe, 1995).

Cultivation was both shifting and sedentary, and the main crops grown were sorghum, Eleusine millet and bananas. Cassava, maize, sweet potatoes and beans from the Americas were brought to East Africa later, when the first colonists arrived in the eighteenth century. The occurrence of trypanosomiasis, carried by the tsetse fly, was a major limitation on pastoralism, so many savanna areas had low levels of livestock. Land and resource management took place through common property regimes (see Section 6.3.2 below), which were governed through traditional authorities. Various systems were in place: Box 6.2 gives the example of the Barabaig people of Tanzania.

6.3.2 Colonisation and post-colonial land use

Most of the East African savannas were colonised by the end of the nineteenth century by the British, with the exception of Ethiopia, which remained independent apart from a brief occupation between 1930 and 1941 by the Italians. The British colonies, notably Kenya, became settler colonies, and by the 1920s, these Europeans formed small but influential communities (Davidson, 1990). One of the main effects of this was the redistribution and selling of land, normally the best land, to white settlers. For example, over two million hectares of the fertile highlands of Kenya were sold or given to European settlers by the British colonial government.

As technological changes were imported into the colonies, large-scale export agriculture began, the principal crops produced being coffee, cotton, sisal and tobacco. Ranches for large-scale cattle-raising were also established.

Infrastructure, principally the railways and communications, was set up in order to aid marketing of produce. In order to get natives to work on the settler land, people were first forcibly rounded up, and later made to pay taxes. In Kenya, for example, Africans were forbidden to grow cash crops on their own land, so that they would not have a share of the market, or earn enough money without having to work for Europeans. Traditional farmers were, in effect, turned into wage-workers. Labour was also drawn from areas with little immediate economic potential of their own, and migrant labour systems were well established by the 1920s (Siddle and Swindle, 1990).

The late nineteenth and early twentieth centuries saw two major events in the region. The first was the great outbreak of rinderpest, which destroyed livestock numbers and saw the decimation of many pastoralist communities, such as the Karagwe and Maasai. The second was the popularisation of the East African savannas as one of the favourite sites for big-game hunting. Famous western travellers, high-ranking colonial officials, politicians and aristocrats all travelled

---

Box 6.2 Traditional land management by the Barabaig of the Hanang Plains, Tanzania

The Barabaig are semi-nomadic pastoralists, numbering more than 30 000 and living in the Hanang district of eastern Tanzania. They mainly herd cattle, approximately six per head, but maize is also a staple food, obtained through trade of livestock, or by shifting cultivation. The climate of the area is semi-arid with periodic droughts, and an average rainfall of 600 mm. The dominant vegetation is *Acacia* and *Commiphora* woodland interspersed with open grassland.

Table 6.5 shows the resource availability and constraints for pastoralism in the landscape. Although there are variations in the environment, together with unpredictable rainfall, the Barabaig system is sustainable over time because it has flexible and opportunistic responses to the changing patterns of resource availability and production constraints. The key element to this management strategy is seasonal grazing rotation (Figure 6.6). Grazing plains are also managed using burning, practised mainly between September and November to burn off dry, dead vegetation, and stimulate new green shoots. Firebreaks around homesteads are usually burned preceding this. Fires are also used to reduce tick populations, a major carrier of livestock diseases in the area.

So as to facilitate the use of resources, the Barabaig have a common property tenure which entails a hierarchy of jural institutions that control access to and use of land. A public assembly of male adults have the ultimate control over land rights, but at the clan and neighbourhood levels, more local problems are governed. Access to water and trees are subject to strict rules, and there are regulations to control degradation, such as the restriction to permanently inhabit the *darorajega* or plains.

Box 6.2 (Continued)

Table 6.5 Resources and constraints for pastoralism in the savanna landscape for the Barabaig (after Lane and Scoones, 1993). Indigenous land type names in italics.

Land type	Geology	Soils	Tree species	Grass species	Constraints
Basotu plains (*darorajand*)	Calcareous volcanic tuff	Mollisols Deep fertile	*Acacia* spp., *Commiphora* spp., *Grewia* spp.	*Pennisetum* spp., *Aristida* spp., *Setaria* spp., *Heteropogon* spp., *Hypparhenia* spp.	Lack of available water in dry season. Grass failure in droughts
Bottomlands (*muhajega*)	Volcanic tuff	Vertisols Deep fertile	Few trees *Acacia drepanclobium*	*Eragrostis superba, Pennisetum mezianum, Cynodon dactylon*	Appropriation for wheat farms. Gully erosion in farms
Barabaig plains (*darorajega*)	Volcanic tuff	Mollisols Deep fertile	*Acacia* spp., *Commiphora* spp.	*Pennisetum* spp., *Aristida* spp.	Intensity of use. Lack of grass in poor rainfall
Bottomlands: river and lake margins (*ghutend*)	Fluvial and lacustrine deposits	Sandy/silty clay loam. Salt pans	Few trees	*Chloris gayana, Cynodon dactylon*	Heavy livestock pressure. Salinity
Hills/mountains escarpments/ ridges (*hayed*)	Volcanic tuff (plus granitic intrusions)	Lithic thin, stony, fertile clay loam	*Acacia* spp.	*Aristida* spp.	Mt Hanang forest reserve. Steep slopes
Miombo woodland (*darabet*)	Granitic	Loamy sands Sandy loams infertile	*Brachystegia spiciformis* and other miombo species	*Aristida* spp., *Brachiaria* spp., *Tragus* spp., *Dactyloctenium* spp.	Tsetse fly. Poor forage quality
Valleys and depressions (*mbuga*)	Granitic	Colluvial heavy fertile	*Cassia* spp., *Commiphora* spp. thickets	*Brachiaria* spp., *Dactyloctenium* spp.	Tsetse fly

Box 6.2   (Continued)

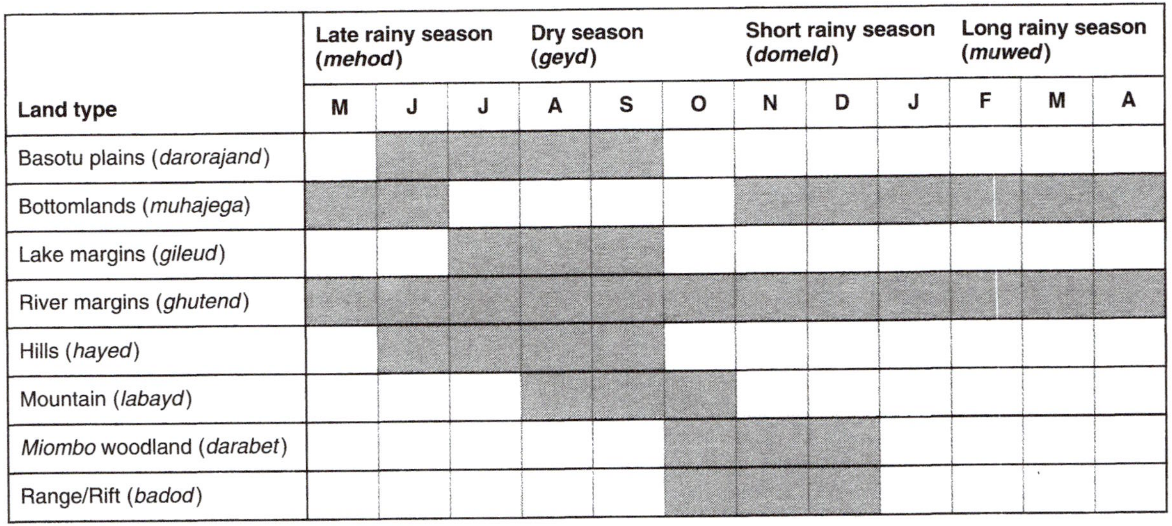

Land type	Late rainy season (mehod)		Dry season (geyd)			Short rainy season (domeld)				Long rainy season (muwed)		
	M	J	J	A	S	O	N	D	J	F	M	A
Basotu plains (darorajand)			▨	▨	▨							
Bottomlands (muhajega)	▨						▨	▨	▨	▨	▨	▨
Lake margins (gileud)				▨	▨							
River margins (ghutend)	▨	▨	▨	▨	▨	▨	▨	▨	▨	▨	▨	▨
Hills (hayed)			▨	▨	▨	▨						
Mountain (labayd)						▨						
Miombo woodland (darabet)							▨	▨				
Range/Rift (badod)						▨	▨	▨				

Figure 6.6   Traditional Barabaig grazing rotation management in different resource areas of the savanna landscape (after Lane and Scoones, 1993). Indigenous names in italics.

to the East African savannas to undertake safari hunting. For instance, between April 1909 and March 1910, the then US President Theodore Roosevelt travelled with over 200 trackers, skinners, porters and gun-bearers, and shot over 3000 specimens of African savanna game (Akama, 1996). Little (1996) comments that the accessibility and uncontrolled hunting by colonial parties proved disastrous for the region's wild animals, especially the large mammals. He goes on to say that by the 1940s, virtually no elephant or buffalo were left in the Baringo District of central Kenya. The ivory trade was also significant at this time. In the mid-nineteenth century, over 385 000 kg of ivory was exported from East Africa, and substantial exploitation continued well into the twentieth century (Spinage, 1973). This period saw a decrease in elephant impact on savannas, and the recovery of many woodland areas.

Colonisation also had the effect of changing demographic patterns in the region. Although there were huge losses of life from conflicts and resistance of natives to the colonial authorities and the forced labour regimes, later, particularly after World War I, population numbers rose. This may be attributed to the spread of roads and railways, allowing food to reach areas that had previously been isolated, and the introduction of western medicines. There was also an influx of immigrants into the region, notably the South Asians into Kenya, Tanzania and Uganda as indented labour to build the railways. As a result, urbanisation was also on the increase. All these factors had a huge influence on savannas and their use, either through direct destruction or through intensification of use.

Land tenure

As mentioned before, land tenure systems prior to colonisation were based on community-property systems, regulated by traditional authorities and values. After colonisation, much land was declared crown land and unequally distributed, with the best land going into the hands of the colonist settlers. African Native Reserves were also established, and restricted indigenous people to smaller and smaller areas for sustaining their livelihoods. In the independence era, the situation has changed variably in East Africa. For example, in Tanzania, the colonial land tenure system was replaced by the socialisation of land-ownership, whereas in Kenya, very little has changed (Bruce *et al.*, 1998b). The tenure systems of East African savanna states in 1996 are shown in Table 6.6.

6.3.3 Present-day land use

Savanna lands in East Africa are utilised for various purposes. Wood is a major source of energy for rural and urban populations, and is consumed for domestic as well as industrial means. Natural resources, in general, are important to local people. The Batemi agropastoralists of north-central Tanzania, for example, use over 90 species of woody plants for purposes including firewood, fencing, construction and cultural artefacts (Smith *et al.*, 1996). Johns *et al.* (1996) identified 53 cultivated food plants, 20 wild fruit species and 18 wild leafy vegetables in the Mara region of East Africa alone. Nevertheless, the

Table 6.6   Land tenure systems in East African savanna countries (after Bruce et al., 1998b)

Country	Official tenure objective	De facto dominant tenure type	Private ownership (freehold)		State leasehold		Community-based tenure		
			Exists	Significant	Exists	Significant	Indigenous, legally recognised	Other, legally recognised	Extensive
Ethiopia	State ownership	Alternative community based	Yes	Yes	Yes	No	Yes	Yes	No
Kenya	Private ownership	Private ownership	Yes	Yes	Yes	Yes	No	No	Yes
Somalia	?	?	No	No	Yes	Yes	No	No	Yes
Sudan	Mixed	?	Yes	Yes	Yes	Yes	?	No	Yes
Tanzania	Unclear	Alternative community based	No	No	Yes	Yes	No	Yes	Yes
Uganda	Mixed	Indigenous community based	Yes	Yes	?	Yes	Yes	No	Yes

economies of the region are dominated by plantation and subsistence agriculture, and pastoralism, although tourism plays a significant role in revenue earning in Kenya. This also probably reflects the long history of wildlife tourism in the country, stemming from the big-game hunters of the early twentieth century, and the establishment of game or safari parks later in the 1940s.

Agriculture and pastoralism

Agricultural production in East African savannas is mainly by small-scale subsistence farmers and pastoralists (Tables 6.7 and 6.8). The main crops grown are maize, wheat, sorghum and millet, although other vegetables and pulses are also important. Agriculture accounts for over 80% of exports, the two dominant crops of the region being tea and coffee. Pastoralism provides meat and dairy products to the region, but in terms of exports, only plays a significant role in Somalia and Sudan.

Tourism

Although tourism in the savanna areas of Tanzania, Uganda and Ethiopia were historically important, political upheavals and economic decline in these countries led to their demise in the tourist industry. However, they are making a recovery. In Tanzania, tourism is currently the fastest growing industry. It expanded by 600% from 1985 to 1990, and in 1992 earned over US$129 million (Neumann, 1995). Nevertheless, Kenya is the most popular destination for visitors wanting to see the big animals of the savannas. In 1995, there were

Table 6.7 Production figures for maize, coffee, tea and cattle for East African savanna states in 1998 (based on FAO, 1999). Crop production figures in million tonnes, and cattle per head.

Country	Maize	Coffee	Tea	Cattle
Ethiopia	2 500 000	204 000	700	29 900 000
Kenya	2 600 000	57 000	280 000	14 116 100
Somalia	121 000	?	?	5 200 000
Sudan	55 000	?	?	33 350 000
Tanzania	2 822 401	34 020	27 000	14 302 000
Uganda	750 000	180 000	23 000	5 370 000

Table 6.8 Total and agricultural populations of East African savanna states in 1997 (based on FAO, 1999). Figures in millions of people.

Country	Total	Agricultural
Ethiopia	60.1	50.3 (84%)
Kenya	28.4	21.8 (77%)
Somalia	10.2	7.4 (73%)
Sudan	28.0	17.7 (63%)
Tanzania	31.5	25.1 (80%)
Uganda	21.0	16.7 (80%)

1 883 900 visitors to national parks and game reserves in Kenya (Central Bureau of Statistics, 1996). In 1994 tourism earned approximately 12% of Kenya's gross domestic product (Economic Survey, 1995). However, this wealth does not come without difficulties, and conflicts over wildlife resource management and pastoralists are ongoing debates. This is discussed more fully in Section 6.4.1.

6.4 Current management problems and strategies

Tourism is one of the most challenging management problems in the East African savannas. Visitors have been attracted to the region since colonial times, and this continues today. The savannas offer diverse activities including eco-tourism, sports holidays and cultural and archaeological interests, but over-whelming, most tourists come to see the diverse wildlife, or 'big game'. However, tourism is linked to problems of conservation, which in turn have influenced land distribution, and conflicts still abound between 'saving' wildlife and sustaining peoples' livelihoods.

6.4.1 Conservation

Conservation, i.e. setting aside land for the preservation of big game, in East African savannas began seriously during colonial times, when it was realised that the big-game hunting was unsustainable, and that large populations of mammals had been significantly reduced. Game reserves were set up, and although hunting was prohibited in these areas, licensed shooting was still allowed. It was not until the 1930s that the effects of over-exploitation were fully realised, prompting the colonial government to search for areas of protection. This led to the development of national parks, the first of which, in Kenya, was set up near Nairobi in 1946 (Kock, 1995).

Although there has been a long history of conservation efforts in East African savannas, it has been directed from colonial, western views of conservation (Akama, 1996), characterised by limited African involvement in state conservation activities. Thus, when East African states became independent, international environmental NGOs, such as the IUCN, now the World Conservation Unit, began to aid de-colonised nations in planning and managing their conservation areas. Neumann (1995) concludes that there was a distinct continuity between colonial and post-colonial situations. Park and wildlife laws have not been significantly changed, boundaries remain mostly unaltered, forced relocations persist, and international agencies still play a critical role.

Conservation and tourism
The delimitation of conservation areas has had varying effects in the savannas. The most obvious benefit has come from the tourist industry in terms of revenue, although as Akama (1996) points out, most revenue from national parks goes entirely to the national government and tour operators, and few financial resources are allocated for local development. The major negative

effect has been the restriction to use resources within protected areas, and the inability of local people to protect themselves and their property from wildlife damage. In Kenya, state law prohibits any form of wildlife killing.

Pastoralists have perhaps been most severely affected by the establishment of conservation boundaries. The following case of the Maasai illustrates this well. It was decided in 1993 that an ecotourism site, the Serena Lodge, funded by the Aga Khan Foundation, would be located at Kimba in the Ngorongoro Conservation Area on land belonging to the Maasai village of Oloirobi (Neumann, 1995). It was stated by the investors that the lodge would provide money for conservation, reduce poaching and promote the welfare of the local communities. However, the site at Kimba has two springs which provide the only sources of reliable water during the dry season, and the lodge was built right on the main western livestock access route in and out of the crater. The Maasai were never consulted or involved in the planning of the project, and they fear that as well as the water sources being relocated to the lodge, their livestock will not be tolerated in the locality (Neumann, 1995).

Poaching

Illegal hunting has been and still is a major conservation issue in East African savannas. In Kenya, although certain laws, such as the Wildlife Act of 1976, were established to help reduce illegal hunting, it continued unabated until the end of the 1980s (Kock, 1995). In Kenya, from the 1970s to 1989, populations of elephant declined by 85% and rhino by 97% (Cumming et al., 1990). The ivory and horn trades were responsible for the huge decline (e.g. Barbier et al., 1992). In recent years, poaching in some countries like Kenya has decreased as a result of the establishment of a more independent Wildlife Service and community wildlife activities. Still, illegal hunting does continue. For example, in the Serengeti National Park, it was estimated that the total annual off-take between 1992 and 1993 was 159 811 wildlife, 44 958 resident and 111 691 migratory mammalian herbivores – the equivalent of 11 950 tons of meat (Hofer et al., 1996). This was mainly wildebeest, zebra and impala.

6.4.2 Desertification and land degradation

Pastoralists in East Africa have a long history of conflict with government authorities. In colonial times, they were seen to be causing overgrazing, and several destocking programmes were initiated in an attempt to reduce grazing pressure (Little and Brokensha, 1987). Many of these were abandoned after the negative and sometimes violent responses of herders. However, in the 1940s and 1950s, efforts were made by the colonial governments to introduce new resource management techniques such as rotational grazing. This only resulted in further alienation of pastoralists from the state. With the independence of savanna states, agricultural production was seen to be the economic driving force, and many pastoral lands were settled by farmers, further reducing their grazing areas.

This behaviour comes from an inherent belief by governments, inherited from colonial notions, of pastoralists being unable to contribute to a national economy. For example, although Tanzania is ranked as one of Africa's highest cattle resource countries, livestock production has failed to satisfy national demands for meat and dairy products, and failed to exploit potential export markets (Lane and Scoones, 1993). It is acknowledged that 99% of livestock is in the hands of traditional pastoralists; yet the measures to utilise this wealth are based on converting traditional pastoralism into private and state ranches, which are seen as being 'modern'. This has led to the appropriation of large areas of land and the destruction of traditional pastoralist lifestyle and land management systems (Lane and Scoones, 1993).

The transformation of traditional pastoral life

The increased land pressure has limited the ability of many traditional pastoralist communities to cope with changing circumstances. For example, the Barabaig of Tanzania (see Box 6.2), are under increasing pressure from encroaching farmers, resulting in the forced year-round use of the previously seasonally protected *darorajand* or plains (Lane and Scoones, 1993). A similar situation has arisen through the establishment of commercial wheat farms in the Basotu Plains, which has removed access to traditional *muhajega* or bottomlands grazing resources. Both these changes have resulted in more intense use of other resources and have led to environmental degradation, including soil loss, gully erosion and siltation (Lane and Scoones, 1993).

For the Maasai of the Great Rift Valley of Kenya and northern Tanzania, their present situation is the result of European settlement and land tenure changes (Siddle and Swindell, 1990) (Figure 6.7). In the early twentieth

Figure 6.7 Maasai pastoralists of the East African savannas.
(Photograph by the author)

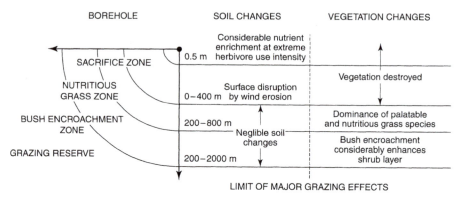

Figure 6.8 The effects of boreholes on soil and vegetation (after Perkins and Thomas, 1993).

century, cultivators began to encroach onto their land when they were contained in government-controlled southern reserves. Additionally, under the veterinary control systems of the time, the Maasai were prevented from passing through European land to the north from where they traditionally acquired stock to upgrade their herds. The outcome has been a loss of land and a poorer quality of herd, resulting in the abandonment of traditional management methods. Although many Maasai still use traditional seasonal grazing rotation management, it is set against a background of declining resources. Many have also changed their dietary habits, with less reliance on their livestock products.

The impact of boreholes

Although traditional pastoralists were nomadic, migrating seasonally for resources (Lamprey, 1983; Sinclair and Fryxell, 1985; Ellis and Swift, 1988), recent government settlement programmes have forced pastoralists to use limited rangeland. Permanent water is provided by watering points or boreholes, and the consequences have been predictable (Andrew, 1988). Transects from undisturbed savanna to boreholes have shown livestock density and dung concentrations increase, trees and palatable perennial herbs decrease, annuals and unpalatable perennials increase, amount of bare ground, soil compaction and erosion increase, and soil nutrient concentrations increase (Tolsma *et al.*, 1987; Barker *et al.*, 1989; Georgiadis and McNaughton, 1990) (Figure 6.8). The importance of and concern with these boreholes lies in the tendency to raise livestock densities throughout the year, leading to local overgrazing and range deterioration (Sinclair and Fryxell, 1985). They create patches of degradation, animal use and nutrient levels.

6.5 Concluding remarks

The East African savannas, determined primarily by plant-available moisture and soil chemistry, are characterised by a plethora of large herbivores dependent on the spatial and temporal heterogeneity of plant communities (Belsky, 1995).

As a consequence, research in the region has been largely focused on herbivory, and its effects on ecosystem dynamics. Land use is dominated by small-scale agriculture and pastoralism, and both have come into direct conflict with wildlife and its conservation.

Throughout the recent history of East African savannas, from colonial times to post-independence, government policies have been to protect wildlife and resettle people. In some countries, such as Tanzania, where tourism is beginning to become profitable again, the trend of coercing people continues, but in others places, such as Kenya, schemes to involve local communities in conservation efforts are beginning to work. These savannas have a great potential for sustaining local economies, and allowing outsiders to experience the natural landscape, but can only succeed if management responsibilities are transferred from powerful stakeholders to the hands of locals.

Key reading

Belsky, A.J. (1995). Spatial and temporal landscape patterns in arid and semi-arid African savannas. In: Hansson, L., Fahrig, L. and Merriam, G. (eds), *Mosaic Landscape and Ecological Processes*. Chapman and Hall, London, pp. 31–56.

Dublin, H.T., Sinclair, A.R.E. and McGlade, J. (1990). Elephants and fire as causes of multiple stable states in the Serengeti–Mara woodlands. *Journal of Animal Ecology*, **59**: 1147–64.

Lane, C. and Scoones, I. (1993). Barabaig natural resource management. In: Young, M.D. and Solbrig, O.T. (eds), *The World's Savannas. Economic Driving Forces, Ecological Constraints and Policy Options for Sustainable Land Use. Man and the Biosphere Vol. 12*. UNESCO, Paris, pp. 93–120.

Sinclair, A.R.E. and Arcese, P. (eds), (1995). *Serengeti II. Dynamics, Management, and Conservation of an Ecosystem*. The University of Chicago Press, Chicago.

Chapter 7

The savannas of southern Africa

7.1 Introduction _____

7.1.1 Distribution, climate and soils

Extending from 34°S, the savanna biome comprises about 1 435 713 km² or 53.7% of southern Africa (Rutherford, 1997). It occurs in east-central Namibia, in almost all of southern Botswana, the southern tip of Zimbabwe, southern and eastern Swaziland, all of southern Mozambique, and approximately a third of South Africa, predominantly in the northern parts of the Cape Province and Transvaal, and the eastern seaboard of Natal and eastern Cape Province (Figure 7.1).

Most of the savanna is on the extensive plains of the Kalahari Basin and the coastal platform of Mozambique. Some is also found on more hilly, rocky topography such as escarpments and valley slopes along the east coast and parts of the southern Cape, varying in altitude from a few hundred metres in the south to 2000 m above sea level in the north. Soils can be calcareous, dystrophic, mesotrophic or eutrophic. Dystrophic soils are highly leached, with low nutrient content, whereas eutrophic soils have been subject to very little leaching and have a high nutrient content. Mesotrophic soils fall between the two types. Mean annual rainfall ranges between 200 and 1000 mm (Schulze, 1997). Most rainfall is exclusively a summer phenomenon, lasting six to over seven months, between October and April (Schulze, 1965). Where rainfall is low, temperatures can drop so low in winter as to cause frosts – mild on a few occasions per year, but severe every couple of decades.

Towards the east and south of the savanna zone, the vegetation grades into the arid shrubland of the Nama-Karoo, which shares many species and genera with the savanna (Gibbs Russell, 1987). Desert grassland becomes dominant west and south-west of the savanna zone, and towards the north, moist savannas grade into woodlands and open deciduous forests. Within the savanna zone, forests can be found along river courses and other areas of enhanced moisture status, such as areas of deeper soils.

7.1.2 Vegetation structure and composition

Ecologists have traditionally subdivided the African savannas into functionally different types along a moisture gradient, also correlated to a soil fertility

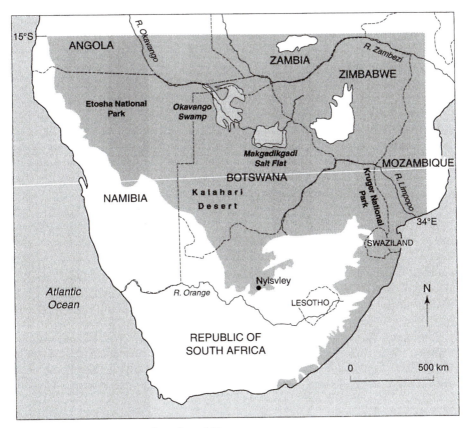

Figure 7.1 The savannas of southern Africa.
(Reproduced by kind permission of Justin Jacyno and Kevin Dobbyn)

gradient. For example, Huntley (1982) proposed that the southern African savannas can be divided into two main types: arid/eutrophic and moist/dystrophic. This distinction may have its origins in the South African tradition of differentiating between 'sweet' and 'sour' veld. The former has a forage protein content sufficient for cattle maintenance throughout the year, while the latter requires protein supplementation during the winter months (Booysen and Tainton, 1978). Sweetveld occurs in lowland, warm, arid areas on fertile soils, whereas sourveld occurs in highland, cool, wet infertile areas. However, this classification can only be applied to cattle-raising, as both sweet- and sourveld support indigenous herbivores throughout the year.

Although the eutrophic/dystrophic distinction appears to be valid (reflected in the many levels of ecosystem structure and function), Scholes (1990) argues that when linked to the dry/moist distinction, it tends to hide the fact that changes caused by soil fertility are both more discrete and fundamental. For example, in areas with extremely homogeneous parent materials, such as the Kalahari sands, geomorphological processes have resulted in the deposition of nutrient-rich areas such as the Mubabe depression and the pans of the central

Kalahari (Scholes, 1990). It may, therefore, be more useful to distinguish between fertile and infertile savannas (Scholes, 1990).

Rich versus poor savanna

Fertile savannas in southern Africa are found on base-rich, calcretous (calcretes) soils derived from Post-African basic igneous rock such as basalt, or fine-grained sediments such as shale (Scholes and Walker, 1993) (Figure 7.2). They are nutrient-rich as a consequence of either active soil formation taking place on those surfaces, or due to their being located on volcanic material of recent volcanic origin. These savannas are typical of hotter, drier lowland valleys, although they can extend into moist savannas on other base-rich substrates such as termitaria. Rainfall occurs in five or six months of the year and ranges from 250 to 650 mm year^{-1}.

Infertile savannas typically occur on the high-lying surface of the African shield, which is extremely old and highly weathered (Scholes and Walker, 1993) (Figure 7.3). The soils are acid and infertile, mostly as a result of prolonged leaching. They can be very deep and lateritic, with clays predominantly kaolinites and oxides of iron and aluminium. These savannas can extend into fertile savanna areas on acidic sands. Precipitation ranges from 500 to 1100 mm year^{-1}, mostly falling during six to nine months of the year.

The main functional distinction between southern African savannas is between the broad-leaved and fine-leaved types (Huntley, 1982; Scholes, 1990). The fine-leaved savannas are in nutrient-rich, arid conditions and the broad-leaved savannas are generally in nutrient-poor, moist environments. In

Figure 7.2 Fertile savannas of southern Africa dominated by fine-leaved species.
(Reproduced by kind permission of Bob Scholes, CSIR, Pretoria, South Africa)

Figure 7.3 Infertile savannas of southern Africa dominated by broad-leaved species.
(Reproduced by kind permission of Bob Scholes, CSIR, Pretoria, South Africa)

the Nylsvley Nature Reserve and Research Station, South Africa, the broad-leaved savannas are characterized by *Burkea africana* in the tree layer, and *Eragrostis pallens* in the grass layer. The fine-leaved savannas are dominated by species of the spinescent genera *Acacia* and *Commiphora* and grasses of the Eragrosteae and Paniceae (Scholes and Walker, 1993). The exception to this pattern are the *Colophospermum mopane* savannas, which, though broad-leaved, are functionally arid/fertile savannas (Scholes, 1997).

Table 7.1 lists the dominant herbaceous and woody species of the two savanna types, and Table 7.2 outlines their main characteristics. Other less common woody plants include geoxylic suffrutices or 'underground trees'. These occur on deep, infertile sands in the broad-leaved savanna, and are usually contra-seasonal, i.e. they produce new leaves and flowers at the height of the dry season.

Plant species richness at a whole biome scale is high: southern African savannas contain 3–14 species m^{-2}, and 40–100 species 0.1 ha^{-1} (Whittaker *et al.*, 1984; Cowling *et al.*, 1989). Gibbs Russell (1987) reports 5788 species in the savanna areas of southern Africa (about 632 000 km^2).

Savanna vegetation in southern Africa underwent considerable changes during the Quaternary period. Data suggest that the vegetation alternated from woodland savanna during warm interglacial phases to cool upland grassland during glacial maxima, and to mesic woodland with *Podocarpus* forest during some intermediate phases (Avery, 1993; Scott, 1995; Scott *et al.*, 1995). At the end of the Pleistocene, about 8000 years BP, there is evidence for a change to semi-arid savanna (similar to the current south-western Kalahari). Broad-leaved

Table 7.1 Common woody and herbaceous species of fertile, fine-leaved and infertile, broad-leaved savannas of southern Africa (after Scholes, 1997)

Savanna type	Dominant woody species	Dominant herbaceous species
Fertile, fine-leaved	*Acacia erioloba, A. hereroensis, A. mellifera, A. nigrescens, A. robusta, A. tortilis, Acacia* spp., *Commiphora pyracanthoides, Dichrostachys cinerea, Euclea divinorum, Grewia flava, Piliostigma thonningii, Sclerocarya birrea, Tarchonanthus camphoratus*	*Aristida* spp., *Bothriochloa insculpta, Dactyloctenium* spp., *Digitaria* (*pentzii*) *nuda, Eragrostis* spp., *Chrysopogon* spp., *Panicum kalaharense, P. maximum, Sporobolus* spp., *Stipagrostis uniplumis, Themeda triandra*
Infertile, broad-leaved	*Afzelia quanzensis, Albizia versicolor, Baikiaea plurijuga, Brachystegia* spp., *Burkea africana, Combretum apiculatum, C. molle, C. zeyheri, Combretum* spp., *Faurea saligna, Isoberlinia* spp., *Julbernardia globiflora, Marquesia macroura, Ochna pulchra, Pterocarpus rotundifolius, Sclerocarya birrea, Terminalia sericea*	Species of Andropogoneae, *Aristida* spp., *Cymbopogon plurinodis, Eragrostis* spp., *Hyparrhenia* spp., *Hyperthelia dissoluta, Schizachyrium jefferysii, Setaria* spp., *Themeda triandra, Trichopteryx* spp.

savannas were established about 7000 years BP, but contemporary savanna structure probably only came about around 1000 years BP (Scott, 1996).

Plant life strategies

Only approximately 2% of the primary productivity by trees in a broad-leaved savanna at Nylsvley is invested in reproductive structures every year (Scholes, 1997). Seeds are typically large, up to a few grams each, and have a low viability due to parasitism by insect larvae. Seeds can remain viable for many years when stored, but there is no dormancy mechanism, other than a hard testa, and so most seeds germinate in the first year. The main controls on recruitment are from fire, herbivory and competition with grasses, but once saplings escape these, they have a competitive advantage over grasses. The life-span of savanna trees varies from a few decades in pioneer species such as *Dichrostachys cinerea*, to several centuries in *Adansonia digitata* (Guy, 1970).

Although there are no formal studies on pollination of the southern African savanna trees, observations suggest that insects are predominant, especially bees, and bird-, moth- and wind-pollination also occurs (Scholes, 1997). Some specialisations include *Kigelia africana* pollinated by nectar-feeding bats, and *Acacia nigrescens* by giraffes (Du Toit, 1990). Tree dispersal mechanisms include wind, ingestion by mammals and birds, and ballistic dispersal, i.e. explosive splitting of pods.

Table 7.2 A comparison of the features of fertile, fine-leaved and infertile, broad-leaved savannas of southern Africa (after Scholes, 1997)

Feature	Fertile, fine-leaved	Infertile, broad-leaved
Mean annual rainfall (mm)	400–800	600–1500
Age of erosional surface	Recent	Ancient
Parent material	Basic igneous lavas, mudstones and siltstones, limestones	Acid crystalline igneous rocks, aeolian sands, sandstone
Cation exchange capacity	High	Low
Phosphorus availability	Moderate	Low
Dominant tree family or subfamily	Mimosoideae (dry), Burseraceae (very dry)	Caesalpinoideae (wet), Combretaceae (dry)
Tree leaf size (cm)	0.1–1	2–10
Mycorrhizal associations and biological nitrogen fixation	Predominantly VAM; moderate nitrogen fixation	Predominantly ECM; low nitrogen fixation
Dominant grass subfamily (and tribe)	Arundinoideae (Chloridoideae, Panicoideae)	Panicoideae (Paniceae, Arundinelleae, Andropogoneae)
Grass growth form	Creeping (stoloniferous)	Bunch (caespitose)
Large mammal herbivory	High (10–50%)	Low (5–10%)
Mean tree nitrogen content at maturity (%)	> 2.5	< 2.5
Grass nitrogen content at senescence (%)	> 1	< 1
Insect herbivory	Seasonally recurrent, mostly of grass by grasshoppers and harvester termites, also episodic by locusts	Episodic, mostly of woody plants by lepidoptera larvae
Main tree anti-herbivore defence mechanism	Structural (mainly thorns)	Chemical (mainly polyphenols, but especially tannins)
Fire fuel load	Low	High
Fire frequency	Quintennial or longer	Annual–triennial

VAM = vesicular-arbuscular mycorrhiza, ECM = ectomycorrhiza.

In the broad-leaved savannas, tall, caespitose (tuft-forming) grass species dominate, whereas in the fine-leaved savannas, there is a higher proportion of stoloniferous (lawn-forming) species (Blackmore, 1992). In more arid areas, wiry grasses with narrow or rolled leaves resistant to desiccation dominate. Most grass species are perennial, with the proportion of annuals increasing with aridity, and on mesic sites in years of drought. Perennial grasses start producing seeds in December, about two months after the onset of the rains, the peak seed set period being March (Veenendal, 1991). There is high seed predation (mainly by ants, rodents and birds), a lifetime in the soil store of one to a few years, and a low rate of establishment success (O'Connor and Pickett, 1992).

Almost all savanna grasses have an innate dormancy mechanism to prevent germination in the first year they are produced (O'Connor and Pickett, 1992). After a delay of several months, exposure of the seeds to a wetting event of about 20 mm will cause germination within six days, but not all the seeds of one species will germinate at the first wetting. This protects the species from any subsequent drought (Veenendal, 1991). Drying and re-wetting will trigger further germination. Seedling establishment and survival depends on micro-site, such as the influence of a tree canopy (Veenendal, 1991). Mortality of mature perennial tufts only comes about through severe drought, defoliation or frequent intense fires. Even though most savanna grass seeds are dispersed by the wind, water, or by attachment to animals, the vast majority are transported only a few metres away from the parent plant (Veenendal, 1991). For example, the mass of dissemules of 11 grass species from Botswana ranged from 0.13 to 5.62 mg, and the average distance they were disseminated by wind was 13 m (Veenendal, 1991).

Plant productivity

Above-ground primary production of the herbaceous layer is linearly related to annual rainfall up to about 900 mm (Scholes, 1997). However, the rate of production is also controlled by nutrient availability (Scholes, 1993). Table 7.3 compares the productivity of the broad- and fine-leaved savannas. Woody plant production is on average 50% higher in fine-leaved savanna than in broad-leaved savanna, and 70% higher in the herbaceous layer. Around 50% of the total primary production occurs below-ground (Scholes and Walker, 1993). This excludes root exudates and mycorrhizal respiration. There is no significant difference between broad- and fine-leaved savannas in the below-ground biomass. Total net primary production has been estimated to be in the range of

Table 7.3 Primary productivity over one year in fine-leaved and broad-leaved savanna at Nylsvley, South Africa (after Scholes and Walker, 1993)

	Fine-leaved		Broad-leaved	
Primary productivity	Mean	SD	Mean	SD
Above-ground production (g m^{-2} year^{-1})				
Tree leaves	399	89	265	38
Tree twigs	81	34	57	85
Grasses and forbs	199	12	120	8
Below-ground production (g m^{-2} year^{-1})				
Fine roots	493	66	497	59

SD = Standard deviation.

$500–1500$ g m^{-2} year^{-1} (Rutherford, 1978). Depending on savanna structure, 5–95% of this can be herbaceous layer production, of which 90% or more is typically produced by grasses.

7.1.3 Fauna

Southern African savannas have a rich and diverse array of wildlife, including invertebrates, rodents, bats and birds. However, it is the big game, carnivorous species such as lions (*Panthera leo*) and cheetahs (*Acinonyx jubatus*), and large herbivores such as the African elephant (*Loxodonta africana*), white rhinoceros (*Ceratotherium simum*), black rhinoceros (*Diceros bicornis*) hippopotamus (*Hippopotamus amphibius*) and African buffalo (*Syncerus caffer*), that are most well known. The long history of wildlife hunting in the region and habitat transformation has meant that some large mammals, such as the roan antelope (*Hippotragus equinus*), wild dog (*Lycaon pictus*) and black rhinoceros, are endangered.

7.2 The main savanna determinants

Aside from geomorphology, mentioned above, what other processes cause the creation and long-term persistence of nutrient patchiness in the southern African savannas? Du Toit (1988) has shown that the feeding movements of mammalian herbivores may lead to nutrient concentration in particular areas. Similarly, termite foraging and mound-building results in localised accumulation of bases (Malaisse, 1978b). Rutherford (1983) suggests that the broad lateral spread of tree roots, which can exceed the canopy area by a factor of seven, results in an accumulation of nutrients in the sub-canopy, which is frequently exploited by a different suite of grasses. Pre-colonial human activities such as cattle penning, firewood collection and iron smelting, were also widespread and important inputs to nutrient build-ups (Blackmore *et al.*, 1991). Nutrient-rich patches of human origin on sandstone in the Northern Transvaal (Fordyce, 1980) and on Kalahari sands in Botswana (Denbow, 1979) have been dated on the basis of archaeological evidence and ^{14}C analysis to be $700–1000$ years old.

7.2.1 Plant-available moisture (PAM)

The annual potential evaporation greatly exceeds the annual precipitation in savannas, so for much of the time, water is a limiting factor. Mean ecosystem potential evaporation rates for southern African savannas peak at about 4 mm day^{-1}, because of the low stomatal conductance (0.3–0.8 cm s^{-1}) and low total leaf area (typically less than 1 cm^2 m^{-2}) (Scholes, 1997). Many plants have adaptations to the varying degrees of water availability, which will also depend on soil characteristics. Perennial grasses survive the water stress of the dry season by dying back down to ground level, whereas the annual grasses and forbs return to the soil as seeds.

Savanna tree adaptations to PAM

Most trees in southern African savannas are deciduous, and of the small proportion that are evergreen (around 5% of species), the majority are strongly sclerophyllous. The deciduous trees are also sclerophyllous, but less so. For example, the leaf area : mass ratio is typically $6-7$ m^2 kg^{-1} (Rutherford, 1979). Blackmore (1992) determined the mean LSM (leaf specific mass, an index of sclerophylly) for abundant species from 15 savanna sites in South Africa, and found that they ranged from 1.0 to 1.7 dm^{-2} for woody plants, and from 0.5 to 1.06 dm^{-2} for grasses. Differences between broad- and fine-leaved savannas were not great. LSM for species from nutrient-rich sites ranged from 1.00 to 1.81 dm^{-2}, and in a nutrient-poor site from 0.96 to 1.65 dm^{-2}. Leaf fall and leaf death (in grasses) is regarded as a drought-avoidance mechanism, although there is not much evidence to support this view. The usual cause appears to be the drying of the soil profile during the dry season, but an early frost can also precipitate leaf fall, as can a prolonged midsummer drought.

One of the consequences of transpiration is the cooling of the leaf. Species such as *Acacia tortilis* have smaller leaves, and these assist in keeping the leaf temperature close to the air temperature when there is insufficient water for transpiration (Bate *et al.*, 1982). This permits water conservation by stomatal closure without the risk of overheating the leaves (Pendle, 1982). Water loss is further controlled by the prevalence of compound leaves, typical of the *Acacia* genus. The clustering of tiny leaflets may weaken the linkage between atmospheric vapour pressure deficit and transpiration rate, so retaining moisture (Jarvis and McNaughton, 1986). In moist savanna species, leaves are characteristically thick, sclerophyllous and hairy, with sunken stomata and well-developed cuticles. For example, the stomatal openings of *Ochna pulchra* are completely covered by a hairy cuticle until the leaf is fully expanded (Dyer, 1980; Ludlow, 1987, cited in Scholes and Walker, 1993).

The below-ground to above-ground biomass ratio is high in the South African savannas, particularly for the fine-leaved (arid, nutrient-rich) savanna, suggesting that growth of fine roots is predominantly an adaptation in response to aridity (Baines, 1989, cited in Scholes and Walker, 1993).

7.2.2 Plant-available nutrients (PAN)

As already mentioned, nutrients play a fundamental role in differentiating between the two main savanna types found in southern Africa (Figure 7.4). If the total nitrogen in all the litter-fall fluxes is measured, the flux of nitrogen and phosphorus through the fine-leaved savanna is substantially higher than through the broad-leaved savanna (Scholes and Walker, 1993) (Figure 7.5). This is because there is a greater mass of litter fall in the fine-leaved savanna, which contains higher mean levels of nitrogen and phosphorus. This is particularly true for grass litter, which even when dead, sustains high nutrient content.

Soil fauna play a significant role in decompositional processes, but probably have more important effects on the spatial redistribution of litter and nutrients.

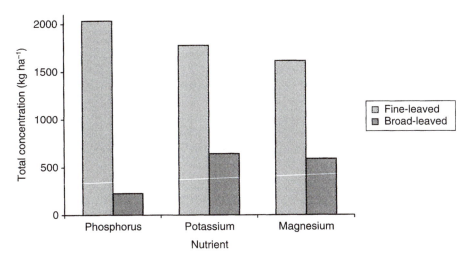

Figure 7.4 The total concentrations (kg ha⁻¹) of some nutrients in fine-leaved and broad-leaved savannas. This graph comprises data from the woody, herbaceous and soil layers (based on Blackmore *et al.*, 1991).

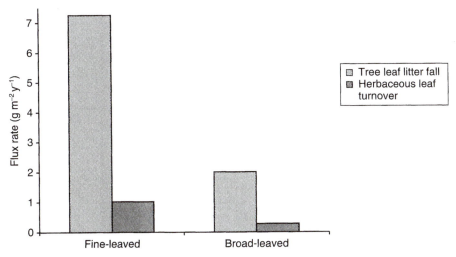

Figure 7.5 The flux rate of nitrogen (g m⁻² year⁻¹) in fine-leaved and broad-leaved savannas (based on Scholes and Walker, 1993).

The main macro-faunal groups include millipedes, dung beetles, coleopteran larvae, ants, cockroaches and termites, the latter being the most prominent.

Termites and nutrient cycling

At least 21 species of termites have been recorded at Nylsvley, of which 68% are soil feeders and 32% litter feeders (Ferrar, 1982a,b). There is a conspicuous difference in termites between the fine-leaved (0.16 g DW m^{-2}) and the broad-leaved (3.17 g DW m^{-2}) savannas, but particularly in the soil-feeding termites (0.06 versus 2.16 g DW m^{-2} respectively) (Ferrar, 1982c). This low litter-feeding biomass in the fine-leaved savanna is comparable to other nutrient-rich African savannas, the reasons for which are unknown (Scholes, 1990). The combined organic matter consumption by soil- and litter-feeding termites at Nylsvley is $9-14$ g m^{-2} year^{-1}.

Termites forage for material, which is then deposited in their nests, causing local changes in soil fertility and vegetation (Wood, 1976; Malaisse, 1978b; Griffioen and O'Connor, 1990). The most commonly reported fertility change is an increase in the concentration of basic cations. Where there are large mound-building termites (e.g. *Microtermes*) part of this enrichment results from the fine mineral particles brought up from the subsoil for nest construction, and cemented with saliva.

Plant adaptations to PAN

Most savanna grasses and trees have mycorrhizae-infected roots (predominantly VAM), which may assist in the acquisition of both nutrients and water (Högberg, 1989). Although both broad- and fine-leaved savannas are dominated by leguminous trees, there is no evidence for nitrogen-fixing symbiotic activity. Almost all of the nitrogen fixation is performed by nodulated leguminous forbs or by free-living bacteria and soil surface crusts of blue-green algae (Zietsman *et al.*, 1988).

Losses of nutrients through nitrification, fires and leaching are inferred to be greater in the fine-leaved than in the broad-leaved savanna. If so, what prevents the nutrient levels in the fine-leaved savannas from 'running down' to the levels sustained by the surrounding broad-leaved savannas? The answer to this lies in the nutrient fluxes resulting from large mammal feeding behaviour.

7.2.3 Herbivory

Mega-herbivores (species exceeding 1000 kg in adult body weight) typically form 50% or more of the total biomass of large mammalian herbivores in savanna communities (Owen-Smith, 1988). These include the African elephant, white rhinoceros, black rhinoceros and hippopotamus. African buffalo are the predominant grazers (consuming 15–35% of the grazing biomass), except where white rhinoceros and hippopotamus are abundant (Owen-Smith and Cumming, 1994). In the drier regions of southern Botswana and northern Namibia, blue wildebeest (*Connochaetes taurinus*) were the most abundant grazers until their numbers were reduced through hunting. Small vertebrates, such as rodents (Korn, 1987), and avifauna including guineafowl and francolins (Tarboton, 1980) may also be important, although their distribution may be more patchy and their abundance vary widely over time.

About 75–90% of consumption by indigenous large herbivores is concentrated on the grass component (Owen-Smith, 1993); the consumption of trees and shrubs is much lower, even when elephants are present. Elephants, however, do exert a major impact on woody vegetation, although severe damage is episodically associated with droughts, fires and severe frost. In fertile savanna, where trees are relatively shallow-rooted, the destruction of trees by elephants, coupled with their suppressant effect on regeneration, can lead to a progressive opening of the woodland. This has been documented for *Acacia–Sclerocarya* parkland in the eastern basaltic region of the Kruger National Park, despite the management policy of restricting elephant damage by annual culling (Viljoen, 1988).

Differences in herbivory between the fine-leaved and broad-leaved savannas

Large ungulate herbivores such as cattle, impala and kudu feed in the fine-leaved savanna in preference to the broad-leaved savanna in all months of the year, except when the latter has been recently burned (O'Connor, 1977; Zimmerman, 1978, cited in Gandar, 1982a; Scholes and Walker, 1993). For example, it was found that impala at Nylsvley increased from 6 to 51 individuals km^{-2} from pre- to post-fire in the broad-leaved savanna, whereas there was a decrease from 39 to 2 individuals km^{-2} from pre- to post-fire in the fine-leaved savanna (Gandar, 1982b). The degree of preference varies according to the season (Figure 7.6), with the highest preferences for fine-leaved savanna occurring at the end of the dry season, and the lowest during the early wet season (Skinner *et al.*, 1984; Scholes and Walker, 1993).

Grazers spend sixfold more time in the fine-leaved patches, although the grass production there is less than twice that in broad-leaved savanna. This preference can be attributed to the forage quality there. However, the herbaceous productivity in the fine-leaved savanna is insufficient to support the herbivore biomass, so it seems that the bulk of energy requirements are derived from the broad-leaved savanna, with nutrient requirements being gained from the fine-leaved savanna. If defecation and urination are proportional to time spent in an area,

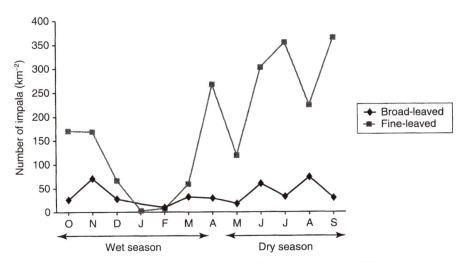

Figure 7.6 Seasonal habitat preferences by impala (based on Scholes and Walker, 1993).

the herbivores provide a mechanism for continued nutrient enrichment of the fine-leaved patches.

Other herbivores include invertebrates (Gandar, 1982c), of which there are grass and tree leaf consumers. The dominant insect herbivores of the herbaceous layer are grasshoppers. In the broad-leaved savanna, they consume 13 g DM m^{-2} year^{-1} (about 6% of the above-ground grass production), while in the fine-leaved savanna they consume 40.6 g DM m^{-2} year^{-1} (Gandar, 1982c). Tree leaf-eating insects are mostly found in the broad-leaved savanna, where species such as *Burkea africana, Ochna pulchra* and *Terminalia sericea* are periodically subject to defoliation by lepidopteran larvae, principally the emperor moth, *Cirina forda* (Scholtz, 1982). There is therefore a predominance of insect over mammalian herbivory on trees in the broad-leaved savanna, and a reversed situation in fine-leaved savannas. This may be because the lower populations of birds and other insect predators in the broad-leaved savanna (reflecting the low unpalatability of leaves) exert less control here than in the fine-leaved savanna (Tarboton, 1980; Ferrar, 1982a,d).

Adaptations to herbivory

Trees in the nutrient-poor broad-leaved savanna tend to be unpalatable because of their chemical composition, while fine-leaved savanna trees have thorns to restrict herbivory to acceptable levels, but whose leaves are still highly sought after. The degree of chemical defense against herbivory could be related to soil fertility. A model by Coley *et al.* (1985) suggests that plants adapted to low-nutrient soils are slow growing, with inherently low rates of photosynthesis and nutrient uptake. They cannot afford to lose leaf tissue or nutrients by herbivory, so invest in defensive mechanisms. Plants growing on fertile substrates, on the other hand, are fast growing and have high rates of photosynthesis and nutrient uptake (Bryant *et al.*, 1989). They can therefore afford to lose some tissue to herbivory rather than invest in defense.

Adaptations to herbivory differ between the broad- and fine-leaved tree types. The former contain high leaf concentrations of digestion-retarding secondary compounds, particularly tannins (Owen-Smith and Cooper, 1987). There are two main forms: the condensed tannins, which are located in the cell wall, and are therefore principally effective against mammals (Furstenburg and Vanhoven, 1994; Cooper and Owen-Smith, 1985); and the hydrolysable tannins, which are found in the cell contents, and are mainly effective against insect herbivores. Toxins such as alkaloids are relatively rare, and mostly confined to some forbs and the 'underground trees' such as *Parinari capensis* (highly unpalatable) and *Dichapetalum cymosum* (extremely toxic) (Cooper, 1985). Fine-leaved trees generally use thorns to deter herbivores, which rather than preventing herbivory, retard the rate of leaf ingestion to a tolerable level (Cooper and Owen-Smith, 1986). Dangerfield *et al.* (1996) found that longer, less spinescent shoots of *Acacia tortilis* are produced in habitats recently protected from wildlife. The evergreen species are the least palatable, but the few that are palatable, e.g. *Boscia albitrunca*, are a key browsing resource and heavily utilised.

The stoloniferous (lawn-forming) grass species are probably the best adapted to herbivory. They have the ability to re-grow after defoliation by virtue of the basal location of the bud, and remain highly productive. When grazing is removed, they revert to a more upright growth form. Some savanna grasses also contain tannins, aromatic oils, and are high in silica bodies, restricting grazers to those with hypsodont dentition (Scholes, 1997). To deter invertebrate grazers, many grasses are hairy and have serrated edges.

7.2.4 Fire

Lightning fires have probably been significant in southern African savannas since mesic conditions allowed the build-up of significant levels of combustible herbaceous fuel. Although *Australopithecines* and other ancestral human forms probably employed fire (Thackeray *et al.*, 1990), it was widely used from the Stone Age, some 1.5 million years BP to historic times (Hall, 1984). The current fire regime is greatly influenced by and is the product of the system of land use being applied in an area, namely, commercial ranching, subsistence ranching or wildlife management.

The causes of fire

According to Komarek (1971), Africa has a unique fire climate in that the probability and occurrence of lightning fires are accentuated because at the end of the dry period, dry lightning storms frequently occur and ignite many fires. The importance of lightning as an ignition source is indicated by Siegfried (1980, cited in Trollope, 1984a), who reported that probably 73% of all fires occurring in Etosha National Park between 1970 and 1979 were caused by lightning.

Natural fires are more frequent in moist savanna than in arid savanna, not only because of the higher rainfall leading to higher herbaceous fuel biomass, but also because most of the herbaceous matter is unpalatable to herbivores. The natural fire frequency in these moist savannas must have been every 1–2 years. The lower rainfall and higher rates of herbivory in the arid savannas means that herbaceous fuel is reduced, and the frequency of fires is determined by the occurrence of exceptionally wet seasons and stocking rates. Fire frequencies in these areas probably varied from 5 to 30 years or more (Scholes, 1997).

Today, people cause the majority of fires in the southern African savannas. These are set for pasture regrowth by both commercial and subsistence ranchers, for clearing land for agriculture, as a prescribed tool in areas of conservation, through poaching and arson, and as a result of accidents.

Fire behaviour

Fires in southern African savannas are largely confined to the dry season from about May to October, at the end of which the herbaceous grass fuel layer is very dry (Table 7.4). This is the time of high-intensity burns, when the moisture content of the fuel is low, and climatic conditions, particularly the low relative humidity, are particularly favourable for fire propagation (Van Wyk, 1971). The various characteristics of the fuel, and its effects on fire behaviour, are obviously extremely important, but little work has been done on this (Trollope, 1984b). Harrison (1978, cited in Trollope, 1984b), for instance, measured soil surface temperatures of less than 260 °C below *Burkea africana* canopies to over 816 °C in a thick, dry litter bed below *Ochna pulchra*. Mean temperatures at the base of grass tufts were around 460 °C. The temperatures above 1 m in tree canopies never exceeded 260 °C. Frost (1985) estimated that 74% of standing dead grass and grass litter was consumed in a fire at Nylsvley in 1983, while 58% of twig and bark litter and 42% of tree leaf litter was consumed.

The effects of fire

Although there have been numerous fire studies in the southern African savannas, the results have been to a large extent confounded because of the presence of herbivory (e.g. Van Wyk, 1971; Robinson *et al.*, 1979). Nevertheless, the frequency of burning and the different types of fire probably affect the degree to which the plants recover during the interval between fires. For example, crown and surface head fires cause the highest death of stems and branches of *Acacia karroo* and other woody plants, as compared with backfires (Trollope, 1984a). This is because of the heat energy in those types of fires being released above the soil surface at levels closer to the terminal buds of the aerial portions of the trees and shrubs.

Table 7.4 The fire regimes in the savannas of southern Africa (based on van Wilgen and Scholes, 1997)

	Fine-leaved, fertile	Broad-leaved, infertile
Fuel characteristics	Erect swards of grass with patchy distribution provide fine fuel to support fire. Trees are unimportant in carrying fire, but provide material for smoldering combustion. Dung adds fuel for combustion. Less efficient fuel consumption than in infertile savannas because of patchiness. Within burned patch, over 90% of fuel is consumed, but burned patches only make up 50% of landscape	Erect swards of grass with continuous distribution provide fine fuel to support fire. Trees are unimportant in carrying fire, but provide material for smoldering combustion. Fuel consumption of grass layer is efficient (70–80%)
Fire frequency	2–11 years, with mean of every 8 years in Kruger National Park, and 3–10 years in Etosha National Park	1–6 years, with mean of every 3 years in Kruger National Park
Fire season	Dry season	Dry season
Fire intensity	Range from < 100 to 4000 kW m^{-1}	Range from < 100 to 6000 kW m^{-1}
Fire type	Grass layer surface fires	Grass layer surface fires

Rutherford (1981) showed that less intense, slow burns result in higher mortality in small plants, e.g. 13% compared to 9% in fast, intense burns. However, in canopy top-kill, the opposite trend was seen: mortality was greater in the fast burn (43.2%) than the slow burn (23.5%). In both cases, over 90% of the canopies killed were below 2 m in height. Six months after the fire, the leaf biomass of dominant species in the burned plots was less than that of the same species in the unburned plot, the reduction depended slightly on the species and strongly on the size of the individual. Therefore, *Ochna pulchra* plants shorter than 1 m showed a 90% reduction in leaf biomass, but individuals between 2.5 and 5 m tall showed only a 26% reduction.

In general, fire frequency has a weak negative long-term effect on total grass basal cover (O'Connor, 1985). In relatively high rainfall areas on infertile soils, grass basal area declined slightly in unburned swards, possibly because of the increase in woody biomass. On fertile soils in dry areas, grass basal area decreased on both annually burned and fire-protected plots.

Yeaton *et al.* (1986) compared the structure, composition and dynamics of the grass layer in three stands of varying fire history at Nylsvley. Species richness was highest on the annually burned stand, lower in the area burned 5 years previously, and lowest on a stand burned 12 years previously. Total basal area increased with time since the last fire, mostly because of the increasing dominance of *Eragrostis pallens* and *Digitaria eriantha*. In *E. pallens*, the plant density increased from the annually burned area to the 5-year stand, but then decreased to the 12-year stand. This may have been a result of self-thinning through competition.

In the higher rainfall areas (above 600 mm year^{-1}), the tree/grass mosaic is maintained with fire alone. This is because even though the woody species coppice, rainfall is sufficient and reliable enough for adequate grass material to accumulate under grazing conditions to support frequent fires, which burn the coppice growth and control woody seedlings (Trollope, 1984a). In drier savannas, rainfall is too low and erratic to support a significant number of fires under grazing conditions. This allows the regeneration of woody plants from coppice and seedling growth. Therefore, in dry savannas fire maintains woody plants at a height and in a state highly acceptable to browsing animals.

Plant adaptations to fire

Plants in the southern African savannas have many different adaptations to fire. Escaping the dry season in the form of seeds, bulbs, corms or by dying back to ground level is common in the herbaceous layer. Woody plants also show seasonal dormancy, their aerial parts dying back each year and the perennating buds protected in underground stems, lignotubers or rootstocks. These include species such as *Lannea edulis*, *Parinari capensis* and *Dichapetalum cymosum* (White, 1976). Protection by thick, corky bark is seen in many species including *Strychnos cocculoides* and various *Acacia* species.

Protection from burning is clearly observed in bush clumps, which are very resistant to combustion even under favourable fire conditions (Frost, 1984). Fires generally skirt around the edges of the bush clumps, leaving the centre unburnt. This may be caused either by a lack of grass fuel or by relatively non-flammable grass species growing under the trees and shrubs of the bush clump. For example, in the eastern Cape, two common grass species occurring under bush clumps are *Panicum maximum* and *Kaerochloa curva*, both of which generally have higher moisture contents than the grasses growing between the bush clumps (Frost, 1984). Practical significance is that once bush encroachment progresses to bush clump stage, then fire is a far less effective management tool than during the initial stages of encroachment.

The impact of fire on animals

Little is known about the effect of fire on vertebrates and birds, other than the various escape mechanisms and post-fire environment exploitation of resources (Frost, 1984). For example, Gandar (1982c) found a marked increase in the numbers and biomass of grasshoppers on unburned patches of savanna immediately after an early spring fire, and a significant decline in the burned area. Arboreal insect populations, on the other hand, were almost completely destroyed, due to the exposed environment in the trees.

Gandar (1982a) observed that the number of impala at Nylsvley on burned areas increased from day 10 after the fire, until day 30, peaking at 0.51 impala ha^{-1}, after which they declined slowly over a period of 3 months to the pre-burn densities of 0.06 ha^{-1}. Impala at Nylsvley show a 14-fold preference for the fine-leaved over the broad-leaved savannas, but this preference is reversed when the broad-leaved savanna is burned because of the increase in food availability.

The interactions between fire and elephants has already been discussed (see Section 6.2.3). However, at Nylsvley, another mechanism to account for tree mortality has been proposed. Porcupines are common here, preferentially eating the basal bark of young individuals of *Dombeya rotundifolia, Burkea africana* and to a lesser extent *Terminalia sericea* (Yeaton, 1988). It has been suggested that once bark is removed by porcupines, a scar forms. When a fire then occurs, the exposed wood is burned a short distance into the tree. Successive fires enlarge the scar, causing lop-sided development of the tree. Eventually a windstorm may cause the stem to snap, which may not necessarily kill the tree, but normally does. In the fire which Yeaton (1988) observed, about 2% of the mature individuals of *Burkea africana* and *Terminalia sericea* (greater than 16 cm trunk diameter) were felled, although they constituted a much greater proportion of the tree basal area. This example indicates how interacting factors can operate over a long period to affect vegetation structure and density.

7.3 Human determinant of the savanna

Southern Africa has been a region of extensive political unrest, first through tribal expansions and warfare, secondly with the arrival of colonists in the nineteenth century, and thirdly due to the installation of apartheid in South Africa at the end of the 1940s.

7.3.1 Pre-colonial land use

Prehistoric people in savanna regions were hunters and gatherers, who probably lived in small populations and had little impact on the environment. Khoikhoi pastoralists, grazing domesticated animals such as sheep and later, cattle, probably migrated from central Africa into southern Africa about 2000 years BP. Although details of subsequent movements are disputed, from around AD 300 (Early Iron Age), mixed subsistence farming had evolved, and these agro-pastoralists survived by 'slash-and-burn' agriculture, hunting and raising domestic sheep and goats (Maggs, 1984; Hall, 1987). Common crops grown were sorghum and millet, as well as other vegetables, and although there was an early reliance on browsing animals such as goats, this changed rapidly as a cattle-based economy was established. Village density was high, with one located every few kilometres, but they were generally small in size (Maggs, 1984).

The success of the Early Iron Age farmers was because of their use of iron for a variety of agricultural and domestic purposes. This time also marks the beginning of clearing savanna woodlands for cropping and grazing, but also for fuel-wood and iron production (Hoffman, 1997). The transition from Early to Late Iron Age (around AD 1000) was marked by a change to an economy based on the association of political power and wealth with cattle, and the development of regional population centres with long-distance trade links (Hall, 1987).

The increasing importance of cattle in the agricultural economy of the Late Iron Age led to a range of ecological problems (Hoffman, 1997). The rise and fall, in the ninth and fourteenth centuries, of a number of well-established economic centres in the Limpopo Basin and eastern Kalahari may have been related to the deterioration of the grazing lands, as well as to the shift in trade networks to more northerly centres in the *miombo* (Denbow, 1984; Maggs, 1984; Hall, 1987). From the middle of the seventeenth century, European tradesmen and farmers were arriving at Cape Town, and their expansion into southern Africa began. However, it was not until the 1830s that the savanna areas were opened. The discovery of large agricultural and mineral resources in the region meant that within a few decades much of the region had been annexed or colonised, though not necessarily controlled, by European powers.

The early nineteenth century also saw military conquests of Shaka Zulu and others, and mass regional displacement and political restructuring, especially in the eastern and northern areas. Although disputed (Hall, 1987), some authors believe that the rise in the eighteenth century of human and cattle populations, and then the fall through a series of droughts, aided the subsequent amalgamation of the people into the broader northern Nguni society (Hoffman, 1997).

7.3.2 Colonisation and post-colonial land use

As the Dutch and English settlers moved into southern Africa, the population of indigenous people, such as the San hunters–gatherers, the Khoikhoi pastoralists and the Bantu-speaking farmers, fell, through their assimilation into other cultures and outright extermination campaigns. In South Africa, there followed a struggle for power, which led to the Anglo-Boer War of 1899–1902, and finally British rule in 1910 (Mentis and Seijas, 1993). The large population of colonist settlers meant that independence was a hard fought battle, and the southern African states were some of the last to gain independence in the 1960s.

Colonisation also saw the decimation of indigenous herbivore populations and their replacement with a few species of domestic animals. Although Iron Age people had traded in ivory and animal skins for centuries, there are no indications that they had a major effect on native herbivores, probably because of their lack of weaponry. The colonists, however, had firearms, and elephants, rhinoceros and other animals were hunted to satisfy the huge international demand for wildlife products, especially ivory. Hall-Martin (1992) estimates an elephant population of more than 100 000 in South Africa before the big game hunting of the late-eighteenth and nineteenth centuries. By 1920, there were fewer than 120 individuals left.

Land tenure

Land distribution inequalities were a major subjugation in most southern African savanna states, but South Africa is a particularly good example. Following the establishment of British rule, the 1913 Native Land Act was proclaimed, under which native people were permitted to own land in only 8% of South Africa, and were prohibited from rental farming (Mentis and Seijas, 1993). The major purpose of this was to cut off the native people's source of independent wealth, and force them into wage labour for the colonists, especially in the mining sector. Native leaders protested at these actions, but failed, and over the next 35 years, more racial laws followed. In 1948, the Nationalist Party came to power and apartheid had officially begun.

In 1959, nine tribal reserves were created, and over the next 20 years more than three million natives were forcibly relocated to the reserves or 'homelands', which were typically remote, undeveloped regions. As a result of these policies, wealthy colonists dominated the commercial agricultural economy, heavily regulated and subsidised by the government, while native people led a subsistence livelihood in tribal areas owned by tribal authorities.

Following independence in savanna countries, such as Namibia, came the redistribution of land. This meant the subdivision and reassignment of what were either freehold or long-term leasehold white settler farms, into smaller holdings for resettlement by Africans, usually retained in state ownership, or on permits or leaseholds from the state (Bruce et al., 1998a). Nevertheless, the displacement of communal area holders still continues in many countries today, especially where land is seen to be in scarce supply. In Botswana the Land Board system has attempted to decentralise the authority of tribal communities over land, while at the same time easing traditional land administration authorities out of control (Bruce et al., 1998a). In South Africa, the Communal Property Associations Act of 1996 is modelled on community land trusts, aimed at delivering land to groups of purchasers, with title held communally, but use rights to at least some of the land held individually (Jensen, 1998).

Table 7.5 Stocks of cattle and goats in three southern African savanna countries in 1998 (based on FAO, 1999)

Country	Number of cattle	Number of goats
Botswana	2 330 000	1 820 000
Namibia	2 192 359	1 710 190
South Africa	13 800 000	7 000 000

7.3.3 Present-day land use

Today, there are more than 50 million people in southern Africa, and most are concentrated in the savanna and grassland biomes. The various land uses include pastoralism, cultivation of crops, mining and afforestation. A wide variety of savanna plants are traditionally used for food, medicine and domestic purposes. Savannas are also the basis of two major industries: cattle-ranching and wildlife-related tourism. In 1989, they had turnover of around US$1000 million (Mentis and Seijas, 1993).

Pastoralism
Extensive livestock ranching is the most common agricultural practice in southern Africa, with 84% of savanna land in South Africa used for this purpose (Grossman and Gandar, 1989) (Table 7.5). The highest livestock concentrations occur in the humid savannas, e.g. in South Africa, where consumption is estimated at 900 kg ha^{-1} year^{-1} (Owen-Smith and Danckwerts, 1997). In the Transvaal savanna region, the overall stocking level of domestic livestock (47 kg ha^{-1}) is 1.5 times the biomass density of wild herbivores in the Kruger National Park (32 kg ha^{-1}). About two-thirds of the cattle and goats in the savannas are raised on large commercial ranches with the objective of meat production. The rest are kept on communally grazed lands where meat production is secondary to asset accumulation in the form of livestock, as well as other services such as milk and draught power (Scholes, 1997).

Cultivation
The total area under cultivation in South Africa in 1988 was around 130 000 km^2 or about 10.6% of the land surface (Anon., 1994, cited in Hoffman, 1997). Data show that there was a steady increase in the area cultivated between 1911 and 1965, but that this has levelled off in the last three decades (Figure 7.7). This suggests that most productive land has already been converted for cultivation. Nearly half of the total land cultivated in South Africa is used for maize production, and most of this is in savanna areas. Sugarcane is also an important crop (Figure 7.7).

Conservation, wildlife and ecotourism
The private wildlife industry in southern Africa, established in the 1960s, is one of the most well known, and far exceeds areas devoted to parks and reserves.

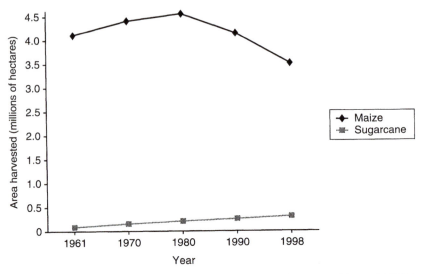

Figure 7.7 The area of maize and sugarcane harvested between 1961 and 1998 in South Africa (based on FAO, 1999).

Operations vary from pure wildlife commercial game ranches and private reserves to mixtures of domestic animals with antelope species. Its success is attributed to the incentives of the ranchers to conserve their wild animals because of the important source of income that can be derived from them. Trophy-hunting and ecotourism are particularly high-income activities (Mentis and Seijas, 1993).

The largest and most well-known parks in southern African, such as Kruger National Park, are in savanna areas. In recent years there has been a trend towards restocking private land with native mammals, primarily for licensed trophy-hunting, and now increasingly for recreation and tourism (Scholes, 1997). Large areas of communal land in Botswana, Zimbabwe and Namibia also have wildlife as their main source of income. However, the ecological promise of wildlife-based meat production systems in savannas has not yet been realised, mainly because of unresolved problems relating to harvesting the animals in a cost-effective manner and marketing the meat products (Scholes, 1997).

Wood harvesting

Fuelwood harvested mainly from savannas is the principal energy source in most southern African households, and for the foreseeable future, demand will remain high. With the exception of the *Baikiea*- and *Pterocarpus*-woodlands of Zimbabwe, Botswana and Namibia, which yield high-value hardwoods, commercial exploitation is minimal, and timber is used mostly in the craft industry (Scholes, 1997).

7.4 Current management problems and strategies _____

The southern African savannas have a great potential for sustaining its long history of wildlife tourism. However, there is no doubt that whether it be eco-tourism or organised trophy-hunting, these activities have to go hand in hand with local people's livelihoods. The current limitations to effective resource management in the southern African savannas are protectionist conservation policies and land tenure, the threat of alien plant invasions, and inadequate fire management. Further land-use conflicts reflect government perceptions of pastoralists causing desertification.

7.4.1 Conservation

About 11.6% of the southern African savanna biome is currently under conservation status (Rebelo, 1997). Botswana has the highest area under conservation at 17.88%, followed by South Africa (9.96%), Namibia (5.74%) and Swaziland (4.96%). Still, this does not mean that all savanna types are adequately conserved, or that these conservation areas contain only savanna. For example, the Etosha National Park in Namibia conserves 4.3% of the savanna biome, yet a third of the area is occupied by the Etosha Pan, a saline pan. It also seems that savanna types shared with other countries are better conserved than endemic types (Rebelo, 1997).

Although private wildlife enterprises have generally been successful, both economically and in terms of conservation, because of the financial incentive to preserve habitats and species, state-run parks have come under criticism due to their inefficiency and inaccessibility (Mentis and Sijas, 1993). In these areas, resettlement, illegal hunting and farmer encroachment has been frequent as a result of restrictions placed on resource use within conservation areas. Community-based conservation programmes, such as the CAMPFIRE programme of Zimbabwe (see Section 4.4.4), have been implemented to give local communities greater authority over their own resources. However, as with other savanna regions of Africa, these are not always as successful as intended. In Botswana, for example, wildlife management area-related policies aim to develop rural people's economies while helping conservation, but have ended up being coercive, and without community participation (Twyman, 1998).

7.4.2 Alien plants

Although a number of plant species were introduced prior to colonisation, none seem to have had any significant impact on the natural vegetation (Richardson et al., 1997). Colonisers, and later settlers, brought with them a vast array of agricultural and horticultural plants from areas such as Europe, the Indian sub-continent and Indonesia. About 65% of the approximately 150 invasive alien species of South African savannas arrived between 1800 and 1900 (Henderson and Wells, 1986).

Most alien invaders in the southern African savannas are from South America (Richardson *et al.*, 1997). Herbs include *Alternanthera pungens, Bidens bipinnata, Chenopodium album, Datura stramonium, Schkuhria pinnata* and *Tagetes minuta.* Several succulents from the Americas are also abundant, and include *Agave sisalana, Cereus jamacaru* and *Opuntia ficus-indica.* Widespread woody invaders are *Caesalpinia decapetala, Jacaranda mimosifolia* and *Sesbania punicea.* In South Africa, *O. ficus-indica* has the most extensive infestation, followed by *Melia azedarach, Lantana camara* and *Opuntia aurantiaca* (Henderson and Wells, 1986).

These plant invasions pose several threats to savannas, including native plant extinction, and the alteration of soils, hydrology, geomorphological processes and fire regimes (Cronk and Fuller, 1995). However, they also have economic consequences. For example, combating the invasion and degradation of grazing land by alien plants, as well as livestock injury, in the case of *Opuntia aurantiaca* in South Africa, has resulted in great employment of time and resources, and has cost millions of rands. Richardson *et al.* (1997) outline some of the methods of control currently in place.

7.4.3 Fire management

Fire prescriptions developed for livestock farming have often been applied to nature reserves to promote the large mammal component. Recently, this 'farming' method has come under criticism, as it may promote some components of the biota at the expense of others. Calls have been made for more variable fire regimes (varied frequencies, seasons, burn intensities) in conservation areas in line with the objective of conserving biotic diversity (Mentis and Bailey, 1990). Where feasible, 'natural' fire regimes have been proposed, meaning in practice that lightning fires be allowed to burn and other fires suppressed (Stander *et al.*, 1993). The idea is that the fire regime under which biota evolved is most likely to ensure its future survival. However, Van Wilgen *et al.* (1990) point out that it is important to know what the management objectives of an area are and whether a 'natural' fire regime will meet those objectives. Also the characteristics of 'natural' fire regimes are unknown, and so the effects may not meet expectations.

The prescribed burning season is early spring, after the first rains. It is argued that burning at this time will cause the most damage to trees but the least damage to grass, cause the least nutritional stress to grazers, and expose the ground surface to erosion for the shortest period. Fire ignition by lightning may have peaked at this time historically, because of the coincidence of 'dry' thunderstorms and dry grass. The current limitation is probably over-restrictive, injurious if applied too rigidly or too late, and makes management more difficult.

In Kruger National Park, between 1980 and 1992, 47% of fires were from controlled burning. This was applied to remove dead and unpalatable herbaceous material, and maintain an adequate tree : grass ratio (Trollope, 1993). Another 23% were caused by refugees fleeing from Mozambique via the park into South Africa. Only 10% were caused by lightning, probably because of the other causes reducing fuel loads. The other 20% arose as a result of various causes including poachers, tourists, arsonists and accidents. The evidence indicates that even if controlled burning was discontinued, the status quo would be maintained because of the various other causes of fire. The fire regime in Kruger National Park is highly variable, depending on climate, fuel and sources of ignition, which may be beneficial in providing a variety of habitat types and promoting biodiversity. However, long-term burning in the park indicates that as fire frequency increases, so does the intensity of grazing by wild ungulates. This increased grazing pressure may lead to eventual reduction in grass biomass to a state that is unproductive in terms of providing forage and with a lower diversity of perennial species. This indicates the necessity of long-term vegetation monitoring.

7.4.4 Overgrazing and bush encroachment

Even though numbers of livestock are at their lowest in 60 years in South Africa, overgrazing by domestic animals is still regarded as a major cause of land degradation. The same is also felt about other savanna countries, particularly Botswana, where controversy exists as to the cause and severity of the problem (e.g. Abel, 1993; Dahlberg, 1993; De Queiroz, 1993; White, 1993; Ringrose et al., 1996; Sefe et al., 1996). This debate is vital to communal farmers, since it impinges directly on their country's livestock development policy (Pearce, 1993). Tenure policies have also been fundamental to pastoralism in these countries. Botswana has advocated 99-year leases of huge ranches to large stakeholders, i.e. ranch privatisation (Bruce et al., 1998a), but to what consequence for traditional nomadic pastoralists?

The historical removal of large native herbivores from savannas, the alteration of fire regimes, and overgrazing have all been blamed for the general 'bush encroachment' problem in the savanna biome today (Grossman and Gandar, 1989) (Figure 7.8). It is estimated that because of bush encroachment, about 2.6% of the savannas in South Africa are unusable, 63% are threatened, and much of the rest has a reduced carrying capacity of up to 50% (Grossman and Gandar, 1989).

Solutions to these problems seem to have two sides. On the one hand, land distribution policies in countries such as South Africa are increasing the demand for pastoral land, and so programmes are being developed to tackle the problem, e.g. increasing surface water resources using pipes in the southern Kalahari. However, as Palmer and van Rooyen (1998) have detected using satellite data, the consequence of artificial surface water is an exacerbation of desertification in the region, and a change from herbaceous to unpalatable woody biomass. Therefore, in this case, the situation seems to be worsening. However, on the other side is the new approach, again in South Africa, of community action in combating desertification at all levels of project planning, decision-making and execution (van Rooyen, 1998). The main limitations of this, from the example of the Mier Rural Area in South Africa, have been political and ideological

Figure 7.8 Bush encroachment in southern African savannas.
(Reproduced by kind permission of Bob Scholes, CSIR, Pretoria, South Africa)

differences within the community, and the perception of desertification as a low priority compared to the basic needs of health care, food and education. Nevertheless, a holistic approach to desertification is vital to ensure any success.

7.5 Concluding remarks

Southern African savannas can be divided into two distinct forms governed by the plant-available nutrients: the nutrient-rich fine-leaved savannas, and the nutrient-poor broad-leaved savannas. There are various differences between the two savanna types, including productivity, herbivory and fire occurrence. However, both are undoubtedly going to come under increasing pressure as population rises and land redistribution programmes are established in the region. In countries such as South Africa, the post-apartheid democratic government faces social and economic problems that will have direct consequences for savanna lands. In the region as a whole, rethinking resource management, particularly in relation to conservation and pastoralism, poses a major challenge.

Key reading

Blackmore, A.C., Mentis, M.T. and Scholes, R.J. (1991). The origin and extent of nutrient-enriched patches within a nutrient-poor savanna in South Africa. In: Werner, P.A. (ed.), *Savanna Ecology and Management. Australian Perspectives and Intercontinental Comparisons*. Blackwell, Oxford, pp. 119–126.

Huntley, B.J. (1982). Southern African savannas. In: Huntley, B.J. and Walker, B.H. (eds), *Ecology of Tropical Savannas*, Ecological Studies 42. Springer-Verlag, Berlin, pp. 101–119.

Mentis, M.T. and Seijas, N. (1993). Rangeland bioeconomics in revolutionary South Africa. In: Young, M.D. and Solbrig, O.T. (eds), *The World's Savannas. Economic Driving Forces, Ecological Constraints and Policy Options for Sustainable Land Use. Man and the Biosphere Vol. 12.* UNESCO, Paris, pp. 179–204.

Scholes, R.J. (1997). Savanna. In: Cowling, R.M., Richardson, D.M. and Pierce, S.M. (eds), *Vegetation of Southern Africa.* Cambridge University Press, Cambridge, pp. 158–277.

Scholes, R.J. and Walker, B.H. (1993). *An African Savanna. Synthesis of the Nylsvley Study,* Cambridge University Press, Cambridge.

Chapter 8

The dry dipterocarp savannas of mainland Southeast Asia

8.1 Introduction

8.1.1 Distribution, climate and soils

Savannas and savanna woodlands extend throughout Asia, and are the predominant formation across mainland Southeast Asia and the Indian subcontinent. Much of the Indian savannas are derived from human disturbance, as described in Section 1.6.5. True lowland open savannas are of limited occurrence in mainland Southeast Asia (Blasco, 1983; Stott, 1984), as for example, at Ban Me Thuot in Vietnam, Mondolkiri in Cambodia, Kanjanaburi, Sakon Nakhon, and the so-called 'bald hills' of the Petchabun range in Thailand (Blasco, 1983). They may originate either because of heavy inundation during the rainy season, as at Sakon Nakhon, although much of this site is now destroyed, or through intense cutting, burning and grazing, as on the Petchabun mountains (Stott, 1991b). Characteristic herbaceous and woody species of this 'grass forest' (*pàa yâa* in Thai) are *Imperata cylindrica, Panicum repens, Saccharum spontaneum, Sorghum halepense, Vitiveria zizanoides, Eupatorium odoratum, Acacia catechu, A. siamensis, Careya arborea* and *Pterocarpus macrocarpus* (Stott, 1991b). These savannas support large browsing and grazing mammals, such as barking deer (*Muntiacus muntjak*), sambar deer (*Cervus unicolor*) and banteng (*Bos javanicus*).

In contrast, savanna woodlands or 'savanna forests' of mainland Southeast Asia are much more widespread, closed in physiognomy, and wooded (Schimper, 1903). Several types of savanna forest have been described by various workers, including the *Diospyros* or *Te* forest, *Than-Dahat* forest, and the *Sha-Dahat* thorn forest of Burma (Stamp, 1925; Richards, 1952), but the most significant and extensive savanna formation in mainland Southeast Asia is undoubtedly the dry dipterocarp forest (Blasco, 1983) (Figure 8.1). Dry dipterocarp forest, beginning in Manipur State, India, is widespread in Burma, Thailand, Cambodia, Laos and Vietnam (Stott, 1976; Blasco, 1983; Mistry and Stott, 1993). Since most of the research on dry dipterocarp savanna forests has been carried out in Thailand, the following description of its ecology, use and conservation will be restricted to this country, although some comparative examples may be given from other parts of Southeast Asia.

All the savanna forests of mainland Southeast Asia have a typical tropical monsoon climate with five to seven distinct months of dry conditions, starting

Figure 8.1 The extent of the dry dipterocarp savanna in mainland Southeast Asia.
(Reproduced by kind permission of Justin Jacyno and Nigel Page)

from October or November, until May or June (Stott, 1991b). Annual precipitation is about 1000–1500 mm, and being essentially a lowland formation, it rarely grows above 100 m above sea level or exceeds the latitude 20°N (Smitinand, 1962, 1977; Stott, 1988a). As a result, daily maximum temperatures are normally above 20 °C throughout the year, with absolute minimum temperatures in winter usually not below 8 °C.

The savanna forests are found mostly on sandstone, but also on granite slopes and ridges, and on old alluvium. Soils are red-yellow oxisols, characteristically well drained, with a low clay fraction, a high proportion of sandy loams, and a substantial amount of stoniness (Bloch, 1958). The soils can also form a laterite hardpan that causes waterlogging during the wet season (Bloch, 1958). The humus content of the soil is very low (< 2–4%), the C/N ratio is about 12.7, the pH between 5 and 6.2, and the total nitrogen content is low at all depths (Aksornkoae, 1971; Sangtongpraow and Dhamanonda, 1973).

8.1.2 Vegetation structure and composition

The dry dipterocarp forest (*kanyin* in Manipuri, *indáing* in Burmese, *pàa ten-grang* in Thai, and *forêt claire à dipterocarpacées* in Indo-China) can be arranged into four main associations: the *Dipterocarpus* association, often considered the true dry dipterocarp forest (e.g. Khemnark *et al.*, 1972; Blasco, 1983; Santisuk,

Table 8.1 Distribution of Dipterocarp and *Shorea* associations across Southeast Asia, and their characteristics

Distribution	Manipur State, India	Burma	Thailand	Indo-China (Laos, Cambodia, Vietnam)
Dipterocarp association	*kanyin* forest (*Dipterocarpus tuberculatus*)	*indáing* forests (*Dipterocarpus tuberculatus*)	*Dipterocarpus tuberculatus–D. obtusifolius* forests	*forêt claire à D. tuberculatus, D. obtusifolius, D. intricatus*
Shorea association	–	semi-*indáing* and *indáing* scrub forests	*Shorea–Pentacme* forests	*forêt claire à Pentacme siamensis*

Characteristics	Dipterocarp association	*Shorea* association
Dominant trees	*Dipterocarpus tuberculatus*, *D. obtusifolius*, *D. intricatus*. *D. tuberculatus* more important in east than west, *D. obtusifolius* found throughout mainland Southeast Asia, and *D. intricatus* generally occurs eastwards from north-east Thailand	*Shorea obtusa* and *Shorea siamensis* (syn. *Pentacme suavis* and *Pentacme siamensis*)
Canopy height (m)	15–40	15–25
Crown cover (%)	Up to 80	below 60
Ground cover	Dense grasses and herbs such as *Imperata cylindrica*, *Apluda mutica*, *Heteropogon* spp., *Themeda* spp. and *Arundinaria* spp. (pygmy bamboo)	Grasses or pygmy bamboo. Dense, tall *Arundinaria* spp. particularly characteristic of more mesic sites, while xeric sites have thin grass cover, although *Apluda mutica* can be dense at some sites
Tree seedlings	Numerous	Inhibited
Epiphytes, orchids and climbers	Abundant	Present
Floristic diversity	Average	Poor
Site characteristics	Found on flat and gently sloping land, between 300 and 900 m. Soils are deep, sandy loams to clay loams, acidic, of the red-lateritic yellow podzolic group	Occur on dry ridges and gentle to steep boulder strewn slopes, between 130–1000 m altitude. Soils are shallow, stony and belong to red-yellow podzolic group. Mostly sandy loams over sandstone, granite, shale or quartzite parent rock

1988); the *Shorea* association; the pine–dipterocarp association; and the mixed dry dipterocarp association. The last two are considered to be ecotonal in character: the pine–dipterocarp forest is found mostly on ridges and mountains over 750–1100 m, combining floristic and physiognomic elements of the high hill evergreen pine forests; while the mixed dry dipterocarp forest is found mainly on limestone in association with monsoon forest (Stott, 1976; Santisuk, 1988). As a result of their prevalence, the following text will deal mostly with the dipterocarp and *Shorea* associations (Table 8.1) (Figure 8.2).

Although there has been much controversy about the origins of the dry dipterocarp savanna forests (Stott, 1991b), evidence based on pollen analyses now strongly suggests that a drier climate prevailed over much of equatorial Southeast Asia during the late Quaternary glaciation (Flenley, 1979, 1982). It is hypothesised that seasonal lowland forests and savannas were much more extensive, and covered a greater part of the Sunda shelf up to 18 000 years BP, when sea level was 180 m lower than the present-day level (Whitmore, 1984). As the ice retreated in higher latitudes around 10 000 years BP and the monsoons re-established, the savanna forests may have contracted back to their edaphic and topographical 'core' communities, probably on steep and stony slopes with shallow soils (Stott, 1984, 1988b). Then, as Neolithic people increased their burning and cutting activities, the dry dipterocarp forest probably spread from their core areas into more fire-sensitive associations such as the dry evergreen forest where the climate still permitted fires to burn in the dry season (Stott, 1988b).

Figure 8.2 Dry deciduous dipterocarp forest (savanna forest) dominated by *Dipterocarpaceae* (e.g. *Shorea obtusa* and *Dipterocarpus intricatus*), with a ground cover of pygmy bamboo (*Arundinaria* spp), north-east Thailand.
(Reproduced by kind permission of Philip Stott, School of Oriental and African Studies, University of London)

Plant life strategies

Grasses generally begin to wither after the last Southwest Monsoon in December or January, and leaf shedding takes place from January or February. This continues until around early or mid-March, when some new leaves then begin to flush. New plants, especially annuals, start to appear in March or shortly after burning in the dry season. Kanjanavanit (1992) divides the ground cover at Khao Nang Rum Research Station, West Thailand, into six main phenological groups based on flowering and fruiting (Table 8.2).

Most trees begin shedding their leaves from January, peaking in February, and continuing until March. Leaf flush then occurs in March and April, and flowering and fruiting can also take place at this time (Figure 8.3). Pollinators may include large moths, butterflies and, to a lesser degree, birds (Ghazoul, 1997). Fruit dispersal can vary among species, but many trees drop their fruits directly below the canopy and the seeds germinate right after the fire season, along with the early rains (Sukwong *et al.*, 1975; Santisuk, 1988), when the nutrients released from fires are readily available. Once young seedlings of the dominant trees, such as *Dipterocarpus tuberculatus*, emerge, they can remain in a low, almost under-shrub stage for decades if they are burnt back and shaded by an overhead canopy. However, once light is admitted, the seedlings send up new straight shoots which are renewed until the buds are out of reach of fires.

Table 8.2 Six phenological groups of the ground cover in dry dipterocarp savanna forest in Khao Nang Rum Research Station, West Thailand (based on Kanjanavanit, 1992)

Phenological group	Examples of species
Flower and fruit more or less simultaneously from early or mid-rainy season (April–June) to early dry season (November–January)	Grasses such as *Setaria pallide-fusca, Panicum auritum, Coelorachis mollicoma*, all members of Cyperaceae and most low shrubs such as *Flemingia* sp., *Leea indica, Helicteres* sp.
Flower during the rainy season (from May), fruit at the end of the Southwest Monsoon rainy season or in the early dry season (October–January)	Geophytes such as *Cyanotis barbata, Zingiber zerumbet, Globba obscura, Kaempferia pulchra*
Flower and fruit in the early dry season (October/November–January)	Most common pattern in herbaceous layer, e.g. *Apluda mutica, Sorghum nitidum, Heteropogon triticeus, Eupatorium odoratum, Elephantopus scaber, Desmodium motorium*
Flower and fruit during the rainy season	Geophytes such as *Curcuma parviflora, Globba obscura, Kaempferia pulchra* and ground orchid, *Habenaria linguella*
Sporadic flowering and fruiting throughout the rainy season and early dry season	Shrub *Pluchea polygonata* and herb *Hedyotis* sp.
Flower and fruit after fire or other disturbances	Grass *Imperata cylindrica*, herb *Elephantopus scaber*, geophyte *Eurycles* sp.

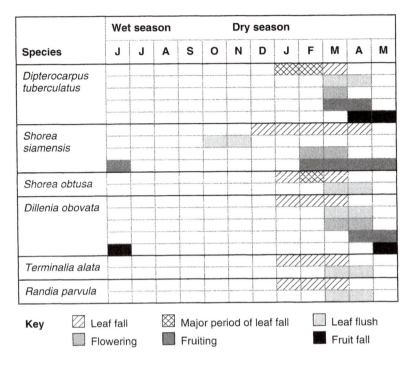

Figure 8.3 Some phenological characteristics of five dominant trees from dry dipterocarp savanna forest in Khao Nang Rum Research Station, West Thailand (based on Kanjanavanit, 1992 and Stott, 1988a).

8.1.3 Fauna

There have been few studies on the faunal component of the dry dipterocarp savanna forests. Table 8.3 lists some of the characteristic mammals found in Thailand. The largest animal associated with *indáing* in Burma is the *siang* or Burmese wild ox (*Bos sondaicus*) (Barrington, 1931), but it is now rare. The banteng (*B. banteng*), common in Burma and eastwards, did occur in Manipur State, but may have disappeared through heavy slaughter by forest tribes (Seshadri, 1969). The more frequently recorded mammals are not just confined to savanna forests, but are spread over a number of vegetation types. This probably reflects the reliance of animals on various vegetation types for resources (Rabinowitz, 1990). Animal diversity in the savanna forests is limited: small mammals (notably nocturnals) and birds are the least diverse, though there is a richness of herpetofauna (Stott, 1984).

8.2 The main savanna determinants

Workers in Southeast Asian savanna forests have had varied and opposing views on which factors determine the vegetation, although moisture, nutrients and

Table 8.3 Characteristic mammals of the Thai dry dipterocarp savanna forests (based on Stott, 1984)

Species	Common names and description
Insectivora	
Tupaia glis	Common tree shrew – forages in secondary forest for insects, spiders, fruit, seeds, buds, lizards and small rodents.
Talpa micrura	Eastern mole – found in slash-and-burn clearings, usual diet insects, but also lizards, snakes, mice and small birds.
Suncus murinus	House shrew – introduced, commensal with humans.
Suncus etruscus	Savi's pygmy shrew
Crocidura dracula	Dracula shrew – river valleys and foothills.
Pholidota	
Manis javanica	Malayan pangolin – preys on insects.
Proboscidea	
Elephas maximus	Indian elephant – formerly common in dry dipterocarp forest.
Artiodactyla	
Muntiacus muntjak	Barking deer – browser, fond of fallen fruit (*Gardenia, Terminalia* species), mainly nocturnal.
Cervus unicolor	Sambar – browser of young leaves, soft buds, new shoots of grass and fallen fruit, crepuscular and nocturnal.
Bos javanicus	Banteng – formerly common in dry dipterocarp forest.
Lagomorpha	
Lepus siamensis	Siamese hare – hides in bushes and grass during the day, feeds on grass, bark and twigs in the night.
Rodentia	
Tamiops macclellandi	Himalayan striped squirrel – usually over 700 m altitude, diet of insects and fruit (*Dillenia, Mangifera* species).
Menetes berdmorei	Berdmore's squirrel – forages on ground, known to enter rice and corn fields.
Atherurus macrourus	Asiatic brush-tailed porcupine – feeds on roots, tubers, fruit, bark and carrion.
Hystrix brachyura	Malayan porcupine – eats roots, tubers, bark and fallen fruit.
Rattus surifer	Yellow rajah rat – the predominant rat in the dry dipterocarp forests.
Mus cervicolor	Fawn-coloured mouse – typical of dry dipterocarp forests.
Mus shortridgei	Shortridges's mouse – in the tall grass and pygmy bamboo beneath forest.

fire were recognised by all as key players. The French, for example, working in Indo-China, regarded the savanna forest as a climatic sub-climax, which is 'stabilised' by fire (e.g. Rollet, 1953; Vidal, 1972). The British, on the other hand, working largely in Burma, saw the savanna forest as essentially an edaphic or topographical climax, although they acknowledged that fire may be important in some areas (e.g. Barrington, 1931; Champion, 1936). More recent researchers view the savanna forests of mainland Southeast Asia not as an edaphic, climatic or fire climax (Santisuk, 1988; Stott, 1988a), but as being governed by the four determinants: plant-available moisture, plant-available nutrients, herbivory and fire (Stott, 1988b).

8.2.1 Plant-available moisture (PAM)

The dry season in the dipterocarp savanna forests creates a prolonged period of moisture stress, during which the rate of evapotranspiration exceeds the rate of precipitation, and vegetation growth is limited (Stott, 1984). This dryness is further exacerbated by the low water-holding capacity of most savanna forest soils (Figure 8.4). Many trees and shrubs have, however, developed xeromorphic characteristics as survival strategies during the long dry season. Mechanisms to minimise water loss include densely tomentose leaves and terminal buds, thick bark and leaf shedding. *Shorea obtusa*, for example, has tomentose leaves on one side in its most common association, but in xeric sites has hairs on both sides, and in mesic sites is virtually glabrous (Smitinand, 1968; Stott, 1976). Similarly, the degree of deciduousness and the period of

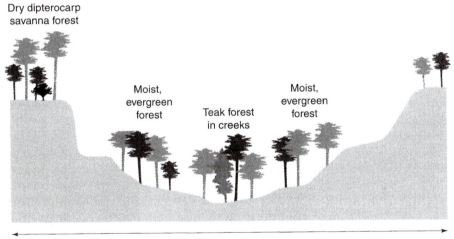

Figure 8.4 A diagram to show the ecology of 'waik' formations in the dry dipterocarp savanna region of Manipur State, India. These 'waiks' were described by Bruce in 1903, and illustrate the role of topography and soil moisture in controlling savanna forest (after Mistry and Stott, 1993).

leaflessness of dry dipterocarp forest trees varies considerably with soil moisture content (Nalamphun *et al.*, 1969; Santisuk, 1988; Kanjanavanit, 1992). For example, leaf shedding is generally one month later in wetter west Thailand, compared to drier north-east Thailand (Kanjanavanit, 1992). The crevices in the thick bark of many trees, such as *Shorea obtusa*, increase the interception rate of moisture, and decrease throughfall and stemflow, thereby additionally reducing soil erosion (Chunkao *et al.*, 1971).

8.2.2 Plant–available nutrients (PAN)

Plant-available nutrients are low in all dipterocarp savanna forests. Paovongsar (1976), in a study of litter fall and the mineral nutrient content of litter in savanna forests, found that leaf litter alone accounted for approximately 74% of annual litter production. The leaf litter and herbaceous ground cover, when burnt, contribute various nutrients to the soil. Most important of these are nitrogen, phosphorus, potassium, calcium and magnesium.

Kutintara (1975) found that the distribution of the dominant species is controlled by the heterogeneous distribution of plant-available nutrients, as well as soil characteristics determining plant-available moisture. So, for example, *Dipterocarpus tuberculatus* is found on soils of various depths with a high potassium but low phosphorus content, at high elevations and a medium slope. *Dipterocarpus obtusifolius* has a range that largely overlaps with *D. tuberculatus* but is common at higher elevations on soils with low potassium and phosphorus content, and on gentle slopes. *Shorea siamensis* occurs at low elevations on steep slopes, and on soils of high phosphorus and low to medium potassium content. *Shorea obtusa* occurs throughout the range, and has the widest tolerance.

8.2.3 Herbivory

Unfortunately little is known about herbivory of native animals, although wild herbivores, such as barking deer, banteng, gaur, the Indian elephant, Siamese hare, sambar deer and the very rare kouprey, frequent the savanna forest for food. Many of these herbivore populations have been reduced through activities such as hunting and war, and it can be hypothesised that they may not be as significant as they were in the past. Invertebrates, such as termites, play a role in removing leaf and other biomass in other world savanna formations, and may therefore also contribute to herbivory in dry dipterocarp savanna forests. However, both mammalian and invertebrate herbivory are probably determining factors only at the local scale.

8.2.4 Fire

Fires in the dry dipterocarp savanna forest have been a common ecological phenomenon during the dry season: first from natural causes such as lightning; and then more frequently and widespread with the advent of Neolithic communities (Schüle, 1990). In recent times, fire has taken an unprecedented leap in frequency, affecting savanna forest resources in closer and closer proximity to human habitations (Goldammer, 1988).

In Thailand, for example, data collected from 1984 to 1986 show that fire affected 3.1 million ha year^{-1} or 20.92% of forested areas (Royal Forest Department, 1988). Fires in the dry dipterocarp savanna forests mainly occur during the dry season, which generally begins in December and lasts until early to mid-April or later, although a Northeast Monsoon normally brings a small amount of rain in February.

Causes of fire

Natural fires are believed to be rare in Thailand although local people claim to witness lightning strikes igniting fires during the onset of the early Southwest Monsoon. Dry thunderstorms are common during this time, and numbers of past dry thunder days extrapolated by Kanjanavanit (1992) indicate that wildfire caused by lightning strike is very probable in Thailand, albeit uncommon.

Most fires in the dry dipterocarp savanna forests are human-induced (Goldammer, 1987). Table 8.4 lists the main sources of fire in some savanna forests of Thailand.

Table 8.4 Causes of fire in savanna forests in northern and western Thailand (based on Kanjanavanit, 1992)

North Thailand (Chiang Mai Province)		West Thailand (Huai Kha Khaeng Wildlife Sanctuary)	
Unknown	44%	Path clearing	Highest frequency
Swidden fire	26%	Unknown	
Minor forest product collection	9%	Swidden/field burning	
Fire-break burning	6%	Forestry officials	
Arson in plantations	5%	Road-side fire	
Hunting	1%	Hunting	
Tourist development (elephant camp)	1%	Camping fire	
Illegal timber cutters	1%	Arson (e.g. land-use conflicts)	
To induce grazing	0.3%	For fun	
		Military purposes	
		Induce young growth (other than for grazing)	
		Nutrient wash from hill ridges to fertilise fields	
		Cigarette ends	Lowest frequency

Factors affecting fire

Climate

Climate is important in affecting fire directly at the time of the burn, but it also controls the fuel moisture content through a range of factors such as rainfall, air temperature and humidity (Mather, 1978a). Burns which occur early in the dry season are generally less intense because of the high fuel moisture content and low daily temperatures, than later in the season, when the fuel is dry, relative humidity low, air temperature near its highest, and wind velocity at a maximum (Mather, 1978a,b; Ruangpanit and Pongumphai, 1983).

Wind direction in relation to the fire can have some important consequences for fire. Kanjanavanit (1992) measured the highest temperatures and shortest temperature duration (600 °C lasting 5 seconds) at ground level in fast-moving, well-sustained head fires. On the other hand, back fires showed a tendency to yield a long duration of maximum temperature, with a maximum temperature above 270 °C for as long as 55 seconds.

Fuels

As well as the fuel moisture content, the actual fuel type and quantity is also important. In dry dipterocarp forest, fuel may be divided into three main categories: grasses; non-grass ground cover, which includes all evergreen or deciduous herbs, forbs and seedlings; and fallen leaves, flowers and fruits (litter) (Kanjanavanit, 1992). Leaf-shedding generally begins in November and peaks in March (Santisuk, 1988), by the end of which the forest floor is carpeted to at least five layers of leaves deep (Kanjanavanit, 1992). As well as increasing the fuel load, the opening of the canopy allows direct sunlight to dry and heat up the leaf litter and undergrowth, giving rise to peak periods of fires (Sukwong and Dhamanitayakul, 1977; Stott, 1984, 1988b).

Where there is a significant, continuous cover of 50–100% of grasses and other non-grass herbaceous vegetation, 'ground cover burns' occur, and where the ground cover is thin but a good carpet of leaf litter is present, 'litter burns' occur (Stott, 1986). With litter burns, Stott (1986) found that temperatures are always highest at ground level, ranging from 250 °C (two to three leaves deep) to 700 °C (over six leaves deep), with a mean of 400 °C. Ground cover burns attain their highest temperatures at 0.5 m, with a mean of 300 °C. Where there are pure stands of grasses and pygmy bamboo, with a cover of 95% and a height of 3 m, fires can reach 900 °C at 0.5–1 m above the ground, and spread at a rate of 1.8 m min^{-1} (Stott, 1986). Kanjanavanit (1992) found similar results. However, she recognises three major fuel-fire regimes based on her experiments around Khao Nang Rum Research Station, Thailand: a mixed ground cover burn; a homogeneous grass cover burn; and a leaf litter burn (Figure 8.5).

Fuel arrangements can also affect the fire regime. Kanjanavanit (1992) found that loosely packed collapsed grass stalks tend to provide better conditions for a burn by allowing more oxygen to circulate the material and reach the combustion zone, while being dense enough for efficient heat transfer. She goes on to show that different grass species form different fuel arrangements as they lose

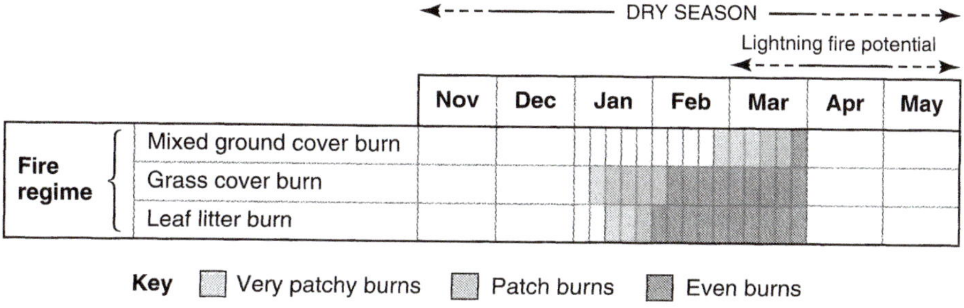

		Nov	Dec	Jan	Feb	Mar	Apr	May
Fire regime	Mixed ground cover burn							
	Grass cover burn							
	Leaf litter burn							

◄--- ———— DRY SEASON ———— - - - - - - - -►

Lightning fire potential
◄- - - ———— - - -►

Key ☐ Very patchy burns   ☐ Patch burns   ☐ Even burns

Figure 8.5 Three proposed fire regimes for Khao Nang Rum Research Station, western Thailand (after Kanjanavanit, 1992).

moisture through the dry season. For example, in her experiments *Apluda mutica* stalks remained upright throughout the dry season, but about half of *Sorghum nitidum* stalks and most of *Heteropogon triticeus* collapsed, mainly after the rains of the Northeast Monsoon in early February (Kanjanavanit, 1992). Moreover, whether the fuel is in patches or is evenly spread influences the homogeneity of the burn, although this does also depend on the moisture content of the fuel (Kanjanavanit, 1992).

The different times of burns in one year can greatly influence the accumulated combustible fuel for the following year (Kanjanavanit, 1992). Burns during or after the main period of leaf shedding leave only a thin litter layer, which cannot sustain fire until enough new litter has accumulated at the end of the next dry season. An unburned area, on the other hand, has a good potential to sustain fire early in the following year, with as much as 95% litter cover, four layers of leaves deep, accumulated on the ground at the beginning of the dry season (Kanjanavanit, 1992).

The effect of fire on vegetation

The vegetation normally recovers quickly after fire, often within two weeks, and even before the onset of the Southwest Monsoon (Stott, 1988b). The release of surface nutrients by the fire induces growth, leaf flushing, and in some species such as *Imperata cylindrica*, also flowering (Kanjanavanit, 1992).

Regeneration of tree species can occur through seeds or underground root buds. To ensure a significant percentage of regenerative success, large quantities of seeds are produced. Of these, the dominant species tend to drop their fruits during and after the peak fire period (February–March) so that germination then takes place on the burnt soil at the beginning of the Southwest Monsoon (Sukwong *et al.*, 1975). Conversely, trees with thick seed coats, such as *Irvingia malayana* and *Sindora siamensis*, tend to drop their seeds before the fires (November–December) (Sukwong *et al.*, 1975), and some workers think that fire may be beneficial to their germination (e.g. Stott, 1988b).

Sprouting from root buds is common, but smaller seedlings are prone to die-back from fire. Santongpraow (1985) found that *Shorea siamensis* and *S.*

obtusa seedling sprouts must reach over 170 cm tall, with a girth of at least 10 cm and a bark thickness of 0.6 cm, in order to survive the fiercest of burns. Whether regeneration is from seed or sprout, most dry dipterocarp savanna forest seedlings take an average of 3–4 years to reach over 170 cm without disturbance from fire (Wacharakitti *et al.*, 1971). This implies that a fire-free interval of more than one year is needed to ensure successful regeneration, although growth, albeit slow, does occur in frequently burned areas (Suthivanit, 1989), and the mosaic nature of many ground fires allows some individuals to escape fire injury (Kanjanavanit, 1992). Kanjanavanit (1992) found that after two years of fire protection, her *Dipterocarpus* savanna forest sites comprised tall, dense seedlings, 30–40% grasses, and a high proportion of non-grass growth. Two years later, very little grass coverage remained (around 10%).

A study by Suthivanit (1989) found that species diversity was maintained in all burned plots, with protected areas showing an increase in the abundance of tree seedlings and herbaceous non-grass species, and a decrease in grass density. Wolseley and Aguirre-Hudson (1997) and Wolseley (1997) have shown that fire-protected dry dipterocarp savanna forest has the highest diversity of lichen species (Figure 8.6), and that fire results in a more fire-tolerant lichen community. These lichen species can withstand increased irradiation and temperatures, and most species use vegetative means of propagation rather than sexual reproduction, thereby ensuring their success in the new habitats created in the burnt environment.

The effect of fire on soils

Fires have little direct effect on the forest soils, in that a surface temperature as high as 700 °C cannot raise the soil temperature at 5 cm depth above 35 °C (Sukwong *et al.*, 1975; Stott, 1986). This is probably a result of the humus

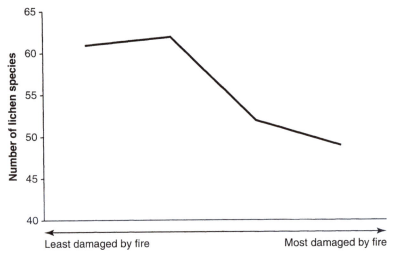

Figure 8.6 Lichen diversity decreases with an increase in fire frequency in dry dipterocarp savanna forest in Thailand (based on Wolseley and Aguirre-Hudson, 1997).

content of the soil, which although less than 2–4% (Sangtongpraow and Dhamanonda, 1973; Sukwong, 1982), has an insulating effect. However, fires, particularly in more frequently burned areas, can reduce the soil nutrient status (Boonplian, 1985; Sukwong and Dhamanitayakul, 1977) and increase soil erosion (Chunkao, 1969). Komkris *et al.* (1969) found that repeated burning increased erosion by up to 3–32 times, especially on steep slopes of > 7 °, while a rare one-off burn can have very little effect.

The effect of fire on animals

As with other savanna regions of the world, little is known about the effects of fire on indigenous animal populations in the dry dipterocarp savanna forest (Nakhasathien and Stewart-Cox, 1990). Some animals do benefit from fire, through the creation of new feeding habitats. Prayurasiddhi *et al.* (1988), for example, found that of the 600 species recorded in the Huai Kha Khaeng Wildlife Sanctuary, Thailand, only 51 (8.5%) profited from fire. These included characteristic dry dipterocarp species such as the butterfly lizard, vultures and falcons. Davies (1997) found greater numbers of termite species in areas subjected to fire compared to fire-protected dipterocarp savanna forest.

Adaptations to fire

The thick bark of dry dipterocarp forest trees is a good insulator against fire. Stott (1986) demonstrated that bark surface temperatures of over 350 °C fail to significantly raise the temperature in the underlying living cambium layer of trees such as *Dipterocarpus intricatus* and *Shorea obtusa*.

Herbaceous layer plants have various mechanisms to avoid the effects of fire. Annual species simply survive as seeds in the soil, or are dispersed away from the area. Dominant grasses, pygmy bamboos and geophytes, such as ground orchids and ginger species, have their perennating buds either underground or protected at the soil surface by a dense mat of dead matter. Some large perennials, including dwarf palms (*Phoenix* spp.) and cycads (*Cycas siamensis*), have either short stems with overlapping leaf bases for protection, or possess underground stems (e.g. *Phoenix acaulis*) (Kanjanavanit, 1992).

Many animals in the dry dipterocarp forest escape fire by taking refuge underground, e.g. burrow-nesting birds, snakes and rodents, or in adjacent moister formations, e.g. large herbivores, or by out-running the fire, e.g. common wild boar and barking deer. Others avoid fire by nesting in branches above the fire zone, some 15–20 m high, although species such as the common golden woodpecker (*Dinopium javanense*) and the black drongo (*Dicrurus macrocercus*) nest remarkably low in the trees, at about 6–8 m, while the fire burns directly below (Kanjanavanit, 1992).

8.3 Human determinant of the savanna

In the first century AD, much of Southeast Asia's population was probably located in scattered groups of shifting cultivators and hunters and gatherers in upland forests (Box 8.1). Communal management was common, especially

Box 8.1 Savanna forest products

The dry dipterocarp savanna forests were, and still are, economically important habitats (Sangtongpraow and Sukwong, 1981) used mainly for timber, but also to extract other minor products. A study of 18 villages on the Phuwiang plateau of Khon Kaen province found 229 species of flora and fauna, and 160 medicinal herbs used by the local people (Chantawong *et al.*, 1992) (Table 8.5). Table 8.6 lists the uses of the main four *indáing* species. Other uses include the extraction of wood oils from *Dipterocarpus* species, Burma lacquer varnish from *Melanorrhoea usitata*, resins and turpentines, and strychnine from *Strychnos nuxvomica* (Stott, 1991b). Forest foods such as insects, wild meat, fungi, leaves and fruit are collected, as are orchids and wild birds (Stott, 1991b).

Table 8.5 Uses of dipterocarp savanna forest by the Sogkhumpoon village, Khon Kaen province, in north-east Thailand (based on Chantawong *et al.*, 1992)

Uses	Types	Scientific name
House construction	Trees	*Dipterocarpus tuberculatus, D. alatus, Lagerstroemia* spp.
Fuelwood	Dry fallen branches	Various
Rituals and beliefs	Trees	*Oroxylum indicum* – used in rites after childbirth, *Morinda* spp. and *Annatto* spp. – leaves used in ceremony to bring back guardian spirit of a person
Food	Vegetables Flowers Shoots Fruits Mushrooms	*Cratoxylum formosum, Careya arborea* *Sesbania* spp., *Cosod* spp. Rattan and bamboo Simarubaceae spp. Various species growing on logs, soil and underground
Medicine	For example, gastric ulcers, cough and sore throat, diarrhoea, enhance blood circulation, constipation	*Tinospora tuberculata*
	Bladder stones; animal illnesses, e.g. milk production in cattle, intestinal worms, diarrhoea	*Melodorum fruticosum* (stem), *Dipterocarpus* spp. (leaves)
Dyes	Plant substances	Various

Box 8.1 (*Continued*)

Table 8.6 Uses of the dominant trees of *indáing* forests (based on Martin, 1971)

Species	Timber uses	Other uses
Dipterocarpus tuberculatus	In construction of house floors, walls, roof rafters, beams and columns. For making agricultural tools and bridge construction. As firewood.	Sap discharged from leaf veins in treating illnesses of the skin. Leaves replace grass (straw) as thatching material in poor regions.
Dipterocarpus obtusifolius	In construction of houses, as before. For making agricultural tools, bridges, coffins and slates for beds.	Leaves sometimes used as thatching material for huts in rice plantations. The flowers are consumed as vegetables.
Shorea siamensis	In construction of houses, as before.	Young leaves consumed as vegetables. Also used for preventing evil spells.
Shorea obtusa	In construction of houses, as before. Excellent charcoal.	Solid resin discharged from trunk employed for sterilising cuts. Reduced to dust and mixed with oil resin, a product prepared for restoring rot-proofing. Making incense sticks.

for resources such as undisturbed forests used for hunting and gathering, or lakes and streams for fishing. Forest cleared for cultivation was often temporarily used by residential or kin-based groups through the agricultural cycle.

Many Southeast Asian cultures traditionally perceived their land as something held in trust for their ancestors and descendants. As such, the management of land, water and forest resources was inherent in the indigenous cultures of savanna areas. This was regulated by an ethno-ecological knowledge system, where different parts of the forest were given different values, and violation of these regulations was thought to bring about spiritual anger, shown through changes in natural resources (Poffenberger, 1990; Pragtong and Thomas, 1990). For example, the law of King Mangrai, dating back almost 700 years, forbids lowland northern Thai from violating 'Pa Sua Ban', or forests dedicated to guardian spirits of each community (Shalardchai *et al.*, 1993).

8.3.1 The role of the state

The state's role in controlling land resources probably began during the first century AD as sedentary farming communities in lowland areas spread

(Poffenberger, 1990). However, territorial expansion and control was largely in river plains and coastal areas, and rarely in the forested uplands and interior. Most farming communities were probably relatively autonomous, especially as distance from the court increased, and central authority weakened. However, these kingdoms were influential in establishing the concept of state domain.

8.3.2 The influence of Europeans

Most of the region was under forest cover until accelerated deforestation in the nineteenth century. This came about because of the increased role played by Europeans in the development of forest management concepts and practices (Figure 8.7). These included well-known foresters such as Kurz (1877), Barrington (1931) and Champion (1936). In Burma (British), Vietnam, Cambodia and Laos (French), colonisation meant the direct control of resources by colonial powers. However, even in Thailand, which remained independent throughout the colonial period, traditional management practices gradually declined during the reign of King Mongkut and King Chulalongkorn in the last third of the nineteenth century (Pragtong and Thomas, 1990). Western, scientific knowledge systems began to be adopted as a means of modernisation in response to pressures from the expanding imperialist Europe.

In order to centralise power, the Thai government gradually took control of all natural resources, with the establishment of the Royal Forestry Department (RFD) in 1896 marking the beginning of state ownership of all forests

Figure 8.7 The influence of European forestry techniques in Southeast Asia: the advent of timber haulage by light tramway.
(Courtesy of the Public Record Office, photograph reference WO 251/611.)

(Pragtong and Thomas, 1990). The aims of the RFD were to regulate harvests of valuable tree species, to capture a portion of the benefits of tree harvest for the central government through royalties and taxes, and to assist in the consolidation of central authority over regional nobility. A Forest Department was also set up in Burma in 1902 for similar purposes.

The colonial states
Colonial rule in Burma, Vietnam, Cambodia and Laos brought about numerous changes (SarDesai, 1997). In Burma (considered a province of India), for example, large tracts of land were brought under rice cultivation, and exports of grain, minerals and timber increased phenomenally. However, this rise in economic activity only benefited the colonist rulers and their helpers, and indigenous people were alienated from land they had owned for centuries. Village life was destroyed: for example, in Vietnam the relative autonomy of the village council was replaced by elected individuals who lacked the traditional following or influence among the peasants. Administrative roles were also allocated to outsiders. In the case of Burma, authority was given to Indians, or minority tribes such as the Karen, who had converted to Christianity. Not surprisingly, nationalist movements began soon after colonisation. However, it was not until anti-colonial movements in India that the independence activities in Burma were accelerated. The British withdrew from Burma in 1948, and the French from Indo-China in the mid-1950s.

8.3.3 The recent history of resource use in Thailand

By the 1950s, deforestation was proceeding rapidly in Thailand, as in other parts of Southeast Asia, and this increased during the 1960s and 1970s because of loggers and migrant settlers (see Sections 8.4.1 and 8.4.3). This prompted several important policies and legislative acts to be formulated, including a revised Forest Act targeting 40% of the land to be covered by forest (Pragtong and Thomas, 1990). Wildlife Conservation and National Parks Acts established the first national park in 1962, followed later by wildlife sanctuaries and watershed conservation units. After 1964, there was a rapid push to gazette large areas of reserved forest land, a time also marked by the emergence of conflicting land-ownership claims between communities and central government. The government's push for economic development brought about the 1968 decision to extend long-term harvest concession leases throughout the country, together with mandatory tree planting requirements in 1970 (Pragtong and Thomas, 1990).

The 1970s and 1980s saw further forest loss and conflict come about. In 1975, the RFD established the Forest Village Programme, where communities living on state lands were resettled in selected forest areas to supply labour for forest production. This required high capital investment to build community infrastructure, and so the programme moved slowly and affected few communities. However, after the military coup in 1976, political stability began to take place, and the flow of migrants into reserved forests increased sharply, encouraged by the military's village registration scheme entitling them to government service programmes. This, as well as illegal logging and upland crop production, was exacerbated by the construction of strategic roads by the government which often penetrated remote forests. By 1980, reserved forests and parks covered 42% of the land, but deforestation had accelerated, leaving only 32% of total land under forest cover (Pragtong and Thomas, 1990).

Government leaders began to recognise the magnitude of land problems, and so, in 1982, a land certification (STK) programme was established to recognise forest occupants' rights (Pragtong and Thomas, 1990). Although well intended, land certification is not yet well integrated into mainstream national forest management systems, and has had only limited impact on community–agency conflicts over forest access. Further policy changes led to the National Forest Policy of 1985, with several new directives on forest management, including the reservation of 40% of the national forest land: 15% in protected national parks, wildlife sanctuaries and watershed headlands, while retaining 25% for 'economic' purposes. Table 8.7 shows the main responses by the RFD and other agencies to the national policy. Although extremely positive in outlook, implementation of this policy, and its various clauses, was difficult, and by 1988 the RFD was engulfed in conflict and criticism (Table 8.7).

This culminated in November 1988, when an unusually heavy rainstorm in the South induced a wave of floods and landslides, destroying villages and leaving more than 200 people dead (Leungaramsri and Malapetch, 1992). Extensive media coverage linked the tragedy to deforested watershed headlands resulting from illegal logging and concession abuse. In January 1989, the government responded with a temporary ban on commercial logging in reserved forests, which was then made permanent in May by a bill passed through parliament. To compensate for the commercial logging ban, Thailand reduced log import tariffs and opened all borders to timber imports. The government claims these are temporary measures, but critics fear the action will result in accelerated forest destruction in neighbouring countries.

8.3.4 Present-day land use

Prior to World War II, Thailand's four major exports were rice, teak, tin and rubber, rice being the main cause of deforestation. It grew from a production of 113 000 metric tonnes in 1890 to 921 000 metric tonnes in 1914, and covered 95% of all agricultural land. After World War II, the main commercial crops became kenaf (*Hibiscus cannabinus*), tapioca (cassava), maize and sugarcane. There was an increase in non-rice crops from 2% in 1950 to 25% in 1980, and a change in cash-crop plantation area from 5.8% in 1959 to 35.7% in 1988. The high grass cover in the dry dipterocarp forest is an important source for a wide range of herbivores, including livestock. Herds, mainly cattle, are often left to graze in the savanna forest, where *Arundinaria pusilla* is particularly favoured (Sukwong *et al.*, 1975).

In 1925, Stamp pointed out, with regard to Burma, 'it is hard to find another part of the British Empire where the timber resources are so extensive, so varied and so accessible, and yet so little appreciated'. The situation has certainly

Table 8.7 Actions arising from the National Forest Policy of 1985 and its various criticisms (based on Pragtong and Thomas, 1990)

Responses to national policy

Protected parks and wildlife areas expanded to cover 10% of land.

Pilot management plans developed together with NGOs, particularly in areas bordering important national parks.

Watershed classification system developed; applied nationally.

Reclassification of reserved forest lands.

Major degazetting programme.

Implementation of STK accelerated.

Registration of new villages within reserved forests halted.

Accelerated RFD planting programmes.

Military and private sector tree planting and plantation programmes.

A major rural development programme for the Northeast Region, known as the Green Northeast (Isan Khieo) Project in 1987.

RFD began developing a nation-wide forestry extension programme, to assist in, for example, agroforestry, community wood-lot technologies and management systems.

Growing national environmental movement.

Negative factors

Long-term harvest concessions included areas now designated reserves.

Vast areas of reserves occupied by registered villages receiving services under government programmes.

With about 1.2 million people occupying land inside reserves, it is not possible to relocate more than a small percentage outside.

Villagers advancing claims on forests through elected politicians, newspapers and NGOs.

Swollen ethnic minorities in mountains (from neighbouring communist regimes in Laos and Vietnam), increasing shifting cultivation beyond sustainability.

Opium substitute crops prone to high pesticide use and erosion into watersheds.

Need land for reform, reservoir construction and other projects.

changed since then, as the demand for timber has grown. *Dipterocarpus tuberculatus* is now recognised as having one of the hardest and best bending properties of all the woody savanna forest species (Forest Research Institute, 1976). *Shorea siamensis*, *Dipterocarpus intricatus* and *Pterocarpus parvifolius* are all economically important hardwood species (Sangtongpraow, 1982), and species such as *Shorea obtusa* are sometimes preferable even to teak (Stott, 1991b).

8.4 Current management problems and strategies

As in other parts of the world, the rate of destruction in the dry dipterocarp savanna forests of mainland Southeast Asia has risen sharply in the last few decades, through increased pressure by humans for forest resources. Current conflicts between people and governments stem from centuries of state attempts to gain exclusive control of vast forested areas through legislation and policing measures, giving little or no regard to the rights of indigenous populations (Poffenberger, 1990). In Thailand, for example, 27.4 million ha or 53% of land was covered in forest in 1961. This decreased drastically over 30 years to 14.2

million ha or 28.03% in 1988 and 27.95% in 1989, according to RFD statistics. The rate of habitat loss now stands at around 2.7% per annum (Dinerstein and Wikramanayake, 1993), and actual forest cover may well be less than 20% (Leungaramsri and Rajesh, 1992). Much of this is savanna forest. Deforestation is still a problem, but associated with this, and one of the major causes of savanna forest loss both historically and today, is the lack of land tenure and the resettlement of people (Subhadira *et al.*, 1987).

8.4.1 Resettlement

The forced movement of people, whether directly or indirectly, is a major issue in savanna forests. It has led to problems of land conflict between groups of people, as well as with areas of conservation. For example, the Korat Plateau in north-eastern Thailand originally comprised over 80% savanna forest. However, following the arrival of people in the early 1970s, through displacement from other areas, settlements were established, and together with the development of roads and tracks, illegal logging and hunting, and mining and army incursions, the pressure on forest resources grew and has become increasingly destructive (Subhadira *et al.*, 1987). Box 8.2 illustrates the case for the Dong Mun Forest.

Box 8.2 Dong Mun Forest Area

The Dong Mun Forest is primarily a dry dipterocarp forest. Above 500 m there is some dry forest, at least 30% of which is deciduous. The dry dipterocarp forest covers the lower elevations between 200 and 500 m, is lower in height and almost 90% deciduous. At one time there were valuable stocks of commercial grade hardwoods, diverse and abundant wildlife and a variety of other forest resources including foods and medicine. The topography is dominated by a series of mini watersheds and uplands which cover the upper Lam Pao watershed (see Figure 8.1). Logging concessions were originally granted to two saw mills in 1961, but cancelled in 1965 when the forest was declared a reserve. New 30-year concessions for an area of 590 km^2 were granted in 1975 to the Kalasin Lumber Company.

Until the early 1960s, Dong Mun remained almost entirely undisturbed by human activity, apart from occasional hunting and use by Buddhist monks. But in 1961, logging concessions were granted to the two saw mills, and this precipitated the first phase of settlement in the area. This period is characterised by free occupancy: forest land was claimed by squatters' rights using informal boundaries and traditional land-tenure systems. Eight households from Tha Kan and Chang Kaew established a small hamlet at Nong Saeng and began subsistence agriculture. However, when Dong Mun was declared a reserve in 1964, these people were evicted.

A second phase of more permanent land settlement began around 1970 (Figure 8.8), with an increase in the sale of land and its clearance for

Box 8.2 (*Continued*)

agriculture. In 1971, a family of 19 members from Udorn Thani province reached the site of Phu Hang village. Their migration was prompted by the flooding of their land by the Lam Pao reservoir and the inadequate compensation and resettlement opportunities provided by the dam authorities. The settlers immediately built houses, claimed large tracts of land, and began clearing the forest for swidden agriculture. Over the next 18 months, a further nine households immigrated to Phu Hang, and in 1972, settlers had their first contact with leftist guerrillas who came to assist them with harvesting rice.

By 1973, two commercial logging companies from Nam Phong and Yang Talad in the north began operations in concession areas, and extended the logging road into the forest. Consequently, 20 new households arrived in Phu Hang, influenced by rumours that the guerrillas were allocating forest land. A year later another 50 households came to settle in Phu Hang, while the original settlers returned to Nong Saeng accompanied by 30 additional households. Again land sales between original and new settlers took place. Meanwhile, the guerrillas warned the settlers that all land sales should have their approval and threatened to appropriate all individual land claims.

Land conflicts increased as immigration increased, and by 1975, the

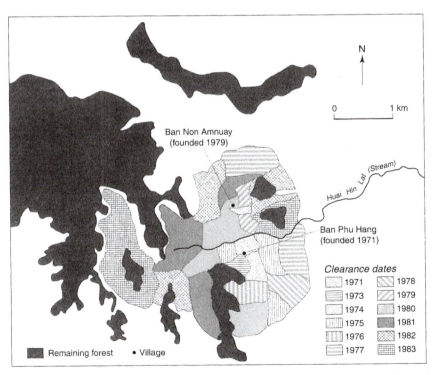

Figure 8.8 The history of forest clearance in the Dong Mun Forest, north-east Thailand (after Subhadira *et al.*, 1987).

Box 8.2 (*Continued*)

guerrillas were controlling all land transactions. At this point, Border Patrol Police were sent in, and fighting intensified over the next years between the guerrillas and the government. During this time, the population in Phu Hang had declined to 60 households and immigration had stopped entirely. It was in the mid-1970s that villagers from Nong Saeng occupied the site of Non Amnuay village, and began to clear land. Throughout the 1970s, the rate and pattern of forest clearance directly reflected the effects of immigration and the level of political stability. This process represented the actions of villagers in cutting timber for domestic use and commercial sale, and the expansion of agricultural activity, rather than timber harvesting by logging companies.

During the 1970s, most of the forest cleared was over lower and middle terraces, east and south of Phu Hang. But after 1980, when fighting had stopped, both villages began to grow from migration. This was the result of: (1) improved accessibility; (2) a 1978 amnesty for illegal residents of reserved forests; (3) implementation of the Forest Village Program (FVP) in Phu Hang in 1978, and the National Forest Allotment Program (STK) in Non Amnuay in 1983; and (4) the end of conflict with leftist guerrillas. As more people arrived in the area, timber began to be cut west of the village and upland swidden was developed. Settlement and clearance began expanding northwards, and around Non Amnuay. By 1985, large areas of the upper slope west of the villages was partially deforested to a distance of 4 km, and the remaining land converted to agriculture. Original dense forest is now only found at the higher elevations and in patches among the irrigated rice and field crops that stretch eastwards towards the Lam Pao reservoir.

Source: Adapted from Subhadira *et al.* (1987).

The management of savanna forest resources has been dominated by traditional forestry policies emphasising commercial production, protection against deforestation, and restrictive legislation to discourage encroachment. In Thailand, the FVP and STK programmes emerged as strategies for resolving forest land problems. However, the case study of Dong Mun Forest indicates that the goals of these programmes have not been realised. The programmes aimed to stabilise the population in national reserved forests and reduce encroachment. However, immigration accelerated after these programmes were introduced, and encroachment has increased. This has been a result of inequalities in the land allocation process, failures to anticipate demands from the continued flow of immigrants, inadequate opportunities for wage labour, and land conflicts embedded in the population's commitment to informal land tenure. For example, in 1983 only 13% of agricultural land in Thailand was titled, and though this has increased in recent years, the lack of recognition of unofficial land tenure by government officials is a major barrier to innovative land management programmes (Anan, 1995).

Community-based management

The increasing problem of land conflict has prompted governments and NGOs to initiate changes towards decentralisation, more people-centred development, and legal recognition over resources (Wiwat, 1993). One of the latest strategies is community-based resource management, where local people manage their own resources using their own knowledge (Seri, 1993). This knowledge is generally based on simple analysis of the environment using physical factors such as terrain, slope, aspect, soil texture, soil taste, colour, temperature and vegetation cover, as well as the observation of signs, e.g. from animals, and human dreams. The communication of indigenous analysis is normally a simple description of land and soils (Table 8.8), and mental mapping of resources, which are commonly reflected by the names given to them.

In terms of land use, there are three different types of forests: watershed forests, sacred forests and village woodlots. Watershed forests (*pa khun nam, pa ton nam, pa sap nam*) provide domestic and irrigation water, and are usually located at the headwaters, near springs or along stream banks. Watershed forest usually consists of evergreen or mixed deciduous forests, and can range in area from 6.4 to 6400 ha, depending on whether it is managed separately or collectively by villages (Shalardchai *et al.*, 1993). These forests are strictly protected and minimal uses are allowed.

Sacred forests are generally smaller than watershed forests, ranging from 2.4 to 32 ha, and may form part of a watershed or woodlot forest. These sacred forests are normally either evergreen forest patches or patches of mixed or dry deciduous forest. They are designated for ceremonial and religious purposes such as *pa pra that* (site of a pagoda containing Buddha relics), *pa aphaiyathan* (Buddhist temple ground forest), *pa sua ban* (forest of village guardian spirits), cremation grounds, and for keeping the belongings of the dead (in Karen culture) (Ganjanapan, 1996). Individual trees may also have significance. For

Table 8.8 Indigenous evaluation of suitable and unsuitable characteristics of land for agriculture (after Ganjanapan, 1996)

Features	Suitable characteristics	Unsuitable characteristics
Terrain	Floodplain, river terrace, mountain top	Steep slopes
Aspect	East	West
Soil colour	Black, reddish	Grey, orange, yellowish
Soil taste	Bitter	Acidic, saline
Soil texture	Clayey loam	Sandy, gravelly
Vegetation	*Dipterocarpus alatus, Dipterocarpus turbinatus, Polyalthia virides,* bamboos, bananas	*Dipterocarpus obtusifolius, Schoutenia hypoleuca, Shorea obtusa*

example, Karen people keep the umbilical cord of a newborn baby on a large tree in the belief that it contains *kwan*, or life essence, of that person.

Village woodlots (*pa chai soi*) range from 16 to 800 ha, and are where grazing, timber cutting, and the collection of woodland products are allowed. It is normally a dry dipterocarp forest covering lateritic soil unsuitable for cultivation. Also included in this category are groups of trees at the edge of paddies and upland crop fields (*pa hua rai plai na*), which many communities protect to allow regeneration.

Community-based management programmes have many problems, and there is a long way to go before communities can really control their own resources. Box 8.3 also illustrates the plight of monks, who have been fundamental in many instances in the fight to recognise local communities as stewards of their natural resources, and the fact that for many local communities, religion plays a large role in their environmental perceptions. Nevertheless, a more people-orientated initiative has meant that in Thailand, the RFD and foresters are beginning to understand the many factors involved in forest management, including local politics, community social structures and organisations, household and community economics, production and marketing microeconomics, communication, and extension methodologies.

8.4.2 Conservation

The destruction of dry dipterocarp forest over large areas has brought about the creation of savanna 'islands'. These are patches of savanna forest with a reduced ecological viability, particularly with respect to fauna, many of which rely on a mixture of tropical semi-evergreen rain forest, monsoon forest and savanna forest for food and shelter. Few such places still remain, and of those, many are under huge pressure from development. The 'Huai Kha Khaeng Wildlife Sanctuary–Thung Yai Wildlife Sanctuary–Kroeng Kavia Non-Hunting Area' complex in Kanchanaburi and Uthai Thani provinces, Thailand, for example, links forests from adjoining Burma to form large stands of savanna forest, which support a complete range of fauna. However, proposals to build the Nam Choan Dam in Thung Yai threatened the ecological integrity of the area (Sricharatchanya, 1988), and though this project has been temporarily postponed, it outlines the continued pressure on savanna forest. Also, apart from Thailand and Vietnam, the other countries in Southeast Asia supporting savanna forest have few or no areas of conservation (Dinerstein and Wikramanayake, 1993).

War in countries such as Burma, Laos, Cambodia and Vietnam has taken a huge toll on the dry dipterocarp forests and their resources (Stott, 1991b). For example, chemicals used during fighting, such as agent orange, have had devastating effects on plants and wildlife. The kouprey (*Bos sauveli*), the world's largest cattle, was once abundant in the savanna and monsoon forest of northern Cambodia. But villagers and guerrillas seeking food, and the animal blowing itself up on mines, has brought it to near extinction (IUCN, 1999). Both the gaur and banteng are classified as internationally threatened, and it

Box 8.3 Monk activists

Several Thai monks have taken an active role in providing a conscious direction for social change and environmental protection. Probably the most well-known case is of Phra Prajak Khuttajitto at the 34 000 ha Dongyai Forest Reserve, Pakham District in the north-eastern province of Buriram. He came to the attention of the nation in 1991, when he took a stand against civil and military bureaucracy and capital interests in the now-failed *Khor jor kor* (a relocation scheme for the poor in degraded forest lands). Although the scheme was supposed to protect reserved degraded state forests from intensified encroachment, it actually actively promoted the establishment of monoculture commercial tree farming. It was the inspiration of government economic policies and the world-wide demand for wood-chip and paper pulp (Lohmann, 1990). Consequently, when Prajak and other monks resisted the military-led evictions, they were accused of anti-state behaviour and hindering national prosperity and development.

Prajak took a religio-political stance of social action informed by reformulated Buddhist truths, and a strategy of conservation for the long-term interests of the people (Taylor, 1993). However, because of the harsh conditions of survival in the area, many settlers did not share his long-term views, were easily bought off, and preferred convenient though environmentally debilitating cash crops such as cassava.

Encamped in Dongyai forest, Prajak's first task, therefore, was to convince the villagers of the importance of the ecosystem and its protection. His task was difficult since the settlers came from different areas, weakening the natural formation of collective action groups. However, he conveyed his messages, using meditation, for example, as a base for building community consciousness and individual awareness, before encouraging involvement in collective action. One activist monk termed this as 'greening the mind' before 'greening the countryside'.

Prajak's nation-wide publicity on conservation and community management in the north-east meant that he lived in constant fear from attack by the government. In fact, the military tried to discredit him, accusing him of being a 'Russian monk' and a 'communist monk', and in early 1994, issued a warrant for his arrest for allegedly felling trees in the reserve to make a new monastery. Prajak claimed the trees were dead, but this action by the government caused him considerable concern. Events became so complex and problematic that in July 1994 he disrobed and disappeared, and his whereabouts are as yet unknown.

Prajak is not the only Buddhist monk who has taken a stance against government policies and tried to help local people. Others such as Luang Phor Naan and Luang Phor Khamkian have also founded active grass-roots organisations. What these all show is that for Buddhist villagers, religion cannot be readily separated from wider social, economic and political concerns, and that Buddhist resistance movements can play a major role in altering social and environmental destruction.

Source: Adapted from Taylor (1996).

is estimated that in Thailand about 915 gaur and 470 banteng remain today, both confined to protected areas (Srikosamatara and Suteethorn, 1995). For these animals, hunting for their horns has been a major threat to their survival.

8.4.3 Deforestation

Although commercial logging has been banned, deforestation continues in Thailand (Figure 8.9). This is due to various reasons (outlined below) but is principally a result of contradictions between forestry policy and other government development policies which in one way or another are related to use of forest resources. In Thailand, for example, there are as many as 19 government agencies involved in land management (Thailand Development Research Institute, 1988).

Causes of continued deforestation in Thailand

Illegal logging is one of the major causes of forest destruction in Thailand (Leungaramsri and Malapetch, 1992). One of the reasons for this is the inability of the Thai government to impose measures on the monitoring of timber factories, woodwork and craft shops, and industries related to timber processing. In fact, these have increased since the logging ban. Also, the government has encouraged logging in neighbouring countries such as Burma and Cambodia, providing opportunities for illegal loggers in Thailand to 'legalise'

Figure 8.9 Deforestation in the dry dipterocarp savanna forests of northern Thailand. A former area of dry deciduous dipterocarp forest and mixed deciduous forest, now cut and burned. (Reproduced by kind permission of Philip Stott, School of Oriental and African Studies, University of London)

their timber by claiming it originated from outside. Clear felling has also been allowed in certain sites marked for infrastructure construction projects such as dams or roads.

Forest destruction is also coming about through commercial plantations of fast-growing trees (Koohacharoen, 1992), deemed as rectifying the deforestation problem. As a projected scenario of the problem in the countryside, the 1985 National Forestry Policy Committee proposed the establishment of monoculture *Eucalyptus* plantations over 61 000 km^2 of reserve forest lands by the year 2020, of which at least two-thirds (though later revised in a policy review to one-third) were to be planted by commercial firms (Lohmann, 1991). There are vested commercial interests in backing reforestation, directed mostly to the pulp and paper industry, which contribute to the rising demands for cheap land, especially that classified as 'degraded' by the RFD. Many of these areas are cleared illegally by reforestation companies, ignoring the 10 million or so rural people who occupy these lands. The government's aim to increase forest cover to 40% has involved reforestation schemes, and has led to problems of land conflict, particularly in the north-east where much of the population live and farm in degraded lands. Widespread opposition to the reforestation policy, and especially the use of *Eucalyptus*, temporarily halted the policy, but lobbying by plantation companies to push for a Plantation Bill and military settlement schemes may bring about further deforestation.

Thailand's export-oriented agriculture policy has encouraged rapid expansion of farmland and forest encroachment, as well as clearance in areas not suitable (Koohacharoen and Paisarnpanichkul, 1992a). For example, hilly terrain cultivation is prohibited, but has expanded, resulting in environmental degradation, especially soil erosion. The policy has also changed local community production patterns of self-reliance to a higher dependence on the external market economy and high inputs. Tourism revenue jumped from 6.5 billion baht in 1979 to 77 billion baht in 1989 (Koohacharoen and Paisarnpanichkul, 1992b). The industry has led to many destructive projects such as luxury hotels, resorts and particularly golf courses, at the expense of forests. There is also the marketing of 'exotic' species of tropical animals through tourism, whether consumed or illegally traded.

Dams, roads and mines are the three most destructive types of infrastructure project affecting forest areas, exacerbated by deforestation in surrounding areas once forests have been opened up (Chantawong, 1992). Although the Thai government plans to increase conservation areas, other government agencies, such as the Electricity Generating Authority of Thailand, are pushing for big projects that threaten forests.

8.4.4 Fire management

Although the dry dipterocarp forests are adapted to moderate levels of dry season fire feeding on limited fuels, more intense and extensive fires caused by the build-up of fuels such as grasses and pygmy bamboo, can be destructive. In some areas, for example, fire protection for 4–5 years or more leads to the development of nearly pure stands of either *Themeda triandra* or *Arundinaria pusilla* grass, with a mean height of 1 m and a ground cover value of around 95%. It can also facilitate tree seedling establishment and development, but what is most likely is the ignition of severe and damaging ground cover burns. In stands of savanna dominated by the grasses *Themeda triandra* or *Arundinaria pusilla*, Stott (1986, 1988a,b) has recorded ground cover burns with an average temperature of 900 °C at 0.5–1.0 m above the ground, moving at speeds greater than 3.0 cm s^{-1} in near windless conditions. These fires have reached fire-line intensities (I) of over 2000 kW m^{-1}, and led to 'spotting', i.e. fire jumping across fire breaks, and to the 'torching' of individual trees.

The result is the death of some tree species as well as larger fauna, the initiation of accelerated erosion on steeper slopes, and even the destruction of property, when wind directional changes can turn flames from the savanna into villages. In northern Thailand, there are so many fires during March and April each year, that smoke covers the entire urban and rural areas of Chiang Mai, Chiang Rai and Lampang provinces (Sangtongpraow, 1986). Nevertheless, the policy in mainland Southeast Asia has been fire protection, and with ineffective policing of savanna areas, many wildfires have occurred.

Stott (1988b) points to the alienation of modern-day society from nature and its elements as an explanation for the negative perception of fire held by many people in Thailand. Perhaps more important is the historic influence of western orthodoxy on forestry training and forest management right across mainland Southeast Asia, which saw fire as detrimental to timber production, and therefore detrimental *per se* (e.g. Macleod, 1971). Present-day savanna forest fire management is still based on these ideas. For example, the Forest Fire Control Unit in northern Thailand concluded that there was a serious need for an educational programme for the 28% of villagers interviewed who thought that fire was a neutral or non-destructive force (Settarak *et al.*, 1986). If, however, wildfire occurrence and excessive burning is to be combated, these perceptions need to change, and prescribed burns employed, which could effectively reduce the fuel load and maintain the savanna forest for the required management objectives.

8.5 Concluding remarks

Although water and nutrients govern the form and functioning of the dry dipterocarp savanna forests, fire is an important force, controlling key processes such as tree regeneration. Since colonial times, the resources of mainland Southeast Asia have been a source of economic wealth, and the savanna forests, as with other forest types, have had a long history of exploitation. It would be reasonable to assume, therefore, that there would be abundant information about these savannas. In fact, the dry dipterocarp savanna forests have been largely ignored in the literature, and although there may be plenty of local research going on, compared to other world savanna formations, little has been recently published. Deforestation continues today, and although some data are available for Thailand, the extent of savanna forest loss in Burma, Cambodia, Vietnam and Laos is impossible to know. The political instability of the region has allowed few researchers to study these savannas – something which is unlikely to change in the near future.

Key reading

Kanjanavanit, S. (1992). Aspects of the temporal pattern of dry season fires in the dry dipterocarp forests of Thailand. PhD Thesis. Department of Geography, School of Oriental and African Studies, London.

Stott, P. (1984). The savanna forests of mainland South East Asia: an ecological survey. *Progress in Physical Geography*, **8**: 315–335.

Stott, P. (1991). Stability and stress in the savanna forests of mainland South-East Asia. In: Werner, P.A. (ed.), *Savanna Ecology and Management. Australian Perspectives and Intercontinental Comparisons.* Blackwell, Oxford, pp. 29–39.

Chapter 9

The savannas of Australia

9.1 Introduction

9.1.1 Distribution, climate and soils

In Australia, savannas are the predominant vegetation in the northern region of the country, and they cover approximately 20% of the continent. They have an arc-like distribution over Queensland, the Northern Territory and the northern part of Western Australia (Walker and Gillison, 1982) (Figure 9.1). In general, Australia is characterised by a long history of weathering cycles and great dissection of the Tertiary planation surfaces. Consequently, many soils of the savanna region have a low nutrient status, particularly in phosphorus, and range from the most widespread red-yellow alfisols and oxisols, to shallow stony soils, deep sands and cracking clays (Mott *et al.*, 1985).

Rainfall varies from < 500 mm year^{-1} to > 1800 mm year^{-1}, and over 90% falls in the wet season between late October and early April (Bureau of Meteorology, 1988). There is a dry season of about 7–8 months. The occurrence of the wet season is predictable in that convection storms and the

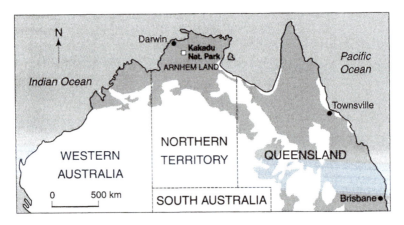

Figure 9.1 Distribution of savannas in northern Australia.
(Reproduced by kind permission of Justin Jacyno and Kevin Dobbyn)

monsoon always deliver the rain each year. However, there is considerable inter-annual variability in the duration and timing of the onset and end of the wet season (Taylor and Tulloch, 1985). There is also a regional difference in rain-fall, peaking in the north-west (1600 mm at Darwin) and decreasing to the east and south (Nix, 1983; McDonald and McAlpine, 1991). Mean monthly tem-peratures vary slightly over the year (mean annual maximum and minimum at Darwin, 32 °C and 23 °C respectively). The greatest variation in mean temper-atures is that associated with the diurnal range which in January is 7 °C and in July is 11 °C. The other main features of the climate are low relative humidity and high rates of evapotranspiration during the dry season (McDonald and McAlpine, 1991).

As a result of the relatively low-lying topography and the strong seasonal rainfall, seasonal inundation of parts of the landscape occurs during the wet season. Tidal swamps and marshes fringe the coast, but where fresh water pre-dominates adjacent to rivers and streams, graminoid marshes occur. Where inundation is less extensive, *Melaleuca* spp. swamps are typical. On higher ground, where inundation is rare or absent, savannas occur.

9.1.2 Vegetation structure and composition

There are a range of savanna types in northern Australia (see below). However, most are characterised by a woody layer dominated by species of *Eucalyptus* and *Acacia*, which range in richness from as low as one species per hectare in inland areas, to more than 80 species per hectare in sub-coastal areas. Approximately 80% of the 503 described *Eucalyptus* species occur in savannas (Gillison, 1994). Other important woody perennial taxa include *Melaleuca* (Myrtaceae), *Banksia* and *Hakea* (Proteaceae), *Albizia* (Mimosaceae), *Bauhinia* (Fabaceae), and *Terminalia* (Combretaceae). The graminoid layer is dominated by bunch grasses. The family Andropogoneae is the major floristic element of the ground layer, and is represented by species of the genera *Bothriochloa*, *Chrysopogon*, *Heteropogon*, *Imperata*, *Sorghum* and *Themeda*.

The most widespread and distinctive *Eucalyptus* community in northern Australia, extending from the Kimberley Region to the Cape York Peninsula, is dominated by *E. tetrodonta* and *E. miniata*, either singly or together (Dunlop and Webb, 1991) (Figure 9.2). Gillison (1994) gives a description of the different types of *Eucalyptus* savanna woodland. The canopy height is generally over 13 m, with crowns almost touching (Story, 1969, 1976), but in wetter areas of Cape York, trees can reach over 30 m (Dunlop and Webb, 1991). Smaller trees in the sub-canopy are patchily distributed, and include *Erythrophleum chlorostachys*, *Terminalia grandiflora*, *Acacia* spp., *Planchonia careya* and *Petalostigma pubescens*. On shallower and rocky soils, stunted *Eucalyptus* com-munities are present. These are less than 7 m in height, have separate crowns, and allow a dense grass layer to develop (Dunlop and Webb, 1991). These com-munities are dominated by *E. foelscheana* and *E. tectifica*. Where there is seasonal inundation, *E. latifolia* and *E. polycarpa* dominate (Dunlop and Webb, 1991).

Palms and cycads are also widespread throughout the region, and can be

Figure 9.2 *Eucalyptus* savanna vegetation. Woodland south of Darwin, the Northern Territory, Australia.
(Reproduced by kind permission of Philip Stott, School of Oriental and African Studies, University of London)

dominant in places. On Cobourg Peninsula and Melville Island, localised populations of the palm *Gronophyllum ramsayi* occur as co-dominants with *E. tetrodonta*, while on the sand plains south-west of Darwin, *Livistona humilis*, normally an understorey to *E. tetrodonta–E. miniata*, forms dense mono-specific stands (Dunlop and Webb, 1991).

Little is known about the climate before about 20 000 years BP (Brockwell *et al.*, 1995). However, charcoal evidence indicates that *Eucalyptus* woodlands probably occupied substantial areas of all present rain forest areas between about 27 000 and 35 000 years BP. From about 20 000–13 000 years BP, the climate was drier than at present (Hiscock and Kershaw, 1992), and palyno-logical, sedimentary and charcoal evidence indicates that *Eucalyptus* woodland dominated much of northern Australia during this late Pleistocene period (Kershaw, 1985, 1989; Walker, 1990; Hopkins *et al.*, 1993). The woodlands probably reached their maximum geographical extent in the period 13 000–8000 years BP.

Aborigines may have also contributed to the expansion of savanna woodland in the late Pleistocene, and affected the rate of rain forest tree recolonisation from the early Holocene to the present. It has been argued that periodic droughts under the influence of the El Niño Southern Oscillation (Box 9.1) would have promoted rain forest fires on a long return frequency throughout the Holocene (Goldammer and Siebert, 1990), and could have helped to keep the vegetation open. Increased rainfall began from about 10 000 years ago (Nott and Price, 1994), resulting in the contraction of woodland and open forest, and the spread of rain forest (Allen and Barton, 1989).

The different savanna types in northern Australia

Monsoon tallgrass (Schizachyrium pastures)

These savannas occur across the northern parts of the continent predominantly on earth and sands of low fertility. *Eucalyptus tetradonta* and *E. dichromophloia* are the common woody elements, with *Melaleuca* spp. occupying poorly drained sites. These trees are interspersed with the leguminous woody species *Erythrophleum chlorostachys* and a variety of woody shrubs including *Atalaya, Capparis* and *Petalostigma* species. The grass layer contains *Themeda triandra, Heteropogon* spp., *Sorghum* spp., *Schizachyrium fragile, Chrysopogon fallax* and *Eriachne* spp.

Tropical and subtropical tallgrass (black spear grass)

These savanna types are located along the eastern coast and divided into northern (tropical) and southern (subtropical) tallgrass types at approximately latitude 21°S. The soils are mostly of low to medium fertility. Tropical tallgrass contains *Eucalyptus crebra, E. alba* and *E. dichromophloia*, while subtropical tallgrass is dominated by *E. crebra, E. melanophloia* and *E. drepanophylla*. The understorey species of both types include *Heteropogon contortus, H. triticeus, Themeda* spp. and *Bothriochloa* spp. Subtropical tallgrass has been extensively modified by people through timber extraction and animal grazing. The latter has caused changes in grass composition from *Themeda* pastures, to *Heteropogon* and *Bothriochloa bladhii*, and more recently from *Heteropogon* to *Bothriochloa pertusa*.

Midgrass (Aristida/Bothriochloa pastures)

These occur in the Northern Territory and Western Australia on infertile earth soils and massive earth soils. In the north, *Eucalyptus microneura* with a herbaceous layer dominated by *Aristida* spp. and *Chrysopogon fallax* occur, while in the south *E. populnea* with *Aristida* spp., *Bothriochloa bladhii, B. decipiens* and *Chloris* spp. predominate.

Midgrass on clay (Brigalow pastures)

Apart from some areas of dense tall shrublands of *Acacia harpophylla* (brigalow) and *A. cambagei*, there are few trees in these grasslands. Native pastures which have developed contain *Dichanthium sericeum, Bothriochloa decipiens, B. bladhii* and *Chloris* spp. These communities occur mostly in Western Australia on cracking clays of moderate to high fertility with good water-holding capacity.

Acacia shrublands (Mulga pastures)

These occur throughout northern Australia mainly on massive earths and stony soils of low fertility and poor water-holding capacity. The woody layer is dominated by *Acacia aneura* (mulga), with the herbaceous stratum comprising *Digitaria* spp., *Monochather paradoxa, Eriachne* spp. and *Aristida* spp.

Gidgee woodlands (Gidgee pastures)

These occur in south-western Queensland and the Northern Territory on fertile clay and clay loams. The *Acacia cambagei* (gidgee) woodlands have been extensively modified and replaced by *Centhrus ciliaris* pastures.

Plant life strategies

In northern Australia, growth is seasonal, with at least 90% of the annual biomass being produced in the summer months (December to April) (Mott *et al.*, 1985). Most herbaceous growth occurs in the wet season, but many of the savanna trees begin active growth late in the dry season (Fensham and Kirkpatrick, 1992; Wilson *et al.*, 1996). The two dominants, *Eucalyptus miniata* and *E. tetradonta*, maintain growth throughout the dry season (Williams *et al.*, 1997).

In the woody layer, there are four main phenological types (Table 9.1; Williams *et al.*, 1997):

(a) evergreen species, which retain full canopy throughout the year;
(b) brevi- or partly-deciduous species, in which the amount of canopy falls significantly, but briefly, during at least one dry season, but to levels not below 50% of full canopy;
(c) semi-deciduous species in which the canopy falls to below 50% of full canopy in each of the dry seasons;
(d) full deciduous species, which lose all leaves during the early to mid-dry season, and remain leafless for at least one month.

About one-quarter of the 49 species studied by Williams *et al.* (1997) in a site near Darwin were evergreen, including the two community dominants, *Eucalyptus miniata* and *E. tetradonta*, and so the foliage cover in this savanna type remains relatively high during the dry season (Duff *et al.*, 1997).

Leaf flushing in most dominant species occurs primarily in the late dry season (August–October), prior to the occurrence of any rain (Bowman *et al.*, 1991; Wilson *et al.*, 1996; Williams *et al.*, 1997). Williams *et al.* (1997) have shown considerable inter-annual, inter-specific, as well as inter-individual variation in leaf fall and leaf flush of deciduous and semi-deciduous species such as *Buchanania*, *Xanthostemon* and *Planchonia* spp., and hypothesise that this asynchrony may reduce the impacts of regular dry season fires.

Leaf flushing in Australian savanna trees is probably determined by changes in the internal water status of the whole plant (Duff *et al.*, 1997; Myers *et al.*, 1997; Williams *et al.*, 1997). Williams *et al.* (1997) found that soil moisture at 1 m did not fall below permanent wilting point during the dry season. Hence reserves of soil water at the end of the dry season are sufficient to support the whole-plant rehydration that precedes leaf flushing in the absence of rain.

Regeneration of the *Eucalyptus* trees can be from seed, or vegetatively from underground organs (Lacey and Whelan, 1976). *Eucalyptus miniata* and *E. tetrodonta*, the two most common overstorey species in northern Australia, begin flowering at the end of the wet season, and ovule development and seed fall are completed within the eight-month dry season (Setterfield and Williams, 1996) (Figure 9.3). Both species have very large seeds (e.g. *E. miniata*, 3–6 × 1.7–5 cm) (Boland *et al.*, 1992).

Table 9.1 Phenology of some woody plants in the savannas of northern Australia (after Williams *et al.*, 1997)

Phenological group and species	Minimum canopy[a]	Growth[b]	Flowering[c]
Evergreen species			
Eucalyptus miniata	Sep–Oct	Jun–Dec	May–Jul
Eucalyptus tetrodonta	Sep–Nov	Jun–Dec	Jun–Sep
Acacia auriculiformis	Sep–Nov	Jun–Sep	May–Jul
Acacia mimula	Sep–Nov	Jun–Oct	May–Jun
Livistona humilis	Sep–Nov	Jul–Nov	Jan–Feb
Maranthes corymbosa	Sep–Nov	Jul–Oct	?
Pandanus spiralis	Sep–Nov	Dec–Feb	?
Brevi-deciduous species			
Eucalyptus porrecta	Jul–Oct	Sep–Nov	Oct–Feb
Acacia aulacocarpa	Jul–Aug	Sep–Nov	May–Aug
Persoonia falcata	Jul–Aug	Jul–Nov	Jul–Nov
Calytrix exstipulata	Aug–Nov	Oct–Dec	Jun–Aug
Eucalyptus bleeseri	Aug–Sep	Oct–Nov	Mar–Aug
Grevillea pteridifolia	Aug–Oct	Nov–Feb	May–Jun
Pittosporum melanospermum	Aug–Nov	Sep–Dec	Jan–Feb
Semi-deciduous			
Erythrophleum chlorostachys	Sep	Sep–Nov	Oct
Xanthostemon paradoxus	Sep–Nov	Oct–Dec	Nov–Feb
Buchanania obovata	Jul–Aug	Sep–Nov	Oct–Nov
Eucalyptus clavigera	Jul–Sep	Aug–Oct	Oct
Alphitonia excelsa	Aug–Nov	Jun–Nov	Feb–Apr
Canarium australianum	Aug–Sep	Oct–Dec	Oct
Timonius timon	Jul–Oct	Oct–Dec	Dec
Fully deciduous species			
Terminalia ferdinandiana	Jun–Nov	Oct–Dec	Nov–Jan
Planchonia careya	Jul–Oct	Sep–Dec	Oct–Dec
Brachychiton diversifolius	May–Oct	Sep–Nov	Aug–Sep
Cycas armstrongii	May–Sep	Jul–Nov	?
Croton arnhemicus	May–Oct	Oct–Dec	Nov–Jan
Cochlospermum fraseri	Jun–Nov	Oct–Dec	Jun–Oct
Ficus scobina	Jul–Oct	Oct–Dec	Jul–Aug

[a] Months of minimum canopy cover, where canopy cover is < 50% for semi- and fully deciduous species.
[b] Months of growth in which new foliage is > 25 % of canopy cover.
[c] Months of peak flowering.

9.1.3 Fauna

Australian savannas have a high biodiversity. For example, in Kakadu National Park, the *Eucalyptus* savannas are one of the richest in animal species (Braithwaite and Werner, 1987). The open forest and woodland habitats support a great diversity of birds, many of which are widespread and highly mobile (Press *et al.*, 1995). Common birds include lorikeets, parrots and

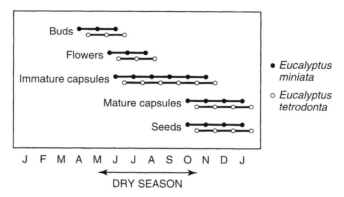

Figure 9.3 The reproductive phenology of *Eucalyptus miniata* and *E. tetrodonta* (after Setterfield, 1997).

cockatoos, numerous honeyeaters, friarbirds, butcherbirds, finches, and raptors such as falcons, kites and hawks. A notable species of these habitats is the partridge pigeon (*Geophaps smithii*). Mammals include the agile wallaby (*Macropus agilis*), antilopine wallaroo (*Macropus antilopinus*), nocturnal mammals such as the northern brown bandicoot (*Isoodon macrourus*), northern quoll (*Dasyurus hallucatus*), brushtail possum (*Trichosurus vulpecula*) and sugar glider (*Petaurus breviceps*), and rodents including the delicate mouse (*Pseudomys delicatulus*) (Press *et al.*, 1995). Common reptiles include the frilledneck lizard (*Chlamydosaurus kingii*), sand goanna (*Varanus gouldii*) and two-lined dragon (*Diporiphora bilineata*). Skinks and snakes are also abundant. The invertebrate fauna is most probably extremely diverse, but information is limited (Press *et al.*, 1995).

9.2 The main savanna determinants

Water, nutrients, fire and, to a lesser extent, herbivory are all important determinants of Australian savannas, and are described below. However, wind may also determine patterns of dominance in Australian savannas, particularly at a small scale. Convective storms are a regular phenomena in some savanna areas, occurring over areas < 1 km^2, but producing gusts over 75 km h^{-1}, and in some cases over 90 km h^{-1} (Gill, 1993). A study by Williams and Douglas (1995) found that a relatively small proportion of trees were damaged following a medium-scale convective storm (9.8% of trees in 3 ha), but that there were clear inter-specific differences in susceptibility to wind damage. Factors causing these differences may include tree height, morphology and architecture, tensile strength of wood, depth of rooting, and resistance to wood-damaging organisms such as fungi and termites. They also found that wind-throw events occurring in the pre-wet season period (November–December), when on average the soils are still relatively dry, may cause proportionally more snapping of trunks, whereas events in the post-wet season period (March–April) may tend to cause more uprooting.

9.2.1 Plant–available moisture (PAM)

The main phenomena controlling climate in northern and eastern Australia are the movement of the inter-tropical convergence zone, the anticyclone (high) pressure belt, tropical cyclones and the Southern Oscillation (Pittock, 1975; Russell, 1988). The latter is the major source of climatic variability in savanna areas of eastern Australia outside seasonal variation. It is a see-saw of atmospheric pressure between the South Pacific and Indian Oceans, with a periodicity of around 2–10 years (Allan, 1988). These extremes in pressure are associated with changes in ocean and atmospheric conditions, termed El Niño, and together comprise the El Niño Southern Oscillation or ENSO (Ramage, 1986; Jin, 1996) (Box 9.1).

Plant adaptations to PAM

The dry season, between April and October, is a period of high moisture stress for most plants. In the Australian savannas, the plants have evolved several mechanisms to cope during this period. Leaf shedding has been observed in periods of severe drought in *Acacia* woodlands, and the degree of leaf loss varies with the season. In the monsoon tallgrass savannas, both *Eucalyptus* and other woody plants, including *Bauhinia*, *Cochlospermum* and *Terminalia*, shed the majority of their leaves at the peak of the dry season (see Table 9.1).

Sclerophylly is well developed in most savanna genera in terms of thickened cuticles, increased glaucousness, rolled margins, dense indumentum, high

---

Box 9.1 The effects of ENSO on Australian savannas

During an ENSO event, eastern Australia usually has lower rainfall. Studies of the influence of the Southern Oscillation on seasonal rainfall in eastern Australia show that it affects rainfall variability and the frequency of extreme events such as droughts and floods (McMahon, 1982; Nicholls *et al.*, 1996). Considering the importance of ENSO events on northern Australian savannas, its effects on community composition and dynamics is now being investigated (Taylor and Tulloch, 1985; Austin and Williams, 1988). For example, Taylor and Tulloch (1985) found that a range of biological processes such as recruitment, distribution and survival, were strongly affected by extreme rainfall events, approximately half of which were associated with extremes of the Southern Oscillation. In north-western Australia, following legislation for increased settlement in 1883, numbers of sheep increased during a period of strongly positive Southern Oscillation. The triple ENSO between 1899 and 1902 resulted in very low rainfall and caused a 60% decrease in animal numbers (McKeon *et al.*, 1991). Therefore, the effects of ENSO are significant for the ecology and economy of the region, and will need to be considered in any prescription for savanna management.

specific leaf weight and an increase in volatile oils. *Eucalyptus* leaves are often pendulous and isobilateral, and both *Acacia* and *Melaleuca* species show marked development of phyllodes. The vertical leaves of *Eucalyptus* trees help to reduce midday heat loads, thereby increasing water use efficiency and hence carbon gain (King, 1997). Some *Acacia* species seem better adapted than *Eucalyptus* species to withstand water stress. *Acacia aneura* (mulga) can withstand leaf water potentials of −12 MPa, and *A. harpophylla* (briglow) can actively photosynthesise at −5.7 MPa. However, their thinner bark and lack of below-ground regenerative capacity suggests *Acacia* are more fire-prone than *Eucalyptus*.

Annual herbs such as *Mitrasacme connata* and *Stackhousia intermedia* survive the dry season in the form of seeds, and perennial grasses and herbs have underground tuberous roots (Dunlop and Webb, 1991). Under moisture stress, the leaves of *Heteropogon contortus* accumulates solutes to make osmotic adjustments (Wilson *et al.*, 1980). Thus these plants can continue to photosynthesise at water deficits of −2 MPa and can survive water potentials as low as −12 MPa.

9.2.2 Plant–available nutrients (PAN)

In most Australian savannas, the soil fertility is low (Mott *et al.*, 1985), and so productivity is dependent upon rapid recycling of nutrients locked up in the plant biomass. Although soil micro-organisms play the major role in the decompositional process, studies show that their activity is strongly dependent on soil moisture and temperature. For example, Holt (1987, 1988) found that in a strongly seasonal savanna woodland near Townsville, north Queensland, 60% of decomposition (measured as rate of CO_2 production) occurred during the four-month wet season. Much lower rates were recorded in the following eight-month dry season, and total decomposition by soil micro-organisms was estimated to be about 2300 kg ha^{-1} year^{-1}. Other losses of nutrients come about through soil erosion, particularly where there has been anthropogenic disturbance or overgrazing (Saunders and Young, 1983), and fire. So to avoid potential losses from fires, some Australian savanna grasses relocate nutrients from the above-ground parts to roots at the end of the summer growing season (Norman, 1963; McIvor, 1981).

The role of termites in nutrient cycling

Termites are active throughout the year (Holt, 1987, 1988). Their mounds are a conspicuous feature of the savanna landscape, particularly in those areas of red and yellow soils where mound densities of 300 per ha^{-1} are common (Lee and Wood, 1971) (Figure 9.4). In Townsville, Holt (1988) found that two mound-building species, *Amitermes laurensis* and *Nasutitermes longipennis*, were responsible for approximately 250 kg ha^{-1} year^{-1} or 10% of carbon turnover at the site. If all termite species were considered, they may be responsible for up to 20% of decomposition, much greater than invertebrate contributions in temperate ecosystems (usually < 10%; Seastedt, 1984).

Figure 9.4 A tall termite mound in the savanna woodlands of the Northern Territory, Australia. (Reproduced by kind permission of Philip Stott, School of Oriental and African Studies, University of London)

Since termites occur in large numbers over semi-arid northern Australian savannas (except on cracking clay soils), their influence on nutrient cycling is widespread (Bonell *et al.*, 1986; Coventry *et al.*, 1988). Termite biomass figures of 25 kg ha^{-1} near Townsville (Holt, 1988) and 40–120 kg ha^{-1} near Charters Towers, north Queensland (Holt and Easey, 1984) confirm this.

There is a strong relationship between fire and termites in savannas, with regular burning causing a decrease in termite populations. Observations of termites near Townsville, north Queensland, show that an increase in mound numbers of *Nasutitermes longipennis* over a period of three years, may in part be due to fire exclusion in the area (Holt and Coventry, 1991).

9.2.3 Herbivory

Australian savannas are considered depauperate in large native herbivorous mammals, with only six species of macropod marsupials qualifying for the title (Freeland, 1991). Up until the Pleistocene, Australia had a megafauna of herbivores (Hope, 1984). The present paucity is thought to have come about as a result of large-scale extinctions (Choquenot and Bowman, 1998). Some workers have suggested that Aborigines may have been responsible for the extinction of many large browsing and, to a lesser extent, grazing marsupials, while others believe they may have simply hastened the process. With the arrival of Europeans, around 150 years ago, came the introduction of large ungulates from Asia, Europe and Africa. Some species are now widely distributed throughout the savanna region (e.g. *Bos taurus, Equus caballu*), some are more localised (e.g. *Bos banteng, Cervus unicolor*), and others are expanding their range (e.g. *Camelus dromedarius*) (Bayliss and Yeomans, 1989). These species are either entirely feral, primarily feral or maintain feral populations together with minimally managed harvested stock (Freeland, 1991). The first two species have resulted in overgrazing and habitat alteration (Braithwaite *et al.*, 1984; Bowman and Panton, 1991).

Mammalian herbivory

Freeland (1991) shows that introduced herbivores have far greater population densities in Australia than in their native habitats. One possible explanation for this is the possible lack of inter-specific competition. Although species richness of Australian herbivores (native and introduced) is comparable to other savanna areas, the combinations of species differ greatly from natural communities, and may influence the type and effects of interactions. Freeland (1991) lists the differences between Australian and other natural herbivore communities as follows: (a) the absence of any extensive period during which the Australian herbivores could have undergone co-evolution resulting in a minimisation of competitive interactions; (b) the absence from Australia of exceedingly large herbivores such as elephants and rhinoceros; and (c) the absence from Australia of browsing species. In fact, because they are not browsed by abundant fauna, Australian *Acacia* species differ from African counterparts in having a low incidence of thorns. The first and last factors, however, lead to an expectation of more intense competition, so further testing of this hypothesis is necessary.

The paucity of potential predators and pathogens seems a more significant factor affecting herbivore populations (Freeland, 1991). Australia lacks larger predators, and relatively few pathogens are capable of infesting the introduced species (Freeland, 1983). A small number of pathogens accompanied the introduction of the herbivores, but virtually none of the major ungulate pathogens, such as trypanosomiasis, are present in Australia (Freeland, 1983). Although savanna plants may have allelochemical/physical defences capable of protecting them against introduced herbivores, the reduced impact from predators and pathogens allows feral species to exceed bounds imposed in native environments.

Insect herbivory

Andersen and Lonsdale (1991) note the lack of literature concerning the role of herbivorous insects in tropical savannas. Fensham (1994) found that sucking insects constituted 79% of all phytophagous insects collected from woody sprouts in the ground layer of a tropical eucalypt forest, and that 21% of species were specialists on single plant species. It is suggested that insect abundance reflects the growth patterns of woody sprouts after regular burning, with plant development tuned to the pressures of insect herbivory (Fensham, 1994).

9.2.4 Fire

Before the arrival of people, lightning was the ignition source for savanna fires on the Australian continent (Kemp, 1981). This changed with the arrival of Aboriginal people in the northern savanna lands, around 20 000 years BP (Russell-Smith and Dunlop, 1987). These people used fire mostly for hunting, access and mobility (Jones, 1969). The extent to which Aborigines may have been responsible for modifying the environment through fire is widely debated (Jones, 1969; Horton, 1982; Clark, 1983; Hallam, 1985). Pollen core and charcoal dating evidence indicates a correlation between an increase in the use of fire since the arrival of the first humans to Australia and the expansion of sclerophyllous vegetation and contraction of fire-sensitive communities (Hopkins et al., 1993). However, these changes may have been climatically induced (Kershaw, 1985).

The causes of fire

Today, most fires are lit by people, commonly by pastoralists (Andrew, 1986), conservation park managers (Press, 1988), and Aborigines continuing traditional burning practices (Haynes, 1985). Fire frequency varies from area to area, but throughout the savanna region there is some annual burning (Braithwaite and Estbergs, 1985). Recent work (Hurst et al., 1994; Cook et al., 1995) has shown that gaseous emissions from Australian savanna fires comprise about 12–17% of global emissions from savanna fires (Cheney et al., 1980), quantitatively similar to those measured from fires in Brazil and the Ivory Coast (Crutzen et al., 1985; Delmas et al., 1991).

The factors affecting fire

Climate

The 'fire climate' in Australia is measured quantitatively, using a fire danger index, which combines the effects of air temperature, relative humidity and wind-speed together with a drought factor for litter fuels and a curing factor (or percentage dead grass) for grassy fuels (see Gill et al., 1996). An index value of 100 represents the worst possible fire weather conditions (Luke and McArthur, 1978). Peak fire danger values vary within the savanna zone. For example, in coastal Darwin, forest fires are most susceptible in July and grassland fires in September; further east and inland, in Kakadu National Park, both forest and grassland fires are more common in September (Gill et al., 1990).

Fuels

Graminoid material is a particularly important fuel in Australian savannas. As the dry season progresses, annual grasses die completely, while perennials die back to the root stock (Gill et al., 1990). Grassy fuels in the region can be as high as 6–8 t ha^{-1} in the first year after a fire, and typically reach 5–10 t ha^{-1}, 4–7 years after fire (Walker, 1981; Mott and Andrew, 1985). Annual *Sorghum* growing in relatively moist, open areas can create fuel loadings up to 14 t ha^{-1} (Rowell and Cheney, 1979). Studies show that these high grass fuel loadings begin to decrease 5–6 years after fire, until they become quite sparse (Figure 9.5). For example, *Sorghum* was found to decline by 24% with 5 years' protection from fire (Gill *et al.*, 1990). During the whole period until the next fire, litter fuel (leaves and twigs of woody plants) may continue to increase (Gill *et al.*, 1990). After 13 years of fire protection in a *Eucalyptus* forest, litter fuels reached an average of 12.1 t ha^{-1}, with a maximum value of 21.5 t ha^{-1}.

The state of the fuel is an important factor affecting ignitability. This is influenced by the weather, primarily humidity (Gill *et al.*, 1996). As the dry season progresses, litter fuels are desiccated in response to the high temperature and low humidity, and grasses are cured: annual grasses dry out first, followed by perennial grasses (Gill *et al.*, 1996). Ignition in *Eucalyptus* savannas is said to be unlikely at a fuel moisture level exceeding 20% of oven-dry weight (Foster, 1976; Cheney, 1981). Although there appears to be little variability in flammability because of mineral content among *Eucalyptus* species, dry leaves of many mesophytic species apparently possess high levels of minerals, and are therefore less flammable (King and Vines, 1969).

Fire behaviour

Early dry season (May–June) fires are characterised by patchiness (Haynes, 1985; Braithwaite, 1987). This is caused by the variety of fuel moisture contents and the different fuel types. For example, annual grasses may be fully cured, whereas perennials such as *Heteropogon triticeus* may be greener. Coarse annual grasses such as *Sorghum intrans* may have bare ground between plants, and in the early dry season conditions of calm winds, moderate humidity and air temperatures, fires can self-extinguish. In fact, many early dry season fires burn in a mosaic fashion and go out at night.

As the dry season progresses, the grasses desiccate further, but deciduous and semi-deciduous woody plants such as *Eucalyptus, Terminalia, Brachychiton* and *Cochlospermum*, lose their leaves, which adds to the fuel content, and also ensures a more continuous fuel distribution. As a consequence, fires become more intense and homogeneous, and at the peak of the dry season can burn through the night (Lewis, 1989; Australian National Parks and Wildlife Service, 1991a, 1991). Temperatures in Australian savanna fires can reach 900 °C (Moore *et al.*,

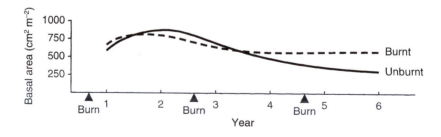

Figure 9.5 The basal area of the savanna grass *Themeda australis* either left unburnt or burnt every second dry season (after Mott and Andrew, 1985).

1995), with a maximum fire intensity measured so far at Kapalga of 20 000 kW m^{-1} (Williams *et al.*, 1999). This is in comparison to maximum values of 30 000 kW m^{-1} in grass fires (Luke and McArthur, 1978) and 100 000 kW m^{-1} in forest fires (Gill and Knight, 1991).

Aboriginal use of fire

Historical records from the time of first contact (Hallam, 1975; Nicholson, 1981) establish that fire was widely used in the savanna landscape. However, it is difficult to ascertain the exact regimes of fire being used at the time. For example, a recent study of the explorers' record in Queensland demonstrated that Aboriginal burning was prevalent in most coastal and sub-coastal open vegetation types, but less so more inland, and that fire frequency was lowest in summer and relatively low in spring (Fensham, 1997).

Haynes (1985) carried out one of the few studies on Aboriginal burning in savanna woodland and forest in north-central Arnhem Land, investigating the Gunei-speaking Aborigines around the Maningrida locality. Seasonal changes are clearly recognised by the Aborigines: they use natural signs such as the first storm of the wet season, or the onset of flowering of a certain species (Haynes, 1985). The Gunei identify six seasons and numerous sub-seasons (other Aboriginal language groups identify different seasons; Jones, 1980), and these are shown in Figure 9.6. Of all the different vegetation types in the area, closed forest, open forest, *Eucalyptus* woodland and open floodplain are the most important in relation to fire management. However, substantial areas of land remain unburnt. Figure 9.6 indicates the times for burning in the four ecosystems.

Fires are controlled almost entirely by timing and placement. Beating out fires with branches is rarely used. For example, fires are set around areas rich in natural resources and hunting grounds in the early dry season, and near camps and walking tracks. Intense fires are set later in the dry season to drive game into hunting grounds. Variations in burning regimes are determined by different tribal customs (Stephenson, 1985; Russell-Smith and Dunlop, 1987). Braithwaite and Estbergs (1985) suggested that Aboriginal burning in *Eucalyptus* forests was mainly in the period May to September with some further burning post-October, while in *Eucalyptus* woodland, burning began in April, peaked in July, was equally prevalent from July to November, and then declined to December.

Fires are lit for many reasons, including for ceremonial purposes, to make walking through the vegetation easier, to keep down mosquitoes, to drive away snakes and for macropod (*Macropus antilopinus* and *M. agilis*) hunting (Haynes, 1985, 1991; Head, 1994). These areas are prepared in the early dry season (*wurrgeng*) for hot hunting fires during the hot weather (*walirr*) by burning firebreaks so that later fires will burn into them. Fires are also set to promote 'green pick' to attract the macropods or to maim or stun smaller animals (Jones, 1980).

Protection of plant food resources is achieved by manipulation of fire (Table 9.2). For example, the maintenance of an unburnt canopy for the

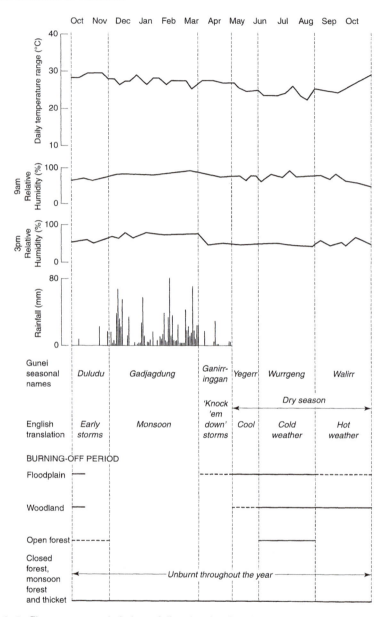

Figure 9.6 The seasons and timing of fires by the Gunei-speaking Aborigines from north-central Arnhem Land (after Haynes, 1985).

preservation of flowers and fruits is recognised as essential by the Gunei. This is a very important reason for burning the open forest only during the 10 week *wurrgeng*. Lastly, there is a general reason to burn, referred to as 'cleaning up the country'. There is no rational reason for this, but it is regarded as being extremely important, and is related to totemic/religious purposes.

Table 9.2 Protection of economic plant resources by manipulation of fire (after Haynes, 1991)

Species	Type of plant and part used	Ecosystem	Means of protection
Dioscorea spp.	Yam, tubers (location indicated by dry stems)	Closed forest margins	Burning around margins of closed forest very early in year
Carpentaria acuminata	Palm, terminal bud	Closed forest	Exclusion of fire (protection of whole plant)
Gronophyllum ramsayi	Palm, terminal bud	Open forest	Burning in early cold weather (protection of plant)
Terminalia carpentariae	Almond tree, seed	Open forest, closed forest margins	Avoid burning until after fruit is ripe
Buchanania obovata	Plum tree, fruit	Open forest	Burning in early cold weather to avoid scorch of developing flower buds
Persoonia falcata	Plum tree, fruit	Open forest	Burning in early cold weather to avoid scorch of developing flower buds
Syzigium suborbicularis	Red apple tree, fruit	Wetter areas of open forest up to closed forest margins	Burning in early cold weather to avoid scorch of developing flower buds
Vitex glabrata	Plum tree, fruit	Closed forest margins	Burning around margins very early in year

Fire and nutrient cycling

Fire in savannas causes rapid mineralisation of nutrients contained in organic matter (Noller *et al.*, 1985; Holt and Coventry, 1991). Most nutrients are deposited as ash, although some are dispersed as fine particles in the atmosphere. Some particles are also lost by volatilisation, especially nitrogen, sulphur and, in hot fires, phosphorus (Cook, 1992). Norman and Wetselaar (1960) found that over 90% (4.5 kg ha^{-1}) of the nitrogen in a native pasture at Katherine, Northern Territory, was lost to the atmosphere as a result of burning. They concluded this to be a small loss offset by annual inputs in the form of rainfall and non-symbiotic fixation in soils.

Cook (1994), working with grassy fuels at the Kapalga Research Station, within Kakadu National Park, found that more than 60% of the total transfer of P, K, Ca and Mg, and more than 50% of the transfer of Cu occurred via transport of entrained ash. Furthermore, the total transfers of these nutrients were relatively low compared with those of the more volatile nutrients N, S and Zn. Rainwater returned greater or approximately equivalent amounts of P, S and Ca to those lost in non-particulate form, but less than 15% of the N lost in fires, and less than 50% of the Mg lost (Cook, 1994). In open woodlands, the losses of K are also well below the inputs from rainfall.

Although symbiotic fixation of N may replace some of the N lost during fires, even the maximum measured rate of fixation in the Kakadu region (1.2 g m^{-2} year^{-1}; Langkamp et al., 1979) could not replace all the N loss. The scant evidence of non-symbiotic fixation of N in Australian savannas suggests that the process is unlikely to be important (Cook and Andrew, 1991). Cook's (1994) study shows that N is the only nutrient for which the losses are likely to exceed the inputs through rainfall and redeposition of particulates, and that the quantity of N lost is substantial compared with the levels of available N in the soils.

Effects of fire on vegetation

Woody plants

Grass fires cause little direct damage to trees, a result of tree adaptations, short fire residence times, and relatively low fuel loads. However, when there are long periods without fire, and fuels become litter based and heavy, late dry season fires can be very damaging. Fire could be the cause of zero recruitment among *Eucalyptus* species in many stands in the Northern Territory, killing young cohorts of seedlings (Hoare *et al.*, 1980; Braithwaite and Estbergs, 1985; Bowman and Panton, 1995). Additionally, fire may be an important determinant of seed supply. Setterfield (1997), working in the Kapalga Research Station in Kakadu National Park, found that intense late dry season fires significantly reduced the fecundity of *Eucalyptus miniata* and *E. tetrodonta*, two dominant savanna trees, and ovule success was affected by early dry season fires. Both late and early dry season fires resulted in a substantial reduction in seed supply compared with unburned areas (Figure 9.7). Fire may therefore have a significant impact on seedling regeneration.

However, the failure of fire protection to result in the recruitment of *E. miniata* juveniles into the canopy suggests that fire may not be the sole cause of the sparse sapling layer in north Australian *Eucalyptus* savannas (Werner, 1986). Fensham and Bowman (1992) have proposed that intra-specific competition is the cause of the suppression of *Eucalyptus* juveniles, demonstrated for other *Eucalyptus* forests throughout Australia (Bowman and Kirkpatrick, 1986; Stoneman *et al.*, 1994). Recent studies in savanna areas suggest that the increase in woody shrubs surrounding patches of monsoon rain forest following the cessation of burning, may facilitate establishment of rain forest species in the long term (e.g. Bowman and Fensham, 1991; Bowman, 1993).

Australian savanna woodlands and forests commonly have a bimodal structure of canopy trees and woody 'sprouts' at the ground layer (Braithwaite and Estbergs, 1985; Fensham and Bowman, 1992). These are maintained in a suppressed state by fire until a fire-free period allows them to grow above the flame height and escape complete defoliation or death (Lacey *et al.*, 1982; Fensham, 1990). Canopy trees suffer an annual mortality rate of at least 1%, rising to 15% after particularly intense fires (Lonsdale and Braithwaite, 1991; Williams,

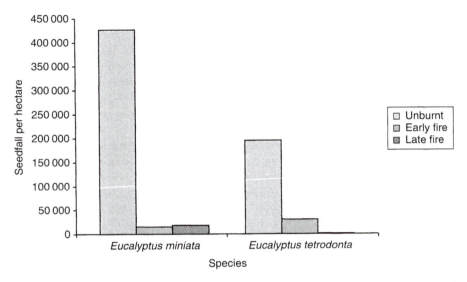

Figure 9.7 The seedfall per hectare of *Eucalyptus miniata* and *E. tetrodonta* in unburnt, early burnt and late burnt savanna (based on Setterfield, 1997).

1995). Height, cover and biomass of trees have been shown to decrease with increasing fire frequency (Bowman and Panton, 1993a, 1995; Wilson and Bowman, 1994), and fire can also influence the dispersion patterns of savanna trees (Lonsdale and Braithwaite, 1991; Hochberg *et al.*, 1994). These effects could be heightened by frequent fires, and could lead to long-term structural degradation (Hoare *et al.*, 1980).

Herbaceous plants
Andrew and Mott (1983) found nearly 400 seeds m^{-2} germinating in stands of annual *Sorghum*, of which 40% survived as seedlings. A similar percentage of seedlings arose from burnt soil, but only 250 seeds m^{-1} were recorded. It seems that direct seed death and bird predation occurs from fires (Gill *et al.*, 1990), but that protective mechanisms allow a crop to be produced the following year. *Sorghum brachypodum* densities were reduced by a factor of 10 after early wet season burning (Lonsdale *et al.*, 1998). A population model based on the demography of unburnt populations predicted that they should recover from a wet season fire, taking 7–16 years to return to normal densities. However, actual field populations did not seem to recover, suggesting that wet season fires not only lower densities, but may also fundamentally change population processes in annual grasses.

Fire protection
A long-term study at Munmarlary, Stage 2 of Kakadu National Park (Bowman *et al.*, 1988; Westoby, 1991), found that five years of fire protection led to long-unburnt *Eucalyptus* savanna (Bowman and Panton, 1995). This was a result of

the release of fire-stunted shrubs and trees such as *Eucalyptus tetrodonta, Acacia* spp., *Grevillea heliosperma, Stenocarpus cunninghamii* and *Erythrophleum chlorostachys* (Bowman and Panton, 1993b), increasing individuals at the 2–8 m tall storey. There has been much debate over this result and the experimental studies conducted. However, a re-analysis of a new set of data from the same sites, this time including only woody species, showed the same results: rain forest tree species do not readily colonise unburnt *Eucalyptus* savanna (Bowman and Panton, 1995).

One possible explanation why propagules have not reached the unburnt plots could be because the plots are too small, too distant from rain forest, or both (Woinarski, 1990; Lonsdale and Braithwaite, 1991). Another is that the savanna environment is unfavourable for rain forest seed establishment because of factors such as grass competition or unsuitable mycorrhizas (Bowman and Fensham, 1991; Bowman, 1993; Bowman and Panton, 1993b). Bowman and Fensham (1991) argue that rain forest seedlings could establish in unburnt savanna once facilitative woody species have ameliorated the environment by increasing the soil fertility, changing the soil micro-flora or changing the micro-climate.

Plant adaptations to fire
Most woody plants have adaptations to withstand fire (Gill *et al.*, 1990). Protection of aerial parts is common, and methods include: a thick bark; elevated and well-separated crowns; obligatory or facultatively deciduousness in the dry season; and a lower volatile leaf oil content in living *Eucalyptus* foliage compared to *Eucalyptus* species of other vegetation types (Lacey *et al.*, 1982). In cases where there has been crown damage, recovery is initiated by shoots from dormant buds in the bark of smaller branches, or if these are damaged, from larger branches or the main trunk (Gill *et al.*, 1990). If all aerial parts are killed, the plants may regenerate by the production of shoots (suckers) from rhizomes and roots (Lacey and Whelan, 1976), or from lignotubers (Lacey *et al.*, 1982). The latter is common in many savanna species, especially *Eucalyptus* species.

In the herbaceous layer, many grasses survive fires by having an annual life cycle, i.e. they escape fires in the form of seeds (Lacey *et al.*, 1982). These seeds have attributes, such as sharply pointed calluses and hygroscopically active awns, which help to bury them in the soil. They also have a period of dormancy, which is normally broken by the beginning of the wet season. Perennial species, on the other hand, have, as with woody plants, mechanisms to protect their growing buds. Grasses, such as *Imperata cylindrica* and *Coelorachis rottboellioides*, protect their buds in deeply buried rhizomes, while *Chrysopogon fallax* and *Themeda australis* have cataphylls which act as shields (Gill *et al.*, 1990).

Effects of fire on soils
Nutrients remaining on the ground in ash and other residue after fire, along with deposited particles, are highly susceptible to erosion through water runoff (Gillon, 1983; Kellman *et al.*, 1985). Fires also destroy cryptogamic crusts and organic cementing materials in the surface soil, thereby reducing surface stabil-

ity (Greene *et al.*, 1990). Therefore, fire also substantially increases the rate of transfer of soil and nutrients into aquatic systems. In northern Australia, the critical period for such erosional losses is during the storms of October and November, before the soil has been stabilised by wet-season production of herbaceous matter. The extent of losses is likely to be determined by fire intensity and fire timing (Andersen *et al.*, 1998). For example, after early season fires, erosion is likely to be low because of incomplete loss of ground layer, and the rapid resprouting of vegetation (Cook, 1992).

Effects of fire on animals

The most important effects of fire on fauna are through the modification of habitat (Andersen *et al.*, 1998). Birds hunt around fires, scavenge immediately afterwards, and many species are also attracted to recently burned areas (Braithwaite and Estbergs, 1987; Woinarski, 1990). These are mostly granivorous, omnivorous and carnivorous species feeding on the ground. Crawford (1979) demonstrated greater species richness of ground, shrub-layer and canopy-using birds in burnt savanna.

Trainor and Woinarski (1994) found that species-rich lizard communities responded variably to a range of experimental fire regimes in a savanna woodland in Kakadu National Park. For example, *Heteronotia binoei* was the only species to show a short-term response to fire, decreasing in abundance directly after early and late burns. On the other hand, species such as *Diporiphora bilineata*, *Carlia amax* and *C. munda* were significantly more abundant in unburned and early-burned sites. Griffiths and Christian (1996) found that many frillneck

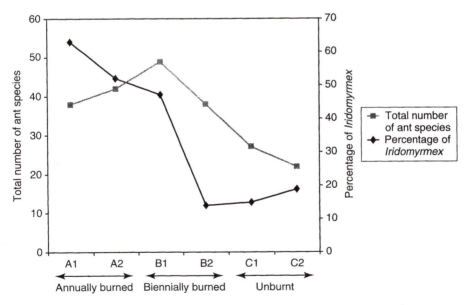

Figure 9.8 The total number of ant species and percentage of *Iridomyrmex* (hot climate specialist) in six plots subject to different fire regimes (based on Andersen, 1991).

lizards (*Chlamydosaurus kingii*) are able to survive early fires by perching in trees, although there is a high level of mortality during late-season burns. The higher abundance of lizards after early-season fires could be attributed to the greater accessibility of food after the removal of ground vegetation.

Whelan and Main (1979) found that acridid grasshoppers in south-western *Eucalyptus* woodland were removed by fire, taking some time to re-establish. The study indicated that the extent of fire was particularly important to grasshopper populations. They recovered more rapidly after burns of small area than after a more extensive fire.

Andersen (1991) analysed the guild structure of ant communities on three fire treatments at Munmarlary after 14 years. He found that annually and biennially burned plots were characterised by a relatively high species diversity, and the presence of the dominant *Iridomyrmex* group (hot climate specialists) (Figure 9.8) and the opportunistic *Rhytidoponera aurata*; this was the reverse of the unburnt plots, which were dominated by other groups, and had a lower species diversity. Differences in the plots were attributed to habitat changes, in particular the effect of fire on the level of litter accumulation and insulation on the ground (Andersen, 1991). These changes affected the ants directly, and also indirectly by influencing the competing effects of the dominant *Iridomyrmex*.

9.3 Human determinant of the savanna

Aboriginal people probably entered Australia from Southeast Asia around 40 000 years BP (Shawcross, 1975), although some authorities now suggest the possibility of it being up to 130 000 years BP (Singh and Geissler, 1985). Radiocarbon dating of artefacts from rock-shelter sites in the savannas of Kakadu National Park have established human occupation in the region to at least 20 000–25 000 years BP (e.g. Allen and Barton 1989; Australian National Parks and Wildlife Service, 1991b). Recent thermoluminescence dates of sediment samples suggest initial occupation between 53 000 and 60 000 years BP, the oldest dates recorded (Roberts *et al.*, 1990a, 1993; ANPWS, 1991b; Roberts and Jones, 1994). However, their validity is currently being debated (e.g. Roberts *et al.*, 1990b,c, 1994; Allen and Holdaway, 1995).

9.3.1 Pre-colonial land use

Prior to colonisation, the Aboriginal economy was mainly land-based, involving gathering and hunting animals such as macropods and emus. Aboriginal resource location and utilisation was, and still is, linked to their perceptions of the annual cycle, seasons and places (Box 9.2). For example, the Gundjeyhmi-speaking Aborigines of Kakadu divide the year into six named seasons (Brockwell *et al.*, 1995). With these seasons come a succession of food resources, or 'bush-tucker', as different plant and animal resources are best harvested at particular times, when they have reached maturity or are in localised abundance.

Box 9.2 Aboriginal cosmology

In the Aboriginal worldview, ancestral figures travelled across a featureless landscape during the original Creation period. Their movements and actions were of profound significance for both the physical and social worlds. A path of travel might create a river channel, an object used and left behind might be transformed into a tree or rock formation, and some aspect of the ancestor's final presence, dormant but still powerful, might be recognised in the shape of a range of hills or the colours of a cliff face. The names of these various 'dreamings' refer to one manifestation of an ancestor that might also be said to have walked around in human form.

The presence of ancestral forms, traces and powers at particular places is an important consideration in managing the landscape. Though the majority of these *djang* or dreaming places can be safely approached, some are affected by behavioural restrictions, while a few, where important and Creator beings are resting, are dangerous under any circumstances.

The relationship between people and sites, and the responsibility of people for sites is considered 'organic' because they come from the same ancestral source. That is, ancestral figures not only caused the physical landscape to be shaped in a particular manner, but disposed social groups and their languages across the landscape in specific patterns.

Source: Adapted from Brockwell *et al.* (1995).

Russell-Smith *et al.* (1997) recorded over 120 species of edible plants and a wide variety of animal foods, including 24 species of fish, 25 reptiles, 51 birds and 18 mammals. While many species are utilised only occasionally (e.g. most waterbirds), or eaten in the form of snacks (e.g. many fruits), the major regional staples comprise 11 plant species and 19 animal species. It is also important to note the wide use of available habitats. Savanna woodlands provide a range of resources (most yam species occur here, and they are abundant in meat and eggs), and are important during the wet season. On the other hand, staple foods are more common in the floodplains and riverine communities, especially during the dry season (Russell-Smith *et al.*, 1997).

In a region such as Kakadu, where there are marked seasonal changes combined with different topographical zones, resource availability is highly uneven over space and time. Therefore, humans organised strategic movements between zones of habitation and exploitation over the course of a year. Most notable was the mid to late dry season concentration on downstream wetlands, where, in addition to edible plants, many animals clustered (Brockwell, 1989). With the approach of the wet season, bands of people began dispersing across the landscape, probably reflecting a broad cultural division that transcended the more localised land-owning groups. The range of movement not only allowed access to varied resources, but opportunities to renew social contacts and ceremonial relations with members of other language groups (Chaloupka, 1981b).

Land tenure

Aboriginal social organisation and land tenure is based on patrilineal-descending land-holding groups, the *gunmogurrgurr*, comprising a number of family units, termed the 'clan' (Rusell-Smith *et al.*, 1997). Each clan owns an estate, their 'country', and the extent of an estate is a function of the distribution of sites owned and looked after by the clan. In general, members from a number of contiguous *gunmogurrgurr* speak a common language, thereby constituting a 'language group'. There are over 580 different Aboriginal languages (Pearce *et al.*, 1996). The Aboriginal cultural geography of the Kakadu region features a distinction between the 'top side', 'stone country' or 'freshwater' groups of escarpment and plateau, and the 'bottom end', 'swamp country' or 'alligator' groups of wetlands and floodplains. This ecological division is complemented by differences in language and ceremonial repertoire (Keen, 1980). The estates of individual land-owning groups were typically neither large nor diverse enough to satisfy all the material requirements of group members. They therefore hunted and foraged over estates of other groups (Hiatt, 1962; Stanner, 1965). This was arranged through relationships established by intermarriage, kinship and ceremonial co-operation, and in western Arnhem Land, through a relationship known as 'company', which could manifest itself in a pattern of close exchange and sharing between contiguous clans.

9.3.2 Colonisation and post–colonial land use

Colonisation had a devastating effect on Aborigines. From an estimated 750 000 people, the Aboriginal population fell to 70 000 by the 1930s (Pearce *et al.*, 1996). The failure of Aborigines to exploit resources for commercial gain meant that their economy was destroyed, and their lands dispossessed by pastoral occupancy (Holmes and Mott, 1993). However, other uses of the savannas by the colonists, such as agriculture, were constrained by several factors. These included the seasonal and variable rainfall, the low nutrient and water-holding capacity of the soils, and the location disadvantage of the savannas (Holmes and Mott, 1993).

Pastoralism

Pastoral activities began in the south-east savanna parts in the 1850s and rapidly expanded to occupy most of the area by 1900. Land-ownership was spread over large companies, families with more than one property, and single family units. Property sizes varied; for example in Queensland they ranged from 130 km^2 to 5180 km^2 (Stewart, 1996). Herds were initially British breeds, and properties contained few fences, yards or equipped watering points; with drought, flood, fire and ticks to contend with also, the life of cattlemen was difficult.

After the Second World War the situation began to change as infrastructure improved, but it was not until the 1960s that big changes occurred. Amongst these was the increasing use of Brahman cattle which have a greater resistance to heat and ticks, and a higher foraging ability and intake (O'Rourke *et al.*, 1992). The initial market for cattle was domestic, but from the 1880s shipments to Europe, particularly the UK, began. In the 1950s and 1960s, the USA entered the market, and became the biggest market for Australian beef, a position now taken over by Japan (Stewart, 1996). Since the 1950s, property sizes have remained static, except in Queensland, where government aims of closer settlement have reduced the areas. The nature of property ownership has also changed, with an increase in single family units and Aboriginal ownership.

Aborigines

Reports of massacres and extreme maltreatment of Aborigines prompted the British government to force the colonial administration to protect them. In response, the colonial administration set aside reserves, which were mostly quite small, with no piped water and little sanitation (Pearce *et al.*, 1996).

These reserves were confined to areas unattractive to pastoralists, such as the arid interior, and the rugged and remote lands of the northern savannas (e.g. northern Kimberley, Arnhem Land, Cape York Peninsula and offshore islands). Hunting and gathering was prevented as land had been taken over for cropping and cattle-raising. This had the effect of replacing the previous varied and healthy diet with one of cheap, portable non-perishables such as flour, tea and sugar. The reserves proved disastrous for the health of the Aborigines, as living habits suitable for a handful of people who moved camp regularly were totally unsuitable for several hundred settled permanently in one area (Pearce *et al.*, 1996). Many Aborigines were 'employed' as labourers, stockriders or domestic servants on large cattle stations. Most pastoralists showed little concern for their workers, and maltreatment was common.

The year 1937 saw the adoption of the assimilation policy which said that part-Aborigines should be assimilated into white communities (regardless of their wishes), detribalised Aborigines were to receive education, and the rest were to stay on the reserves (Pearce *et al.*, 1996). It was not until 1967, however, that the government announced improvements in social welfare, health and housing for Aborigines, much of it coming from Aboriginal pressure for change (Pearce *et al.*, 1996). The period 1972–1976 saw the initial shift of Aborigines to 'outstations' or 'homelands', i.e. small, decentralised communities of close kin established by the movement of Aboriginal people to land of social, cultural and economic significance to them (Blanchard, 1987). At first outstations were established on Aboriginal reserve land, but have now extended into national parks (Kakadu and Cobourg), and onto Aboriginal and non-Aboriginal pastoral stations (see below).

Land tenure

Land tenure arrangements in Australia have been used to encourage land development and rural population growth. They began with closer settlement legislation in the 1860s, and entailed a progression through a series of leases, culminating in the award of a freehold title (Holmes and Mott, 1993).

Although these policies were successful in achieving agricultural development in other parts of Australia (including, to a certain degree, the north-eastern savanna zone), the northern savanna zone has remained frozen at an early stage of the progression, and is incapable of supporting unsubsidised forms of crop production (Holmes and Mott, 1993). Occupied pastoral land in northern Australia is still held under low-rent, fixed-term leases, usually of 1000–4000 km^2, with very modest development and minimum stocking covenants (often not fulfilled), barely able to financially support one family (Holmes and Mott, 1993).

The European experience has indicated the difficulties in savannas providing an adequate economic base, and so many Aboriginal communities will continue to be founded mainly on welfare payments, with additional income from other activities such as tourism (Holmes and Mott, 1993). Yet, there is still an expectation that Aboriginal communities must become economically viable using Eurocentric values. However, for Aborigines, land is critical in sustaining cultural factors, including subsistence foraging and hunting, maintaining complex systems of inheritance, ceremonies and decision-making, and in custodianship and group identification (Coombes et al., 1990).

These Aboriginal values began to be addressed by the Aboriginal Land Rights (Northern Territory) Act of 1976 which issued non-transferable freehold titles to traditional land-owners, reinforced more recently by the Native Title Act of 1993. By 1993, these titles covered about 34% of the Northern Territory, and are leading to the restoration of traditional ways of land use and management. However, this current land legislation is also promoting new conflict between Aboriginal and non-Aboriginal occupancy, and the associated social divisions are strongly felt locally (Holmes, 1996). Resistance towards recognition of further Aboriginal land claims could occur. This situation also precludes multiple land-use options that require shared decision-making between Aboriginal and non-Aboriginal people (Holmes, 1996).

Pastoralists in northern Australia control disproportionately large tracts of land, to which they direct only modest inputs and generate comparably modest outputs. Holmes (1996) proposes several lease tenure reforms to these core pastoral lands including: recognising the ongoing role of pastoralism as the primary use; requiring the pastoralist to adopt sustainable range management; ensuring that other interests and values are adequately recognised within the rights and duties attached to the lease, with prime attention to access and preservation of biodiversity; to ensure that option values for higher uses are adequately addressed by providing mechanisms for the award of additional property rights; and to ensure the lease-holder cannot exercise monopoly control over conversion of land to more intensive uses. Holmes (1996) suggests that leasehold tenure provides an appropriate mechanism for the evolution of multi-value landscape-specific systems of land management, compatible with the primary land use of grazing.

With the decrease in pastoralism, and no alternative large-scale private use for land, the case for private tenures has now disappeared. With a greater public role in resource management, Holmes (1996) emphasises the growing need for strategic regional planning, focusing mainly on marginal regions and urban development regions. Strategic planning needs to be linked to changing land tenures and uses as well as to more informed and sensitive approaches to environmental, economic and social impact assessment.

9.3.3 Present-day land use

Australian savannas are distinctive in that they support very small numbers of people at low densities (0.8% of the Australian population), and over 25% of the savanna population are Aborigines (Holmes, 1996). Compared to other savanna regions of the world which are varyingly subject to increasing population pressure, generating greater demands on their capacity to produce food and other agricultural products, Australian savannas are, for now, under decreasing pressure to engage in these land uses (Holmes, 1996). Although there is some policy attention and research activity directed towards increased agricultural production, most Australian savanna resources are being reassessed in terms of non-commodity outputs, including meeting the aspirations and needs of Aboriginal people, preserving biodiversity and satisfying recreation, tourism and other lifestyle values (Holmes, 1996). Land use in the northern Australian savannas is indicated in Table 9.3. Although forestry is important in other world savannas, it is insignificant in Australian savannas because of the poor soils, variable rainfall, and the susceptibility of the dominant *Eucalyptus* species to termite attack (Haynes, 1978; Lacey, 1979).

Pastoralism

The grazing industry (predominately cattle, although sheep are important for wool) has been a major user of the savanna region for over a century. Today, the savannas support some 7.5 million cattle or 33% of the national herd, and grazing is a year-long activity based mostly on native pastures, and the breeding and some fattening of cattle for beef production (Taylor and Braithwaite, 1996). Although, traditionally, the industry employed many semi-skilled labourers, the labour input has been steadily declining (O'Rourke *et al.*, 1992). The gross value of production is A$1770 million, and the gross value of exports A$1550 million (Taylor and Braithwaite, 1996). The economic prospects for

Table 9.3 Land use in the northern Australian savanna zone (based on Taylor and Braithwaite, 1996)

Land use	% of savanna area
Pastoralism/grazing for beef production	55
Aboriginal use	24
Protected areas/national parks	7
Military use	< 5
Tourism/ecotourism	< 1
Cropping/horticulture	< 1
Mining	< 0.2

the industry are bright, but many pastoral enterprises in marginal areas, such as Kimberley and Cape York, are unviable (Wilcox and Cunningham, 1994) and may be displaced by other uses.

Aboriginal use

Contemporary Aboriginal foraging and hunting for subsistence now combines elements of traditional lifestyles with other elements introduced from non-Aboriginal society. Aborigines now employ both native and introduced flora and fauna, and utilise European technology such as vehicles and guns. However, technology and cash money have not destroyed the cultural significance of foraging and hunting, and enable Aborigines to express their environmental knowledge while reinforcing beliefs in the spiritual value of such knowledge (Pearce et al., 1996).

Presently, about 24% of the savanna zone is under Aboriginal use, and the area of land is likely to continue to increase as an outcome of native title/Mabo claims (Taylor and Braithwaite, 1996) (see Section 9.3.2). There is a desire for Aborigines to be more completely involved in land management issues, one example being the land-use planning exercises in the Cape York Peninsula, known as CYPLUS – the Cape York Peninsula Land-Use Strategy (Baird, 1996). CYPLUS is a three- to five-year planning and land-use study involving an assessment and analysis of natural resources in Cape York, a land-use programme to assess residential expectations and needs, and a final public participation programme, which will attempt to involve communities within the planning process.

Protected areas/national parks

Protected areas and national parks include reserved, dispersed tracts of land varying in size from 1 to 20 000 000 ha (Taylor and Braithwaite, 1996). Many protected areas are theoretically based on conservation, but they are inadequate to ensure maintenance of biodiversity and conservation, and sympathetic management of intervening land or displacement of marginal lands by protected status will be required to achieve these goals (Walker and Nix, 1993). Tourism is becoming increasingly important in these areas, and is seen as the only acceptable commercial use of resources in protected areas (Dingwall, 1992). There is a strong overlap between conservation recognition (e.g. World Heritage status) and attractiveness to tourists (Taylor and Braithwaite, 1996).

Military use

Defence has a significant presence in the savannas of northern Australia, dominated by permanent and temporary deployment of armed units of the Australian Defence Force (ADF) (Barton and McDonald, 1996). The main concentrations are around Townsville, Darwin and Katherine, where the ADF uses Commonwealth land for key training activities (including live firing and detonation, and personnel and vehicle manoeuvring), and accesses other areas of the savanna for large-scale operations such as the triennial Kangaroo exercise (Barton and McDonald, 1996).

These military activities have the potential to destroy plant and animal communities, and cause damage to the physical environment. However, there is a strategic military benefit in minimising the impact of their activities: to conceal their activities from surveillance, including airborne and satellite remote-sensing devices (Barton and McDonald, 1996). As a consequence, the military apply a set of environmental principles, and the aspect of minimising impact is drilled into all ADF personnel. In cases where the ADF want to expand their activities, they have to undergo an environmental impact assessment, as took place in the example of the Townsville Field Training Area (Barton and McDonald, 1996).

Tourism
Tourism in the northern savannas has grown since the 1970s, and now ranks as a major sector in the regional economy. It helps the regional cash economy and increases employment opportunities, and if managed properly, can be compatible with other forms of resource use such as pastoralism, mining, conservation and traditional Aboriginal practices (Holmes and Mott, 1993). Tourism is highly localised in impact, and visitation pressure is greatest during the dry season (Taylor and Braithwaite, 1996).

Surveys show that there is increasing interest in 'outback' rural experiences, sport and adventure, and Aboriginal culture (Collins, 1996), and as such, most tourism in savanna areas is based on natural and cultural features, as is emphasised in areas such as Kakadu National Park (Figure 9.9). Although domestic tourism is more widespread, international tourism has a greater impact on the economy (Harris, 1992). Tourism also generates greater income for local people, compared to other commercial activities such as pastoralism and mining (Dowling, 1993). It produces A$2000 million per annum and is predicted to become increasingly important (ASTEC, 1993). It is the only growth area for the employment of semi-skilled labour, and could be a substitute if extractive or consumptive industries ceased (Wescott and Molinski, 1993).

However, there is still the problem of distance, high travel costs and lack of infrastructure. In the more remote areas, marked seasonality of tourism can devalue economic growth and job security. Furthermore, the remoteness of some areas has led to forms of tourism not beneficial to local communities: capital-intensive resort enclaves, isolated from the locals, with almost all inputs from metropolitan sources; and self-sufficient, safari-style groups, mounted on vehicles, whose main local purchases are fuel and alcohol.

Agriculture
Despite considerable investment in infrastructure and research over the past 30 years, dryland cropping in northern savannas has been unsuccessful. Overall, agriculture is an insignificant land use, highly localised and of limited potential (Taylor and Braithwaite, 1996) (Table 9.4). For example, in the Northern Territory, only 5% of the available land area considered suitable is cropped, while in Queensland this figure falls to 0.5% (Holmes and Mott, 1993). Irrigation projects to aid crop and pasture production in the northern savannas have also failed. For example, the dam on the Ord River in the Kimberley

Figure 9.9 Aboriginal rock paintings as a focus for tourism in the savannas of northern Australia.
(Reproduced by kind permission of Mike Smith, The National Museum of Australia, Canberra, Australia)

District of Western Australia was completed in 1972, but was plagued with high cost burdens and pests, and is now used for recreational purposes (Holmes and Mott, 1993). Failures of irrigation can be accounted for by inexperience, bad management and high costs, but overall they are because of the high costs and low returns associated with a remote, frontier region (Holmes and Mott, 1993).

Dryland crops include sorghum, maize, grain legumes, peanuts and cotton, and sugar is grown under irrigation. However, one growth area could be in horticultural crops: mangoes, bananas, pineapples, melons, avocados, cashews, papaya and lychee are increasingly penetrating temperate markets. Production has a gross value of A$370 million per annum, and exports A$160 million per annum. However, there is community disquiet about the chemicals used and areas cleared, and future production may face tighter controls (Taylor and Braithwaite, 1996).

Mining

The area of savanna affected by mining is small and localised, although mineral exploration is more widespread (Taylor and Braithwaite, 1996). Major operations include the central Queensland coal mines, the bauxite mines at Weipa and Gove, the Mount Isa mines, the gold and base metal operations at Charters Towers, the undeveloped projects in Carpentaria block, and the Ranger uranium, Groote Eylandt manganese and Argyle diamond mines (Ewing, 1996). Remote mines now operate on a fly-in, fly-out basis, so reducing their impact. Mineral exports include iron ore, gold, silver, copper, bauxite/aluminium, manganese, uranium, diamonds, nickel and phosphate, and the gross value of the industry is A$2500 million per year (Taylor and Braithwaite, 1996). Mining accounts for almost 90% of the merchandise exports from northern Australian (ASTEC, 1993).

Environmental considerations are integrated into the mining planning, and the Australian mining industry enjoys a world-wide reputation as a leader in environmental management and rehabilitation. It carries out a significant proportion of environmental research in Australia, and is Australia's largest employer of environmental scientists in specialist fields such as forestry, botany, zoology, hydrology, air quality, meteorology and microbiology (Ewing, 1996). For example, Mount Isa Mines (MIM), in remote north-west Queensland, produces approximately 700 000 tonnes of SO_2 each year as a by-product of processing lead, silver, copper and zinc. MIM is committed to reducing these emissions, and has commissioned a new study on the effect on

Table 9.4 Potential and current land use in northern Australia (after McKeon et al., 1991). Figures in millions of hectares.

Land use	Crop	Sown pasture	Native pasture	Non-agricultural land
Potential	14.6	50.5	236.2	149.0
Current	3.0	4.1	294.4	149.0

the surrounding savanna environment (Griffiths, 1998). Prevailing winds disperse the SO_2 plume from the mine over a 10 000 km^2 area (equivalent to the total Sydney metropolitan area) of savanna woodland north-west of Mount Isa.

9.4 Current management problems and strategies

The continuing emphasis on transferring savanna land to alternative, non-commodity uses, particularly tourism, has increased the need to address conservation issues. Attention is particularly focused on the role of fire in conservation areas, especially the adoption of Aboriginal fire regimes, and the rise in invasive organisms, which potentially threaten the rich diversity of plants and animals in the savannas.

9.4.1 Fire management

With the recognition of Aboriginal land rights, and traditional resource management, fire management policies in the Australian savannas are now aimed at incorporating Aboriginal techniques. At Kakadu National Park, for example, policies for fire management are based on Aboriginal regimes (Table 9.5), where frequent early dry season burning is used to prevent late fires (Gill *et al.*, 1990). These prescribed fires are applied by park staff, but around Aboriginal settlements and areas used for hunting, burning is carried out by Aborigines.

A practical implication of adopting the traditional fire management model is that considerable effort must be given to breaking up grassy fuels in the early to mid dry season, in effect to create a mosaic of burnt and unburnt patches across the landscape (Russell-Smith, 1995). Staff undertake a concentrated programme of burning along roadsides, around campgrounds, and in the vicinity of fire-sensitive habitats as soon as grasses cure sufficiently to carry fire. For

Table 9.5 Fire management objectives of Kakadu National Park Plan of Management (based on Australian National Parks and Wildlife Service, 1991a)

The main objectives in relation to fire management are as follows:

1. to protect life and property within and adjacent to the Park;
2. to maintain, as far as practicable, traditional *bining* (Aboriginal) burning regimes within the Park;
3. to maintain biodiversity;
4. to promote research into the fire sensitivity of environments and species;
5. to provide for the identification and protection of sensitive environments and species;
6. to maintain community education and interpretation programmes covering the role of fire in Kakadu;
7. to minimise the spread of fire from the Park to adjoining land;
8. to minimise the spread of fire into the Park from adjoining land;
9. to monitor the effectiveness of Park fire management programmes.

remote areas, aerial ignition is extensively used employing small helicopters. In most years, burning begins in May and is largely completed for savanna woodlands by mid-July (Russell-Smith, 1995). Floodplain burning, on the other hand, is maintained throughout the dry season.

Implementation of alternative policies is difficult because of the occurrence of unauthorised fires in the late dry season. On the grounds of plant ecology, Hoare *et al.* (1980) have suggested burning every four years in the early dry season, to provide optimal conditions for *Eucalyptus* savanna development. A more recent innovation to aid fire management has been the installation of an expert system at Kakadu (Davies *et al.*, 1986). These decision support systems are discussed more fully in Section 10.3.1.

The nature and practice of Aboriginal burning has relevance for contemporary and future land management. Aboriginal burning has both ecological and social significance, the latter illustrated by the 'cleaning' burns that are set in the early to mid dry season (Russell-Smith *et al.*, 1997). The mosaic of early, late and unburnt patches also has implications for the conservation of fauna. Some recent studies have found that a range of fire regimes are required to adequately conserve groups such as ants (Andersen, 1991), lizards (Braithwaite, 1987; Trainor and Woinarski, 1994), birds (Woinarski, 1990) and biodiversity in general (Braithwaite, 1996). If systematic burning does not take place, there is a real risk of intense late season fires, as has occurred over the vast unmanaged tracts of the western Arnhem Land region (Press, 1988; Russell-Smith, 1995).

9.4.2 Invasive plants and animals

Invasive plants and animals are a particular problem in pastoral areas and in savanna conservation areas, such as Kakadu National Park, where the high biodiversity is threatened (Braithwaite and Werner, 1987). Exotic shrubs, including *Acacia nilotica* (prickly acacia), *Cryptostegia grandiflora* (rubber vine), *Parkinsonia aculeata* (parkinsonia), *Prosopis* spp. (mesquite) and *Ziziphus mauritiana* (chinee apple, Indian jujube) (Grice, 1996), are a particular problem. For example, prickly acacia is a useful tree for fodder and shade, but is now out of control in large areas of Queensland, and likely to spread into the Gulf and the Northern Territory (Radford, 1998; Carter and Cowan, 1993). One of the reasons for its spread dates back to the early 1970s, when a series of floods and out of season rains made ideal conditions for the weed's growth.

Many invasive plants are especially significant to fire management, as they have a high biomass once established and are capable of supporting fires, thereby altering local fire regimes. These include *Lantana* species (Ridley and Gardner, 1961), the grasses *Melinis minutiflora* and *Pennisetum polystachyon* (Skeat, 1986; Tothill and Hacker, 1973), and the shrub *Mimosa pigra* (Lonsdale and Braithwaite, 1988). The latter is a particular problem in Kakadu National Park: it spreads rapidly, stores large quantities of seed in the soil, and can form dense thickets shading out the understorey (Lonsdale and Braithwaite, 1988).

It has been suggested that weed problems in parks may be a result of tourist vehicles acting as a vector for weed seeds. However, Lonsdale and Lane (1994) have shown that there is a relatively low density of weed seeds on tourist cars entering Kakadu National Park. A study by Grice (1996) indicates that management of invasive species needs to involve reducing seed production, and controlling movements of their dispersal vectors. For example, *Cryptostegia grandiflora* can produce 8000 wind-dispersed seeds in a single reproductive phase, and can set seed at least twice a year. A *Ziziphus mauritiana* individual can produce 5000 fruits per year, which are dispersed by mammals such as wallabies and feral pigs, but especially by domestic cattle: propagules pass through the digestive tract of the animals, after which they are more likely to germinate.

One of the most well known and tracked invasive animals is the *Bufo* toad. It is spreading into areas of savannas and could be a threat to local biodiversity. The ease of transfer of parasites between the *Bufo* toads and native amphibians suggests that exhaustive studies need to be undertaken on any potential biological control agent targeted for introduction (Barton, 1997). While studying native parrots, Pell and Tidemann (1997) found that introduced hollow-nesting mynas (*Acridotheres tristis*) and starlings (*Sturnus vulgaris*) dominated the use of the available nest resources in study sites, and the myna particularly demonstrated the potential to reduce the breeding success of the native parrots.

9.5 Concluding remarks

Savanna vegetation in Australia is strongly dependent on the seasonal moisture levels and the availability of nutrients in the soil. Fire is an integral component of savanna functioning, and although native herbivory is considered to be limited, pastoralism, especially cattle-raising, is the main form of land use. Australian savannas are probably the only world savanna formation where pressure to convert land to other forms of use, such as agriculture, is actually decreasing. This is to a large extent the result of the Aboriginal movement, and the redistribution of ancestral land to the indigenous people. Consequently, tourism, together with conservation and Aboriginal land use, is likely to play a greater role in the future of these savanna lands.

Key reading

Ash, A.J. (ed.) (1996). *The Future of Tropical Savannas: An Australian Perspective.* CSIRO, Australia.

Gill, A.M., Hoare, J.R.L. and Cheney, N.P. (1990). Fires and their effects in the wet–dry tropics of Australia. In Goldammer, J.G. (ed.), *Fire in the Tropical Biota: Ecosystem Processes and Global Challenge.* Springer-Verlag, Berlin, pp. 159–178.

Haynes, C.D. (1991). Use and impact of fire. In Haynes, C.D., Ridpath, M.G. and Williams, M.A.J. (eds), *Monsoonal Australia: Landscape, Ecology and Man in the Northern Lowlands.* A.A. Balkema, Rotterdam, pp. 61–71.

Holmes, J.H. and Mott, J.J. (1993). Towards the diversified use of Australia's savannas. In: Young, M.D. and Solbrig, O.T. (eds), *The World's Savannas. Economic Driving Forces, Ecological Constraints and Policy Options for Sustainable Land Use. Man and the Biosphere Series, Volume 12.* UNESCO, Paris, pp. 283–317.

Press, T., Lea, D., Webb, A. and Graham, A. (eds) (1995). *Kakadu. Natural and Cultural Heritage and Management.* Australian Nature Conservation Agency, North Australia Research Unit, The Australian National University, Darwin.

Chapter 10

Savannas in the twenty-first century

10.1 Introduction

Savannas are home to over a billion people. These people come from a range of cultural, social and economic backgrounds, but what unites them all is their dependence on the savannas for their livelihoods. Savannas provide the resources, whether it be land, food or wealth, for people to survive. Yet, despite this reliance from such a huge population, savannas are currently being destroyed and given over to other forms of land use. In the *cerrado*, for example, the rate of agricultural expansion and subsequent habitat destruction presently exceeds that of the Amazon rain forest (Klink *et al.*, 1993) (Figure 10.1).

This loss of savanna is associated with many global problems. War has been, still is, and will undoubtedly continue to be a major impact on savannas. Wars not only cause direct destruction to the savanna ecosystem, but also increase population densities in some areas through displacement. In Chad, for example, civil war during recent decades has caused widespread problems. Militia and others acquired weapons which enabled them to supplement their needs for meat and cash by hunting elephant and rhino in the national parks and fauna reserves (Keith and Plowes, 1997). In addition, nomadic pastoralists were driven out of their normal migration corridors and forced to graze their herds in less suitable sites, thus causing land degradation.

Deforestation is a concern not only in savannas, but in many other ecosystems around the world. In these other tropical ecosystems, deforestation may in fact increase the extent of savannas, albeit 'derived' savannas, as has happened in countries such as India. Linked to deforestation are desertification, savanna burning and climate change. However, the main factor connecting all these problems in savannas is unquestionably population increase. The number of people living in or around savannas is set to rise in the future and, as a consequence, the global issues facing savannas outlined above are only destined to become more prominent. In this final chapter we look at some of these future global trends that will influence savannas. This will be followed by an analysis of some of the new technologies likely to be significant for research and management in savannas in the next century.

Figure 10.1 The large-scale clearance of *cerrado* for agriculture. Here, soybean has been planted. (Photograph by the author)

10.2 Main issues affecting savannas

10.2.1 Population increase

Global estimates of the increase in the world's population over the next century are around 10.2 billion people by the year 2100, nearly a doubling of the present population. Savanna regions of the world are some of the most highly populated places in the tropics, as well as the world, and it is certain that many will see a substantial increase in human populations in the future. With this view in mind, the most likely scenario for savannas seems one of increasing poverty and inevitable land degradation. However, recent studies indicate that this is not always the case.

Obviously some rise in demographic pressure is generally related to land-use intensification and subsequent land degradation, particularly over a short period of time. However, in the long-term, it is too simplistic to link human population increase with environmental degradation. In the savannas of the Machakos District in south-western Kenya, for example, documented evidence suggests that despite a fivefold increase in people between 1930 and 1989, there has been a tenfold increase in the value of output per hectare, and an apparent general 'improvement' in the overall condition of the environment (Tiffen *et al.*, 1994). In heavily populated regions of Kenya and Uganda, small farm sizes have resulted in farmers planting increased numbers of trees (e.g. Holmgren *et al.*, 1994; Place and Otsuka, 1997).

What these studies show is that population increase generally shrinks the amount of land available for subsistence, and so in turn provides the incentive to invest in new technologies, conserve the resource base, and through this, increase production. In other words, some form of land-use intensification takes place. In savannas, where the soils are predominantly PAN-poor, intensification will undoubtedly have to include the use of fertilisers, whether they be inorganic or organic. Obviously organic fertilisers have an important advantage in that they provide a carbon source for microbial utilisation, resulting in the formation of soil organic nitrogen. However, inorganic fertilisers will be necessary to achieve high crop yields, and future strategies to improve their efficiency will be important. In the case of the Machakos District in Kenya, farmers were able to reverse land degradation through the employment of indigenous soil conservation techniques that improved crop and livestock productivity. Promoting these indigenous soil and water conservation methods will be essential in the future if other forms of intensification are to take place (Scoones *et al.*, 1996).

Advances in genetic engineering should allow varieties of crops to be planted in savanna regions in the future which are better adapted to PAM and PAN. In the *cerrado*, for example, there is ongoing research at the *Cerrado* National Research Centre on improving cultivars for their response to water stress, fertilisers and resistance to pests. Although new crop varieties may allow farmers to achieve higher productivity on smaller areas of land, Sanchez *et al.* (1997) also suggest that crops should be diversified so that profitable crops are grown along with basic foods. They point to cultivating indigenous plants using agroforestry techniques, and domesticating indigenous tree species. The domestication of indigenous trees, such as *Uapaca kirkiana* of the *miombo* and *Caryocar brasiliense* of the *cerrado*, they argue, could be used to conserve the genetic resource in living-germplasm banks, and allow tree crops to become higher yielding, more attractive commercially and a means of diversifying diets.

Any changes to intensify land use will have to take place hand in hand with policy improvements. Progress needs to be made in areas such as credit, infrastructure, marketing, research and extension support to farmers (Sanchez *et al.*, 1997). However, land tenure is probably the most urgent issue that needs to be addressed. In the African, Asian and South American savannas particularly, inequality in land distribution and ownership, together with a rise in population, is bound to bring about not only land degradation, but also a decline in social and economic conditions. In some of these savanna regions, changes are beginning to be seen as governments move towards balancing the proportion of land owned and managed by various sections of the population. In many instances, this has been linked to managing community resources, such as wildlife in the African and Asian savannas (see Section 10.2.5).

In other parts of the world, changes are taking place much more slowly, although there is great pressure to redress land tenure issues. For example, the 'Movimento dos Sem Terra' (the Movement of People Without Land) is a well-established and organised pressure group fighting for the millions of landless people in Brazil. It is intensely active in the *cerrado* region, where huge tracts of land still belong to a small number of wealthy people. The problem of land-ownership for pastoralists in African savannas is also one of concern and is certain to get worse in the future. Pastoralists have used certain tracts of savanna land for centuries under traditional property rights, but with the rise in population and expropriation of land for either private or agricultural use, the extent to which their herds can roam has been limited. The changing access of key resources for pastoralists has led to major conflicts between different pastoral groups, as well as between pastoralists and other land users. It is vital that pastoral land rights are recognised in law, so that key resources are protected and that there is a procedural framework for resolving conflicts.

10.2.2 The contribution of savannas to global greenhouse gas emissions and atmospheric ozone

Fires in tropical savannas are a major source of particulate matter and gaseous emissions to the atmosphere (Crutzen and Andreae, 1990; Hao et al., 1990). Some emitted gases such as CO_2, CH_4 and CH_3CL cause atmospheric warming, contributing to the greenhouse effect (Ramanathan et al., 1985), whilst others such as CO and NO_x are involved in complex chemical reactions in the troposphere, causing elevated levels of ozone and acid precipitation (Crutzen, 1988; Crutzen and Andreae, 1990). The high biomass of termites in savannas may also be responsible for considerable emissions of methane (Zimmermann et al., 1982, but see Seiler et al., 1984).

Savanna burning and emissions to the atmosphere

A large amount of work has been carried out on emissions from savanna burning in Africa. The Southern Africa Fire–Atmosphere Research Initiative (SAFARI-92) was set up in Kruger National Park, and the Fire of Savanna/Dynamique et Chimie Atmosphérique en Forêt Equatoriale (FOS/DECAFE-91) at Lamto, Ivory Coast. Both of these experimental sites were subjected to prescribed fires to measure emissions of trace gases and aerosol particles (e.g. Lacaux et al., 1996; Lindesay et al., 1996; Maenhaut et al., 1996; Scholes et al., 1996b). Other studies in African savannas have looked at biogenic nitric oxide (NO) emissions from savanna soils as a function of fire regimes (e.g. Levine et al., 1996; Parsons et al., 1996; Serca et al., 1998). Many of these studies suggest the contribution of savanna fires to global emissions to be smaller than was previously thought. Similar results have been found from the cerrado (e.g. Ward et al., 1992; Poth et al., 1995). In comparison with typical tropical forest, and North American forest burns, results indicate that the cerrado has lower emission factors (Kaufman et al., 1992). However, correlations show a direct link with burning and increased levels of ozone (Kirchoff and Alvala, 1996).

The chemical precursors necessary for ozone formation are believed to be provided by biomass burning, and layers of ozone concentration enhancements have been observed over cerrado regions during the dry season (Kirchoff and Marinho, 1994). Further evidence comes from studies of the large ozone bulge in the South Atlantic Ocean, near the African coast. Preliminary studies at Natal, Brazil (Kirchoff and Nobre, 1986) and later in the South Atlantic (Fishman *et al.*, 1990, 1991; Watson *et al.*, 1990) indicated the bulge to be the result of transport and chemistry inducement by biomass burning in nearby South America and African regions. Trajectory analyses also show that it is the Brazilian sources which contribute to the South Atlantic tropical portion of the ozone bulge at higher levels (above 500 hPa), whereas lower portions of the troposphere over the South Atlantic, from coast to coast, receive burning products from Africa (Fuelberg *et al.*, 1996). Thus, the constituents of savanna burning may in fact help to build up and conserve ozone in the atmosphere.

The role of the underground biomass in savannas

Many of the studies investigating emissions from savanna burning have relied on measurements taken from prescribed fires, where most of the gases released, such as carbon, may be mostly from the above-ground biomass. However, studies in the *llanos* (Sarmiento and Vera, 1979), Ivory Coast (Menaut and César, 1979) and *cerrado* (De Castro and Kauffman, 1998) indicate high root : shoot ratios in savannas (Table 10.1) compared to other vegetation types such as tropical rain forest (e.g. 0.1; Nepstad, 1989, cited in De Castro and Kauffman, 1998) and tropical dry forest (e.g. 0.42; Castellanos *et al.*, 1991). Because of this high below-ground biomass, the *cerrado*, as well as other savannas, may be a significant global carbon pool (De Castro and Kauffman, 1998). Although 'natural' fires may not disrupt the carbon balance of savannas, activities which kill woody plants, such as frequent burning, deforestation and large-scale crop conversion, may shift the functional role of savanna regions from carbon sinks to carbon sources. In fact, San José *et al.* (1998), working with carbon stocks and fluxes in the Orinoco *llanos*, found that protection of savanna, as well as semi-deciduous forest, could sequester significant quantities

Table 10.1 Tree density (number ha^{-1}), shoot biomass (Mg ha^{-1}), root biomass (Mg ha^{-1}), and the root : shoot ratio (R : S) for three different savannas (after De Castro and Kauffman, 1998)

Savanna	Tree density	Shoot biomass	Root biomass	R : S
Cerrado (Brazil)				
Campo limpo	–	2.9	16.3	5.6
Campo sujo	12	3.9	30.1	7.7
Cerrado aberto	1064	17.6	46.6	2.6
Cerrado denso	1000	18.4	53.0	2.9
Llanos (Venezuela)				
Grassland	–	6.0	11.5	1.9
Woody savanna	100	5.3	19.0	3.6
Ivory Coast (Africa)				
Grassland	–	3.5	19.0	5.4
Savanna woodland	800	61.3	37.1	0.6

of CO_2 from the atmosphere into the sub-soil to help compensate for CO_2 released from other land uses.

10.2.3 Climate change in savannas

Most of the models used to predict future climate change agree that an increase in CO_2 and other greenhouse gases will cause a rise in mean global temperatures. Recent simulations suggest that a doubling of atmospheric CO_2 will result in an average increase of 1.76 °C in tropical areas (Hulme and Viner, 1998). These models also predict that there will be a change in the amount and distribution patterns of rainfall, but this is more difficult to assess, and varies regionally. In savannas, where PAM is the primary determinant of the vegetation, changes in precipitation will be crucial. Some savanna regions, such as East Africa and India, are expected to have an increase in rainfall, but for most other savanna regions the climate will probably become drier (Goldammer and Price, 1998; IPCC, 1999). It is also expected that extreme events, such as droughts and the El Niño Southern Oscillation, will become more pronounced. In view of these climatic changes, how will savanna species respond in the future?

Looking at the palaeoecological record, evidence shows that climatic fluctuations have not caused mass extinctions, but that species have responded by changing their ranges. Different species in a community will have different responses under higher CO_2 levels, so species composition will change as different species responses are played out (Bazzaz *et al.*, 1985). Since CO_2 is the primary component of photosynthesis, elevated concentrations of CO_2 will enhance photosynthesis, at least in the short term. Savannas are composed of C_3 plants dominating the woody layer, and C_4 plants mainly in the herbaceous layer (see Section 1.3.1). We may expect C_4 plants to flourish under increasing moisture stress and CO_2 levels, but one of their main advantages, the ability to uptake CO_2 under low concentrations, may be lost, and this could give an advantage to C_3 woody plants. In fact, studies have shown that plants with C_3 photosynthesis are generally more responsive to elevated CO_2 than C_4 plants (e.g. Poorter, 1993).

This could have a number of consequences in savannas. The rise in C_3 woody plants could lead to an overall increase in canopy cover in savannas, and the development of forest formations in some areas. This may be of benefit in areas suffering from wood shortages, but may be problematic in the large expanses of pastoral land where people rely on the tree/grass mosaic for their livelihoods. Agriculture may also change. Many savanna crops, such as maize, are C_4 plants, which explains their success in savanna environments. However, these may actually be at a disadvantage in future climates, and may have to be substituted by C_3 crops, such as wheat (Figure 10.2). This could have direct economic as well as social effects, as peoples' subsistence and nutrition changes.

Other changes in species composition will be in relation to temperature and rainfall. The direct effect of an increase in temperature may be an increase in the water use efficiency of savanna plants, i.e. the increased CO_2 will enable plants

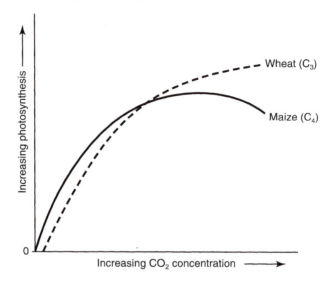

Figure 10.2 Photosynthesis of the crops maize and wheat under increasing CO_2 concentrations at constant light conditions. Note that as CO_2 increases, maize becomes saturated, while wheat continues to increase in photosynthesis (after Moore *et al.*, 1996).

to reduce the transpirational loss without a corresponding reduction in carbon uptake, therefore being able to tolerate water stress at a higher threshold (Walker, 1991). The lengthening of the dry season could have major consequences on savanna plant phenology, promoting earlier leaf drop and delaying leaf flush, and speeding up flowering. Deciduous trees may become more prominent in the woody layer, and annual plants with short life cycles in the herbaceous layer. These shifts could also separate pollinators and dispersers from their host-specific plants, and cause local extinctions.

Changes in plant phenology may have significant repercussions for herbivory in savannas, as the quantity of food may decrease because of the shortening of the growing season. However, food quality may also change. Because of the prevalence of carbon in leaves under increased CO_2 levels, it is hypothesised that there may be a decrease in nitrogen concentration in leaves, and an increase in carbon-based chemical defences (Körner and Bazzaz, 1996). This could mean an overall fall in food available for herbivores, particularly for more specialist feeders, and a switch of generalist herbivores to more palatable hosts, thereby increasing the load on these species.

Higher temperatures in savannas may lead to quicker drying of fuels, thus potentially more frequent and intense fires. However, it is the change in rainfall that will be crucial (Goldammer and Price, 1998). In regions where rainfall is predicted to decrease, and the length of the dry season to increase, savannas will be subject to a higher fire risk. On the other hand, these areas may have an overall decrease in net primary production because of the shorter growing season, and if grazing is taking place, there may be a lower fuel load, therefore decreasing the risk of fire. In areas where rainfall may increase, the greater net primary production will in fact lead to the build-up of fuels, even if grazing is present, and so increase the risk of fire. The invasive characteristics of many alien plants in savannas are intrinsically related to fire, and so it may be expected that they will be favoured, especially with the increased CO_2 allowing them rapid initial growth.

All the factors above will determine species distribution in the future. However, humans are also part of the biogeographical picture, and one of main factors limiting migration of species will be habitat fragmentation, i.e. the creation of huge barriers. We have seen in the previous chapters that most savanna regions of the world are under increasing pressure to convert natural savanna to other forms of land use, especially agriculture. These areas are potential barriers to species dispersal, and may lead to species extinction.

10.2.4 Desertification and the fuelwood crisis

Desertification and the fuelwood crisis have been topics of considerable debate in recent times, not only in terms of their effects on savanna dynamics but also in respect to management implications and policy-making. The majority of work has been carried out in African savannas, where the problem is seen to be most serious. There is an extensive literature on these subjects, most of which is beyond the scope of this book to discuss in detail. You are therefore referred to the following texts and articles which give a basis for further discussion: Leach and Mearns (1988, 1996), Cline-Cole *et al.* (1990), Behnke *et al.* (1993), Thomas and Middleton (1994), Scoones (1995), LeHouérou (1996), Warren *et al.* (1996), Thomas (1997), Darkoh (1998), Mainguet and de Silva (1998), Mortimore (1998) and Puigdefábregas (1998). In this section, the current paradigms will be presented, and new management strategies will be discussed.

The creation of deserts in savannas

Desertification in savanna lands has mainly been attributed to the communal ownership of land. Hardin (1968) termed this 'the tragedy of the commons', assuming that in communal areas nobody managed land sustainably as they were not personally responsible for it. These ideas led to management applications of reducing livestock numbers and changing land tenure patterns in order to return to the 'equilibrium' state of the system (see Section 1.2.5). However, Ellis and Swift (1988) challenged this conventional paradigm of pastoral ecosystems, and in their work in the Turkana region of Kenya, showed that in fact pastoral communities exist in a non-equilibrium environment, and they show great persistence using particular adaptive strategies. They concluded that any successful interventions should be designed to take account of environmental fluctuations, rather than maintaining system equilibrium. Effective management of pastoral savanna lands also needs to address individual case studies rather than using results to generalise about other areas (Oba, 1998).

Following this, other studies carried out in arid and semi-arid savannas found similar results, and have criticised the notion that rampant desertification is occurring (e.g. Behnke and Abel, 1996; Sullivan, 1996; Ward *et al.*, 1998).

These authors agree that rainfall in these dry regions is the major driving force and has the ability to 'recharge' a system that suffers heavy grazing pressure. However, these results do not exclude the possibility that slow, long-term degradation may be occurring, and recently monitoring studies have been established to assess this. For example, Ringrose *et al.* (1996) used satellite data and a geographical information system to determine the causes of vegetation change in south-eastern Botswana. They found that in 1994 most natural vegetation cover had been removed within walking distance (2–3 km) of villages, and adjacent to boreholes. Also, extensive areas of woody weeds were replacing natural woodland. On a larger scale, Prince *et al.* (1998), also using remotely sensed data, found that subcontinental desertification in the Sahelian savanna zone of West Africa was not taking place (Figure 10.3), although some areas contained within the region did show signs of degradation.

Most authors agree that in order to prevent desertification, small-scale participatory approaches need to be advocated (e.g. Kasusya, 1998; Palmer and van Rooyen, 1998; van Rooyen, 1998), and as such, successful solutions in one location may not be applicable elsewhere. The International Arid Lands Consortium was established in 1990 to promote research, education and training for the development, management, restoration and reclamation of arid and semi-arid lands throughout the world. Ffolliott *et al.* (1998) describe demonstration projects supported by the Consortium that demonstrate the applicability of recently acquired research information and technology to management situations. Seeley (1998) emphasises the role of indigenous knowledge as a solution to problems of desertification, but warns that science and communities can

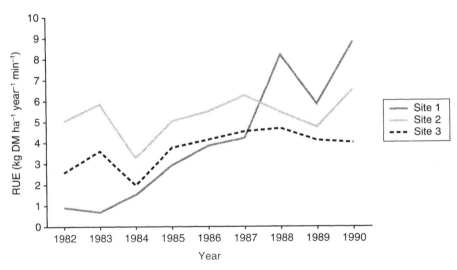

Figure 10.3 Annual rain-use efficiencies (RUE) for three sites in the Sahelian savannas from 1982 to 1990. Sites 1 to 3 have increasing rainfall. RUE is the ratio of annual net primary production to annual rainfall. If extensive desertification was taking place, there would be a decline in RUE with time. In fact, the graph shows an overall increase in RUE (based on Prince *et al.*, 1998).

only work effectively together if the institutional frameworks are conducive to these interactions.

The loss of trees in savannas

Many savanna countries rely on fuelwood, whether it be as firewood or charcoal, for their energy requirements (Figure 10.4). The 'fuelwood crisis' or 'fuelwood orthodoxy' (Cline-Cole *et al.*, 1990) is built on the assumption that general and widespread deforestation is caused by household consumption. It is closely related to desertification, in that tree removal leads to desertification. This 'fuelwood orthodoxy' assumed that fuelwood and deforestation were linearly related to population increase, deforestation proceeding in the form of ripples spreading outwards from urban consuming centres. The expansion of agriculture and conversion of land reduced fuelwood supplies, mostly because farmers did not undertake any management strategies (Cline-Cole *et al.*, 1990).

In recent years, the fuelwood orthodoxy has come under heavy criticism, as it does not reflect many rural African communities (Leach and Mearns, 1988). People respond to population increase and wood shortages by planting trees, using fuel more economically, switching to more abundant crop residues, or intensifying efforts to encourage the natural regeneration of woody vegetation (Bradley *et al.*, 1985; Benjaminsen, 1993; Tiffen *et al.*, 1994). Dewees (1989) describes the management of woody biomass in Kenyan savannas specifically to produce low quality woodfuel, used in times of scarcity. For example, *Tithonia diversifolia* is found on verges and along paths and produces large quantities of fuelwood. The bushes are cut just above ground level and left to dry for a few days before collection.

Figure 10.4 Bags of charcoal made from *miombo* wood for sale at a market in the city.
(Reproduced by kind permission of Paul Desanker, University of Virginia)

The patterns of woodfuel consumption are varied and defy generalisations (Cline-Cole *et al.*, 1990), and most fundamental, it is agricultural land clearance that is the principal cause of deforestation in many areas, not woodfuel consumption (Silviconsult Ltd., 1991; Mearns, 1995; Cline-Cole, 1996). The new woodfuel discourse emphasises resiliency of traditional practices, and community-level action to manage wood resources. Reforestation, though an option, may not be economically viable. For example, in Malawi, a recent review concluded that around 800 000 ha of fast-growing trees would have to be planted to meet estimated deficits, at a cost of well over US$360 million (French, 1986). Plantations cause certain environmental problems of their own, particularly on water balance (Kessler and Bremen, 1991).

10.2.5 Wildlife conservation: the way forward?

Pimbert and Pretty (1997) argue that conservation in savannas remains problematic because of the high dependence on centralised bureaucratic organisations for planning and implementing. Currently, none of the IUCN conservation area categories guarantee the recognition of indigenous people's rights to self-determination and development. Pimbert and Pretty (1997) recommend two immediate priorities: to reform protected area categories and land-use schemes to embody the concepts of local rights and territory in everyday management practice; and to strengthen local control over the access and end uses of biological resources, knowledge and informal innovations.

Because of the lack of policies, the creation of protected areas has frequently promoted hostile attitudes among local communities, which may have been displaced, or restricted from exploiting resources they had long been using. In some regions of the world, the protectionist ideologies of colonial times have meant that conservation is carried out with almost military action, with guards being armed, and the majority of expenditure devoted to law enforcement and public relations. The main ways this could be tackled are through community-based resource management, tourism, and the commercialisation of wildlife, i.e. game viewing, sport hunting and exploitation for commodities such as meat and hides.

Community-based resource management

In recent years, a large number of community-based management projects, aimed at involving local people to manage and conserve their own environments, have been established. The CAMPFIRE programme, probably the most well-known, is described in Section 4.4.4. Although there have been some positive results, many have failed to achieve their objectives, and have ended up being dominated by local elites rather than community members. For example, in 1989, the Kenya Wildlife Service began implementing community wildlife projects in areas around Amboseli National Park and the Maasai Mara National Reserve. However, these wildlife associations ended up being dominated by local élites who monopolised most of the tourism revenue (Sindiga, 1995).

Additional problems with community-based management programmes have been caused by the application of projects set up in one country or area, to another country or region. These community-based management programmes are not always transferable. For example, in the Serengeti National Park, the wildlife is dominated by migratory herbivores, exacerbating the problem of assigning unambiguous ownership of wildlife outside the protected area to a given local community – a pre-condition for any successful privatisation or commercialisation scheme. Also, if the future community conservation strategies are focused on communities currently benefiting most from illegal exploitation, i.e. those adjacent to the Park, then they are likely to attract more people closer to the Park (Hofer *et al.*, 1996).

One of the main inherent problems in all these projects is the failure of policies and legislation to delegate responsibility and authority for tourism development and wildlife conservation from powerful stakeholders such as the state and local élites, to local rural communities. Consequently, mechanisms need to be put in place which encourage local participation in the design as well as the implementation and management of projects. This will also ensure that projects are sensitive to indigenous cultures and the local environment.

Tourism

Nature-based tourism in savannas may instigate sustainable land use in the future. Akama (1996), interviewing visitors to Kenya in 1990, found that over 70% indicated that the main reason why they had decided to visit Kenya was to view wildlife in pristine habitats. However, tourism is not without its shortcomings. In some areas, such as the African savannas, tourism has been instated for many decades, but unfortunately has not always been very 'nature friendly'. Scenes of queues of camper vans and cameras flashing brought about major concern for the welfare of wildlife, and range degradation. There was also the question of conflict over scarce water resources between tourist establishment and local communities (see Section 6.4.1). Political instability also affects tourism potential, as has recently been seen in many parts of Africa. Neumann (1995) comments that in Tanzania, where tourism is a major growing revenue earner, the level of investment, particularly from private and foreign sources, is altering control over land use, and local communities, especially pastoralists, are being evicted from land they have occupied for centuries.

It is obvious that tourism ventures need to be established in co-operation with local people. Proposals to designate the Karanambu Ranch in the northern Rupununi savannas of Guyana, a protected area, led to concern over potential conflict with indigenous land rights and the effect the protected area would have on traditional resource-management practices (Shackley, 1998). It was envisaged that the protected area could be financed by expanding the, as yet, small ecotourism operation at the ranch, most visitors coming to see the rich bird-life and the endangered giant otters.

Following an assessment of the proposed delimitation of the ranch, Shackley (1998) recommended the establishment of a research station and a small nature reserve at Karanambu. These would be a focus for research, training and education, and develop new ecotourism opportunities and facilities. These centres would also help the development of sustainable resource utilisation, including the farming of certain wildlife species such as iguana by local communities. The Karanambu project will be managed by a field biologist and a local trust representing the interested parties, and the aim is to be financially self-reliant by sustainably combining scientific research, ecotourism and development.

Commercialisation of wildlife

The commercialisation of wildlife can be in the form of game ranching, which is the use of rangelands for the maintenance of wild animal populations for subsequent cropping and sale, or as game farming, i.e. the intensive management of usually a single species for breeding and sale for slaughter or other purposes (Kock, 1995). These forms of sustainable wildlife utilisation can occur together with, or as an alternative to domestic animal ranching, farming or other land uses, and have some advantages. It can lead to a more economic use of land and so reduce degradation. However, probably most importantly, it can help the conservation of species which would otherwise be removed from areas where they are in competition with livestock, so reducing the risks of extinction. The latter case is exemplified in the example of capybara and caiman hunting in the *llanos* (Moreira and Macdonald, 1996) (see Section 3.4.2).

In the African savannas, consumptive wildlife utilisation has been developed in southern Africa (there are 3500 game ranches in South Africa alone; Grossman *et al.*, 1992), but only recently has this aspect of conservation been introduced in areas such as East Africa. Kock (1995) describes the success of the Lewa Downs Ranch in Kenya in terms of economic benefits and conservation of rare species, and the increase in ostrich farms in the country. Hunting is also an important activity. Taylor (1992) estimated that over 85% of the revenue gained from wildlife utilisation on communal land in the Nyaminyami District of Zimbabwe was from hunting concessions. Baker (1997) gives a comprehensive review of the tourist hunting industries of six savanna countries in Africa. However, there are major constraints on the development of these wildlife utilisation activities in savannas, including the legal protection of wildlife, illegal hunting, the current economic situation, land tenure, traditional uses and perceptions of wildlife, the industrial capacity for processing products, support services, and the risk of disease (Kock, 1995). As with the community-based management schemes, wildfire commercialisation will not succeed until local communities are given the responsibility for the resource, and legislation to support utilisation.

10.3 New technologies in savanna research and management _____

10.3.1 Decision support systems

Decision support systems (DSSs) are tools which help people to make decisions. All have the common goal of improving decision-making and providing users with the means to assess alternative outcomes more objectively and comprehensively than could be done before (Stuth and Stafford Smith, 1993). DSSs also aim to focus on understanding which decisions are important to different users by identifying objectives, options and perceptions of the decision-making process (Norton and Walker, 1985). In savannas, DSSs have mainly been developed for pastoral (e.g. RANGEPACK: Stafford Smith and Foran, 1991) and fire (e.g. FIRES: Davies et al., 1986; FIRETOOL: Pivello, 1992; Pivello and Norton, 1996) management, or a combination of both (e.g. SHRUBKILL: Ludwig, 1991; MacLeod and Ludwig, 1991). Notably, they have only been developed to date in the technologically advanced savanna countries such as Australia, the USA and South Africa.

Although DSSs may be composed of spatial landscape analysis e.g. geographical information systems, simulation models and expert systems, because of their ease of construction, the latter have been most popular. Expert systems undertake the solution of difficult tasks by using the knowledge of and mimicking the solution methods of human experts in a problem domain. FIRETOOL, the first expert system for managing prescribed burning in the *cerrado* (Pivello, 1992), was constructed using literature and knowledge from various *cerrado* scientists and park managers. It is composed of four subsystems which give instructions to the user on whether a burn is necessary, the type of burn to be used, how to design a short-term burning plan, how to assess the risk of a wildfire in the site, and how to carry out basic precautionary measures. Testing FIRETOOL against the appraisal of an independent *cerrado* fire expert indicated that the expert system had a 70% reliability.

Expert systems, and other decision support systems, are very useful tools in many respects. They can supply information and training for land managers, and help to structure complex problems by identifying the different components. The reliability of the systems may be questionable at the present time, as illustrated by the FIRETOOL example, but as more knowledge becomes available and computer technology evolves, this reliability will undoubtedly increase in the future. Indigenous and local people should also be included as 'experts' in the future development of decision support systems. They have a wealth of knowledge and experience, and although this may bring about issues of knowledge ownership and its safeguard from commercial exploitation, their participation in all forms of land management will be essential.

10.3.2 Modelling

The use of simulation models to predict vegetation change are being developed to help understand savanna functioning. For example, workers at Lamto in West Africa have been using models to explore the effects of tree demography, fire-induced mortality and seed dispersal on the spatial spread of trees (Menaut et al., 1991; Hochberg et al., 1994) and nutrient dynamics and primary productivity (Jacques Gignoux, pers. comm., 1999; LeRoux et al., 1997). Toxopeus (1996) has developed a model for the Amboseli Biosphere Reserve in Kenya, for the management of wildlife and livestock grazing. Modelling is also being used in the savannas of the southern Kalahari to simulate long-term vegetation responses to large-scale disturbances, small-scale heterogeneities and shrub encroachment (Jeltsch et al., 1996, 1997a,b, 1998; Weber et al., 1998). These models are based on a spatially cellular automaton system, and competition between individuals is based on neighbourhood effects, e.g. distance between individuals. Other savanna models are based on patch dynamics, i.e. it is assumed that the maximum effect will be determined by the spacing of the largest individuals (Botkin, 1993). MIOMBO is a patch dynamics model developed for the *miombo*, in particular, for woodlands in Malawi (Desanker and Prentice, 1994), and incorporates the effects of moisture and fire on tree growth, establishment and mortality.

A limited, but growing number of models are being developed to predict changes to climate. Desanker (1996) ran the MIOMBO model for climate change scenarios from four different global circulation models for a doubling of carbon dioxide concentrations. He found varying results from the different GCMs, but in general less moisture would be available for plant growth. This would be disastrous for the species that begin to leaf in anticipation of rainfall, as is common in the *miombo*. Desanker (1996) acknowledges the variations in the results and emphasises the need for further validation to be done to test the model, as well as consideration of the underground competition for water and nutrients.

10.3.3 Remote sensing and spatial analysis

With the advancement of technology and its wider dissemination, satellite imagery and spatial analysis techniques will probably be an important tool for savanna management in the future. They are already being used to monitor changes in vegetation phenology (Franca and Setzer, 1998) and land use, particular with regards to burning (e.g. Scholes et al., 1996a; Ehrlich et al., 1997). Dwyer et al. (1998) analysed vegetation fires from all over the globe using satellite imagery and found that the largest numbers of fires were in the African continent and mainly in savanna woodlands.

Remote sensing is also being used to address problems of land-use conflicts. For example, in the East Kimberley region of the Australian savannas, fire is employed by both Aborigines and pastoralists, but the nature of the land tenure means that they are often competing to burn the same land, resulting in conflict over the appropriate burning regime. O'Neill et al. (1993) used remote sensing to delineate fire patterns in the area over three years, and a geographical information system to integrate land use, land cover and other spatial maps. The study concluded that the Aboriginal burning had not significantly impacted the region, and that their pattern of burning throughout the dry season (in contrast to pastoralists burning during the late dry season) may in fact help reduce fuel loads and the incidence of high-intensity late dry season fires.

Although data derived from these methods are appropriate for mapping and categorising land cover types and burn types, they may not be ideal for identifying the ignition source (O'Neill et al., 1993). The resolution of the remotely sensed data is also an important consideration, although this is improving and some satellites may have pixels of 5 m^2.

10.3.4 Stable isotopes

An increasingly important technique being used in savanna ecology is the use of stable isotopes. Tissues from C_3 and C_4 plants have distinctive $^{13}C/^{12}C$ stable isotope ratios, expressed as $\delta^{13}C$ (Polley *et al.*, 1992). Additionally, these $\delta^{13}C$ ratios in soil organic matter, generally reflect the relative contribution of C_3 and C_4 plants to site productivity over a long period of time. The use of this technique in savannas, therefore, is extremely relevant where natural $\delta^{13}C$ abundance measurements allow discrimination between trees (C_3) and grasses (C_4). For example, Mordelet *et al.* (1997) used $\delta^{13}C$ to show that there was no spatial

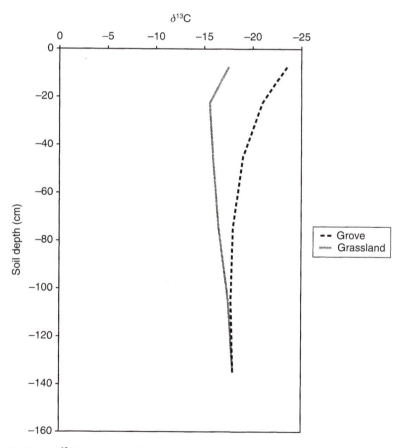

Figure 10.5 Mean $\delta^{13}C$ values of soil organic carbon from groves and grassland in subtropical savanna parklands of southern Texas. Note that as depth increases, the grove soil shows a greater C_4 signature, i.e. $\delta^{13}C$ values are below −20 (after Boutton *et al.*, 1998).

or temporal partitioning between tree and grass roots in the humid savannas of West Africa.

Perhaps more significant is the use of stable isotopes to document vegetation change. A large body of work using this technique has been carried out in the southern Texan savannas to assess interactions between trees and grasses (see Section 1.3.3 for a description of the vegetation dynamics in the Texan savannas) (e.g. Archer, 1991, 1995; Boutton *et al.*, 1998). C_3 woody species have a $\delta^{13}C$ range of -27% to -32%, and C_4 grasses have a range between -13% and -17%. Studies have found that soils beneath present grass-dominated areas show a strong C_4 signature, with $\delta^{13}C$ values between -14% and -18%. However, soils beneath *Prosopis* woody clusters also show a C_4 signature, with $\delta^{13}C$ values between -21% and -23%. The strength of the C_4 signature increased as soils from lower depths beneath the woody clusters were sampled (Figure 10.5). These data give direct evidence that woodlands, groves and shrub clusters dominated almost exclusively by C_3 plants now occupy sites once dominated by C_4 grasses.

10.4 Concluding remarks

Savannas are heterogeneous ecosystems composed of a diverse range of plants and animals. They support a huge proportion of the world's population that will increase substantially in the future. Yet savanna regions of the world are relatively unknown to most people. Having been introduced to people and commenting that I work in Brazil, I have often been asked, 'How is the rain forest?' What is a little worrying is that some academics have asked the same question! People are normally surprised to discover that over 20% of Brazil is in fact savanna! This bias particularly towards rain forest in the tropics is also reflected in the media. Scenes of rain forest biodiversity and its destruction are common in newspapers and on the television, and if savannas are ever referred to, it is normally in the context of lions and wildebeest in East Africa or people in the Sahel dying of hunger. There is no mention of the fact that a fifth of the world's population depends on savannas, or that in many parts of the world, savannas are being destroyed faster than other ecosystems such as rain forest. There is, therefore, a real need to highlight the role savannas play, so that people are aware of their importance and current status, and so that more funding for research and development is directed into savanna ecology and management.

Although PAM, PAN, fire and herbivory are the inherent ecological forces determining savannas, the future of these ecosystems will be largely governed by humans, and their willingness to modify or sustain the environment. The greatest players in this game are governments. They will be responsible for constraining or driving social, economic and land-use policies. This will be important for savanna lands within individual countries, but also over whole regions, where co-operation between countries could bring about better solutions for savannas. The trans-national need for collaboration is highlighted by the example of the Trans Border Conservation Area (TBCA) or 'superpark' planned for the area where South Africa, Mozambique and Zimbabwe meet

(Figure 10.6). This would reopen old wildlife migration routes and help conserve genetic diversity, as well as provide an economic boost, through tourism, to local people in the region. However, although the ecological and economic basis of the superpark has been agreed, political factors on the part of the Zimbabwean state have slowed its implementation (Duffy, 1997). Political instability and civil unrest, which are increasingly becoming characteristic of many African states, will be a major determinant of savanna conservation and management in the future.

Savannas are intrinsically dynamic and variable. Humans have responded to this by modifying behavioural and land-use patterns, and this will be vital in the future to ensure that savannas have an adequate resource base to support people's survival. The challenge for savanna governments is great in that major policy, institutional and infrastructural changes will need to be made. Scientists also face the task of understanding savanna functioning and dynamics better, and closing the gaps in our knowledge about individual savanna formations (Table 10.2). I believe savannas will be places of enterprise and

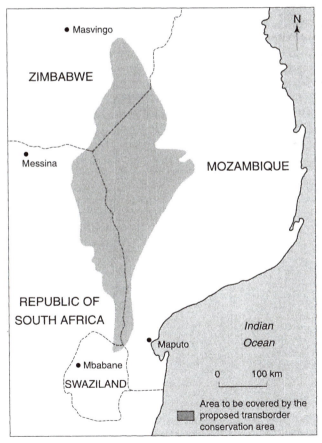

Figure 10.6 The location of the proposed superpark in southern Africa (after Duffy, 1997).

Table 10.2 The major gaps in our knowledge about different savanna formations

Savanna	Some major knowledge gaps
Cerrado	Faunal population dynamics, and interactions with savanna functioning. Their responses to fire. Micro-organisms, mychorrhizal associations. Herbivory, especially invertebrate. Indigenous peoples' savanna management. Effect of land conversion to agriculture on savanna species. The ecology of invasive grasses.
Llanos	The ecology of woody plants. Faunal population dynamics, and interactions with savanna functioning. Nutrient cycling and adaptations to PAN. Herbivory. Factors affecting fire and fire behaviour. Indigenous peoples' savanna management. The effect of dikes on savanna ecology. The ecology of invasive grasses.
Miombo	Faunal population dynamics, and interactions with savanna functioning. The role of termites. Herbivory by invertebrates and mammals other than elephants. Fire behaviour.
West African savannas	The ecology of the three different savanna vegetation types, e.g. phenology. Faunal population dynamics, and interactions with savanna functioning. Invertebrate herbivory. Factors affecting fire and fire behaviour. Indigenous people's savanna management. The effect of hunting on wildlife populations. Wood consumption.
East African savannas	The ecology of savanna vegetation, e.g. grass phenology. Insect herbivory. Fire ecology, e.g. factors affecting fire, fire behaviour, effects of vegetation and animals. The effect of tourism on savanna ecology.
Southern African savannas	Plant phenology. The role of soil fauna such as termites in nutrient cycling. Factors affecting fire such as fuels. Effect of fire on faunal populations. The effects of alien plants. The ecology of woody encroachment.
Dry dipterocarp savannas of Southeast Asia	General ecology of savanna vegetation. Faunal population dynamics, and interactions with savanna functioning. Their responses to fire. Nutrient cycling. Herbivory. Rates of deforestation, effects on savanna ecology.
Australian savannas	Faunal population dynamics, and interactions with savanna functioning. Their responses to fire. The role of the El Niño Southern Oscillation on savanna dynamics. Invertebrate herbivory. The effect of tourism on savanna ecology. The effect of invasive plants and animals.

innovation in the future, and I for one will be looking very closely to see what happens.

Key reading

Behnke, R.H. Jr., Scoones, I. and Kerven, C. (eds) (1993). *Range Ecology at Disequilibrium. New Models of Natural Variability and Pastoral Adaptation in African Savannas.* ODI, London.

Boutton, T.W., Archer, S., Midwood, A.J., Zitzer, S.F. and Bol, R. (1998). $^{13}C^{12}C$ values of soil organic carbon and their use in documenting vegetation change in a subtropical savanna ecosystem. *Geoderma*, **85**: 5–41.

Cline-Cole, R.A., Main, H.A.C. and Nichol, J.E. (1990). On fuelwood consumption, population dynamics and deforestation in Africa. *World Development*, **18**(4): 513–517.

Crutzen, P.J. and Andreae, M.O. (1990). Biomass burning in the tropics: impacts on atmospheric chemistry and biogeochemical cycles. *Science*, **250**: 1669–1678.

Desanker, P.V. and Prentice, I.C. (1994). MIOMBO – a vegetation dynamics model for the miombo woodlands of Zambezian Africa. *Forest Ecology and Management*, **69**: 87–95.

Kock, R.A. (1995). Wildlife utilisation: use it or lose it – a Kenyan perspective. *Biodiversity and Conservation*, **4**: 241–256.

Palmer, A.R. and van Rooyen, A.F. (1998). Detecting vegetation change in the southern Kalahari using Landsat TM data. *Journal of Arid Environments*, **39**: 143–153.

Pivello, V.R. and Norton, G.A. (1996). FIRETOOL: an expert system for the use of prescribed fires in Brazilian savannas. *Journal of Applied Ecology*, **33**: 348–356.

Tiffen, M., Mortimore, M. and Gichuki, F. (1994). *More People, Less Erosion. Environmental Recovery in Kenya.* John Wiley, New York.

Walker, B.H. (1991). Ecological consequences of atmospheric and climatic change. *Climatic Change*, **18**: 301–316.

Appendix

Savanna and related Internet sites of interest

CAMPFIRE Homepage: http://campfire-zimbabwe.org

Centre for Investigating the Ecology of the Tropical Andes:
http://www.ciens.ula.ve/~cielat

Cerrado National Research Centre (CPAC):   http://www.cpac.embrapa.br

East Africa Natural History Society:
http://www.museums.or.ke/textpage/releanhs.html

Food and Agriculture Organisation of the United Nations:
http://www.fao.org

Global Fire Monitoring, NASA:
http://modarch.gsfc.nasa.gov/fire_atlas/fires.html

International Centre for the Study and Development of the *Llanos de Moxos*:
http://www.cris.com/~Moxos/english.shtml

Jay Mistry's homepage – general information on savannas:
http://cedarweb.gg.rhbnc.ac.uk/Jay

Miombo Network: http://miombo.gecp.virginia.edu

Philip Stott's Homepage – general information on savannas:
http://ourworld.compuserve.com/homepages/stott2

South Africa National Parks Homepage: http://www.parks-sa.co.za

Steve Archer Homepage, information on Texan savannas:
http://cnrit.tamu.edu/rlem/faculty/archer/sarcher.htm

The Intergovernmental Panel on Climate Change (IPCC):
http://www.ipcc.ch

The International Fire Information network: http://www.csu.edu.au/firenet

Tropical Savannas CRC: http://savannah.ntu.edu.au

United Nations Convention to Combat Desertification:
http://www.unccd.ch/lite

Glossary of terms

Alfisol Relatively young, acid soils characterised by a clay-rich B horizon, common beneath deciduous forest in temperate and sub-tropical climates.

Alkaloid Any group of nitrogenous organic bases found in plants with toxic or medicinal properties.

Alluvial, alluvium Material deposited by running water.

Annual plant Plant that completes its life cycle in a year.

Apical meristem Tip of shoot or root where growth occurs.

Asymmetric competition Where species A has a strong negative impact on species B, but species B has little or no effect on species A. This is opposed to symmetrical competition, where A adversely affects B, and B has a similar adverse effect on A.

Awn Stiff, bristle-like projection.

Back fire Fire that travels against the wind.

Basal leaf One of the leaves produced near the base of the stem.

Bioindicator An organism that responds to environmental change, and thereby can provide information on both the change and the extent of change.

Browser An animal that feeds on plant materials, especially on the woody parts of trees and shrubs.

C_3 or Calvin photosynthesis Photosynthesis where the first product of CO_2 fixation is a 3-carbon acid. Occurs in most temperate plants and tropical woody flora.

C_4 photosynthesis Photosynthesis where the first product of CO_2 fixation is a 4-carbon acid. This is temporary and followed by C_3 photosynthesis. Its advantages are that it can operate more efficiently at low CO_2 concentrations, high temperature and light intensities, and under considerable water stress.

Caespitose Having low, closely matted stems.

Callus Small hard swelling or outgrowth.

Cataphyll A simple form of leaf on lower parts of plants.

Catena A sequence of soil types repeated in a corresponding sequence on topographic sites.

Cation exchange capacity The extent to which exchangeable cations (positively charged ions, e.g. K^+, N^+, Ca^{2+}) can be held in a soil.

Cauline leaf Leaf growing on upper part of stem.

Chlorosis Abnormal condition caused by lack of green pigment in plants, or nutrient deficiency.

Coleopteran   Large order of insects commonly called beetles.

Compound leaf   Leaf made up of several leaflets.

Coppice   Management technique where woody plants are cut at base to allow regrowth of several stems.

Corticolous   Growing on the bark of trees or shrubs.

Cretaceous   *See* Timeline.

Culmed   Grasses and sedges with a stem.

Cuticle (of plants)   Waxy layer on outside of leaf making it fairy impermeable to water.

Deciduous   Trees having leaves that all fall at a certain time of the year, usually in response to the onset of drought or cold.

Dystrophic   Inhibiting adequate nutrition.

El Niño Southern Oscillation (ENSO)   A periodic phenomena in the Pacific Ocean climatic system that affects ocean currents, wind movements and rainfall regimes across the whole region. It may also have repercussions in much farther regions of the world.

Endemic   Restricted to a certain region or part of a region.

Ephemeral   Short-lived.

Epigeous   Borne above ground.

Eutrophic   Rich in nutrients.

Evapotranspiration   Loss of water from the soil surface by evaporation and by transpiration from plants growing thereon.

Evergreen   Vascular plants that do not shed all their leaves at the same time, and therefore appear green all year round.

Feral   Wild, or escaped from domestication and reverted to wild state.

Glabrous   With a smooth even surface, no hairs.

Glacial   A period of time when temperatures were cold enough to allow the expansion of glaciers in high-latitude and high-altitude areas.

Glaucous   Bluish green.

Gondwanaland   Pre-Mesozoic southern land mass composed of South America, Africa, India, Australia and Antarctica.

Graminoid   Of grasses.

Grazer   An animal that feeds on plant materials, especially on grasses.

Hardpan   *See* Laterite.

Head fire   Fire that travels with the wind.

Herbaceous   Seed plants with non-woody green stems.

Herbivore   An animal that feeds exclusively on plant material.

Holocene   *See* Timeline.

Homopteran   Group of insects including aphids, cicadas and scale insects.

Hydrolysable   Can be broken by a chemical reaction with water.

Hygroscopic   Sensitive to moisture.

Hyperseasonal savannas   Savannas with extremely contrasting seasons. There is a prolonged wet season which causes excess water to accumulate, followed by a dry season where there is water shortage.

Hypsodont teeth   Teeth of grazing mammals such as cows.

Imbricated   Overlapping like scales.

Indumentum   A hairy covering.

Interglacial   A period of time when temperatures were warm enough to allow the retreat of glaciers in high-latitude and high-altitude areas, and the development of vegetation.

Invertebrate   Animal lacking backbone.

Isobilateral   Where a structure is divisible in two planes at right angles.

Jurassic   *See* Timeline.

Laterite   A material that develops gradually over exceptionally long time periods in places where water-table fluctuations lead to gradual accumulation of iron oxides. With time, such accumulations or mottles can grow to become hardened nodules, and coalesce to form an indurated deposit (hardpan).

Latosols   A lateritic soil.

Leaf area index (LAI)   A measure of the photosynthetic area over a given area of ground.

Lepidopteran   Order of insects commonly known as moths and butterflies.

Lichen   An organism formed from the symbiotic association of certain fungi and a green alga or cyanobacterium, forming a simple body.

Lignin   A hard material found in plant cell walls.

Lignotuber   Thickened, lignified fleshy storage underground root.

Macropod   With long feet.

Marsupial   Group of mammals where the young are born immature and migrate to the pouch where they are suckled until mature, e.g. kangaroos, opossums.

Mesic, mesophytic   Conditioned by moist climate.

Mesotrophic   Moderate quantity of nutrients.

Metamorphic rock   Sedimentary rock subject to pressure and heat.

Mineralisation   Conversion of an element from an organic to an inorganic form, e.g. organic material decomposed by micro-organisms.

Miocene   *See* Timeline.

Mutualism   Symbiosis in which both partners benefit from association.

Mycorrhizae, ecto-, vesticular arbuscular   A symbiotic association between plant roots and certain fungi. In ecto-mycorrhizae the fungal tissue encloses the smallest rootlets, with the fungal strands penetrating between the plant cell walls. In vesticular arbuscular mycorrhizae, the external fungal sheath and the fungal strands penetrating plant cells are lacking. Here, vesicles (swellings on invading fungal strands) and arbuscules (discrete masses of branched fungal strands) are present in the plant roots.

Neolithic   Period characterised by the use of polished stone tools and the appearance of settled cultivation.

Neotropics   A phytogeographical area including tropical and subtropical regions of America.

New World   South America, Australia, New Zealand.

Old World   Africa, India and Southeast Asia

Oligotrophic   Poor in nutrients.

Oxisols Highly weathered soil type which, together with ultisols, may be regarded as the 'modal' soil types of the tropics. Have good structural properties, but low nutrient content.

Pantropical Distributed throughout the tropics.

Perennial Persists for several years.

Phanerogams All seed-bearing plants.

Phenology Study of periodic biological events such a flowering, breeding, etc.

Phosphorylation The addition of a phosphate group to a molecule. Important reaction in enzyme activity.

Phyllode Winged leaf stalk with flattened surfaces, acting as leaf.

Phytophagous insects Insects that feed on plants.

Pleistocene *See* Timeline.

Pollard Management technique where woody plants are cut at breast height or above to allow regrowth of several stems.

Precambrian *See* Timeline.

Primary production Fixation of inorganic carbon into organic matter.

Pyrophytic Adapted to fire.

Quaternary *See* Timeline.

Rhizome Thick horizontal stem usually underground, sending out shoots above and roots below.

Rhizosphere, rhizopheric Area of soil immediately surrounding and influenced by plant roots.

Sclerophylly Condition of having hard leaves, which are resistant to drought through having a thick cuticle.

Seasonal savannas Savannas with a wet season with sufficient moisture in the upper soil layers, but always below saturation levels, alternating with another season with marked soil water deficit.

Sodicity The influence of sodium in soils.

Stipe Stalk.

Stoloniferous Bearing a creeping plant stem or runner capable of producing rootlets and stem and ultimately a new individual.

Stomata Minute openings in plant leaf cell wall, especially on underside through which air and water travels.

Suberisation Thickening of plant cell walls due to the deposition of suberin, a waxy substance characteristic of corky tissues.

Symbiosis (adj. **symbiotic**) A long-lasting association between two or more different species of organisms.

Tannins Complex aromatic compounds.

Tertiary *See* Timeline.

Timeline The geological time-scale with the approximate date of commencement in million of years BP (*see* Table on p. 275).

Tomentose leaf Leaf closely covered with matted hairs.

Ultisols Similar to oxisols, but characterised by the presence of a clay-enriched B horizon which can lead to impeded drainage and lateral soil water movement in wet season, and impeded root development.

Ungulate   Hoofed mammal.

Vascular plants   Plants with vessels adapted for carrying or circulating fluids.

Vertebrate   Animal with backbone, e.g. fish, amphibians, reptiles, birds and mammals.

Volatisation   The reaction turning solids into gases.

Water potential (measured in MPa)   The capacity of a plant cell to take up water by osmosis i.e. diffusion of water through a semi-permeable membrane from a dilute to a concentrated solution.

Woody plant   Any perennial plant with secondary lignified tissue in stem.

Xeric, **xerophytic**   Adapted to dry conditions.

Xylopodia   Hardened foot-like stem structure.

Timeline table

Era	Period	Epoch	Approximate date of commencement in million of years BP
Cenozoic	Quaternary	Holocene	0.01
		Pleistocene	2.4
	Tertiary	Pliocene	5
		Miocene	23
		Oligocene	36.5
		Eocene	53
		Paleocene	65
Mesozoic	Cretaceous		135
	Jurassic		205
	Triassic		250
Palaeozoic	Permian		290
	Carboniferous		355
	Devonian		410
	Silurian		435
	Ordovician		510
	Cambrian		570
	Precambrian		650

Bibliography

Abbadie, L. (1984). Evolution saisonnière du stock d'azote dans la strate herbacée d'une savane soumise au feu en Côte d'Ivoire. *Acta Oecologica, Oecologia Plantarum,* **5**: 321–334.

Abbadie, L. and Lensi, R. (1990). Carbon and nitrogen mineralisation and denitrification in a humid savanna of West Africa (Lamto, Côte d'Ivoire). *Acta Oecologica,* **11**: 717–728.

Abbadie, L. and Lepage, M. (1989). The role of subterranean termite fungus-comb chambers (*Isoptera,* Macrotermitinae) in soil nitrogen cycling in a preforest savanna (Côte d'Ivoire). *Soil Biology and Biochemistry,* **21**: 1067–1071.

Abbadie, L., Mariotti, A. and Menaut, J.C. (1992). Independence of savanna grasses from soil organic matter for their nitrogen supply. *Ecology,* **73**(2): 608–613.

Abbadie, L., Lepage, M. and Menaut, J.C. (1996). Paradoxes d'une savane africaine. *La Recherche,* **287**: 36–38.

Abbate, E., Albianelli, A., Azzaroli, A., Benvenuti, M., Tesfamariam, B., Bruni, P., Cipriani, N., Clarke, R.J., Ficcarelli, G., Macchiarelli, R., Napoleone, G., Papini, M., Rook, L., Sagri, M., Tecle, T.M., Torre, D. and Villa, I. (1998). A one-million-year-old Homo cranium from the Danakil (Afar) Depression of Eritrea. *Nature,* **393**(6684): 458–460.

Abel, N. (1993). *Carrying capacity, rangeland degradation and livestock development policy for the communal rangelands of Botswana.* Overseas Development Institute, Pastoral Development Network Paper 35c, pp. 1–9.

Ab'Saber, A.N. (1971). A organização natural das paisagens inter e subtropicais brasileiras. In: Ferri, M.G. (ed.), *III Simpósio sobre o Cerrado.* Editora da Universidade de São Paulo, São Paulo, pp. 1–14.

Ab'Saber, A.N. (1983). O dominio dos *Cerrados*: uma introdução ao conhecimento. *Revista do Serviço Público,* **40**: 41–55.

Adámoli, J., Sennhauser, E., Acero, J.M. and Rescia, A. (1991). Stress and disturbance: vegetation dynamics in the dry Chaco region of Argentina. In: Werner, P.A. (ed.), *Savanna ecology and management. Australian perspectives and intercontinental comparisons.* Blackwell, Oxford, pp. 147–156.

Afolayan, T.A. (1978). Savanna burning in Kainji Lake National Park, Nigeria. *East African Wildlife Journal,* **16**: 245–255.

Afolayan, T.A. and Fafunsho, M. (1978). Seasonal variation in the protein content and the grazing of some tropical savanna grasses. *East African Wildlife Journal,* **16**: 97–104.

Agnew, A.D.O. (1968). Observations on the changing vegetation of Tsavo National Park (East). *East African Wildlife Journal,* **6**: 75–80.

Ajara, C. (1989). População. In: Duarte, A.C. (ed.), *Geografia do Brasil, região*

Centro-Oeste (vol. 1). Fundação Instituto Brasileiro de Geografia e Estatistica, Rio de Janeiro, pp. 123–148.

Akama, J.S. (1996). Western environmental values and nature-based tourism in Kenya. *Tourism Management,* **17**(8): 567–574.

Akpo, L.E. (1993). *Influence du couvert ligneux sur la structure et le fonctionnement de la strate herbacée en milieu sahélien.* ORSTOM, Paris.

Aksornkoae, S. (1971). *A comparison of N contents and bulk densities in a dry evergreen forest and a dry dipterocarp forest at Sakaerat, Pakthongchai Nakhonratchasima.* Forest Research Bulletin 15, Faculty of Forestry, Kasetsart University, Bangkok.

Alencar, G. de (1979). O programa de desenvolvimento de região dos *cerrados*. In: Marchetti, D. and Machado, A.D. (eds), *Cerrados: uso e manejo. V Simpósio sobre o cerrado.* Empresa Brasileira da Pesquisa Agropecuária, CPAC-CNPq, Brasilia, DF, pp. 37–58.

Alho, C.J.R. (1981). Mata cilia como refúgio da fauna do *cerrado* em caso de fogo? In: Alho, C.J.R. (ed.). *Resumos das Comunicaçoes Cientificas do VIII Congresso Brasileiro de Zoologia.* Universidade de Brasilia, Brasilia, DF, pp. 175–184.

Alho, C.J.R. and Martins, E.S. (1995). *Little by little, the cerrado loses space.* Discussion paper, WWF/PRO-CER, Brazil.

Allan, J.C. (1965). *The African husbandman.* Oliver and Boyd, Edinburgh.

Allan, R.J. (1988). El Niño Southern Oscillation influences in the Australasian region. *Progress in Physical Geography,* **12**: 313-348.

Allen, J. and Barton, G. (1989). *Ngarradj Warde Djobkeng: white cockatoo dreaming and the prehistory of Kakadu.* Oceania Monograph 37, University of Sydney.

Allen, J. and Holdaway, S. (1995). The contamination of Pleistocene radiocarbon determinations in Australia. *Antiquity,* **69**: 101–112.

Almeida, S.P. and Silva, J.C.S. (1989). *Influência do fogo sobre aspectos fenológicos de gramineas nativas dos cerrados.* Pesquisa em Andamento No. 28, EMBRAPA/CPAC.

Anan, G. (1995). Social change and potential of the Thai community potential. In: Chalong, S. (ed.), *Critiques of Thai society.* The Social Science Association of Thailand, Amarin Printing and Publishing, Bangkok, pp. 151–191 (in Thai).

Andersen, A.N. (1991). Responses of ground-foraging ant communities to three experimental fire regimes in a savanna forest of tropical Australia. *Biotropica,* **23**: 575–585.

Andersen, A.N. (1996). Fire ecology and management. In: Finlayson, C.M. and von Oertzen, I. (eds), *Landscape and vegetation ecology of the Kakadu region, northern Australia.* Kluwer Academic Publishers, Holland, pp. 179–195.

Andersen, A.N. and Lonsdale, W.M. (1991). Herbivory by insects in Australian tropical savannas: a review. In: Werner, P.A. (ed.). *Savanna ecology and management. Australian perspectives and intercontinental comparisons.* Blackwell, Oxford, pp. 80–100.

Andersen, A.N. and Sparling, G.P. (1997). Ants as indicators of restoration success: relationship with soil microbial biomass in the Australian seasonal tropics. *Restoration Ecology,* **5**(2): 109–114.

Andersen, A.N., Braithwaite, R.W., Cook, G.D., Corbett, L.K., Williams, R.J., Douglas, M.M., Gill, A.M., Setterfield, S.A. and Muller, W.J. (1998). Fire research for conservation management in tropical savannas: introducing the Kapalga fire experiment. *Australian Journal of Ecology,* **23**: 95–110.

Anderson, A.B. and Posey, D.A. (1985). Manejo de *cerrado* pelos indios Kayapó. *Boletim do Museu Paraense Emilio Goeldi Botânica,* **2**(1): 77–98.

Anderson, A.B. and Posey, D.A. (1989). Management of a tropical scrub savanna by the Gorotire Kayapó of Brazil. *Advances in Economic Botany,* **7**: 159–173.

Anderson, G.D. and Talbot, L.M. (1965). Soil factors affecting the distribution of the grassland types and their utilisation by wild animals on the Serengeti Plains, Tanganika. *Journal of Ecology*, **53**: 33–56.

Anderson, G.D. and Walker, B.H. (1974). Vegetation composition and elephant damage in the Sengwa Wildlife Research Area, Rhodesia. *Journal of South African Wildlife Management*, **4**: 1–14.

Andrew, M.H. (1986). Use of fire for spelling monsoon tallgrass grazed by cattle. *Tropical Grasslands*, **20**: 69–78.

Andrew, M.H. (1988). Grazing impact in relation to livestock watering points. *Trends in Ecology and Evolution*, **3**: 336–339.

Andrew, M.H. and Mott, J.J. (1983). Annuals with transient seed banks: the population biology of indigenous *Sorghum* species of tropical north-west Australia. *Australian Journal of Ecology*, **8**: 265–276.

Anon. (1994). *Abstract of Agricultural Statistics*. Department of Agriculture, Pretoria.

Aragão, L.T. (1994). Ocupação humana de Brasilia. In: Novaes-Pinto, M. (ed.), *Cerrado, Caracterização, Ocupação e Perspectivas* (2nd edition). Editora Universidade de Brasilia, Brasilia DF, pp. 171–187.

Araki, S. (1992). The role of the *miombo* woodland ecosystem in *chitemene* shifting cultivation in northern Zambia. *Japan Information MAB*, **11**: 8–15.

Araki, S. (1993). Effect on soil organic matter and soil fertility of the *chitemene* slash-and-burn practice used in northern Zambia. In: Mulongoy, K. and Merckx, R. (eds), *Soil organic matter dynamics and sustainability of tropical agriculture*. IITA/K.U. Leuven, Belgium, pp. 367–375.

Araujo, F.B., Costa, E.M.M., Oliveira, R.F., Ferrari, K., Simori, M.F. and Pires-Junior, O.R. (1996). Efeitos de queimadas na fauna de largartos do Distrito Federal. In Miranda, H.S., Saito, C.H. and Dias, B.F. de S. (eds), *Impactos de queimadas em áreas de cerrado e restinga*. Universidade de Brasilia, Brasilia, DF, pp. 148–160.

Araújo, G.M. de (1984). *Comparação de estudo nutricional de dois cerradãos em solos distróficos e mesotróficos no planalto do Brasil.* Tese de mestrado, Universidade de Brasilia, Brasilia, DF.

Archer, S. (1989). Have southern Texas savannas been converted to woodlands in recent history? *American Naturalist*, **134**: 545–561.

Archer, S. (1991). Development and stability of grass/woody mosaics in a subtropical savanna parkland, Texas, U.S.A. In: Werner, P.A. (ed.), *Savanna ecology and management. Australian perspectives and intercontinental comparisons*. Blackwell, Oxford, pp. 109–118.

Archer, S. (1995). Tree–grass dynamics in a *Prosopis*–thornscrub savanna parkland: reconstructing the past and predicting the future. *Ecoscience*, **2**: 83–99.

Archer, S., Scifres, C., Bassham, C.R. and Maggio, R. (1988). Autogenic succession in a subtropical savanna: conversion of grassland to thorn woodland. *Ecological Monographs*, **58**: 111–127.

Aristeguieta, L. and Medina, E. (1966). Proteccion y quema de la sabana llanera. *Boletin Sociedad Venezolana Ciencias Naturales*, **26**: 129–139.

Askew, G.P., Moffatt, D.J., Montgomery, R.F. and Searl, P.L. (1970). Interrelationships of soil and vegetation in the savanna–forest boundary zone of north-eastern Mato Grosso. *Geographical Journal*, **136**: 370–376.

Athias, F., Josens, G. and Lavelle, P. (1975). *Influence du feu de brousse annuel sur le peuplement endogé de la savane de Lamto (Côte d'Ivoire)*. In: Proc. 5th Int. Coll. Soil Zoology, Prague, 1973, pp. 389–397.

Attwell, C.A.M., Campbell, B.M., du Toit, R.F. and Lynam, T.J.P. (1989). *Patterns of fuelwood utilisation in Harare, Zimbabwe.* A report prepared for the Forestry Commission of Zimbabwe and the World Bank, Forestry Commission, Harare.

Austin, M.P. and Williams, O.B. (1988). Influence of climate and community composition on the population demography of pasture species in semi-arid Australia. *Vegetatio,* 77: 43–49.

Australian National Parks and Wildlife Service (1991a). *Kakadu National Park Plan of Management.* Australian Government Publishing Service, Canberra.

Australian National Parks and Wildlife Service (1991b). *Nomination of Kakadu National Park by the Government of Australia for Inscription in the World Heritage List.* Unpublished Report, ANPWS, DASETT, Canberra.

Australian Science and Technology Council (ASTEC) (1993). *Research and technology in tropical Australia and their application to the development of the region.* AGPS, Canberra.

Avery, D.M. (1993). Last interglacial and Holocene altithermal environments in South Africa and Namibia: micromammalian evidence. *Palaeogeography, Palaeoclimatology, Palaeoecology,* 101: 221–228.

Baines, K.A. (1989). *The water use, growth and phenology of four species of savanna grasses in response to changing soil water availability.* MSc Thesis, University of Witwatersrand, Johannesburg.

Baird, M. (1996). A 2020 vision for Cape York Peninsula: a story of 40,000 years plus 200. In: Ash, A.J. (ed.), *The future of tropical savannas: an Australian perspective.* CSIRO, Australia, pp. 159–164.

Baker, J.E. (1997). Development of a model system for touristic hunting revenue collection and allocation. *Tourism Management,* 18(5): 273–286.

Banda, A.S.M. and De Boerr, H. (1993). Honey for sale. In: Kemf, E. (ed.), *Indigenous peoples and protected areas. The law of mother earth.* Earthscan, London.

Banda, M. (1988). *Epicormic and sucker shoots in miombo woodland management in Zambia.* MSc Thesis, University of North Wales, Bangor.

Banyikwa, F.F., Feoli, E. and Zuccarello, V. (1990). Fuzzy set ordination and classification of Serengeti short grasslands, Tanzania. *Journal of Vegetation Science,* 1: 97–106.

Barbier, E.B., Burgess, J.C., Swanson, T.M. and Pearce, D.W. (1992). *Elephants, economics and ivory.* Earthscan, London.

Barker, J.R., Herlocker, D.J. and Young, S.A. (1989). Vegetal dynamics along a grazing gradient within the coastal grassland of central Somalia. *African Journal of Ecology,* 27: 283–289.

Barnes, R.F.W. (1982). Elephant feeding behaviour in Ruaha National Park, Tanzania. *African Journal of Ecology,* 20: 123–136.

Barnes, R.F.W. (1985). Woodland changes in Ruaha National Park (Tanzania) between 1976 and 1982. *African Journal of Ecology,* 23: 215–221.

Barrington, A.H.M. (1931). *Forest soil and vegetation in the Hlaing Forest Circle, Burma.* Burma Forest Bulletin No. 25, Ecology Series No. 1, Government Printing and Stationery, Rangoon.

Barros, C.J.G. (1994). Caracterização geológica e hidrogeológica. In: Novaes-Pinto, M. (ed.), *Cerrado, Caracterização, Ocupação e Perspectivas* (2nd edition). Editora Universidade de Brasilia, Brasilia DF, pp. 265–284.

Barton, A.L. and McDonald, J.N. (1996). The Australian defence force and the future of tropical savannas. In: Ash, A.J. (ed.), *The future of tropical savannas: an Australian perspective.* CSIRO, Australia, pp. 68–79.

Barton, D.P. (1997). Introduced animals and their parasites: the cane toad, *Bufo marinus*, in Australia. *Australian Journal of Ecology*, **22**: 316–324.

Baruch, Z., Ludlow, M.M. and Davies, R. (1985). Photosynthetic responses of native and introduced C_4 grasses from Venezuelan grasses. *Oecologia (Berlin)*, **67**: 388–399.

Baruch, Z., Hernández, A.B. and Montilla, M.G. (1989). Dinamica del crecimiento, fenologia y reparticion de biomasa en gramineas nativas e introducidas de una sabana Neotropical. *Ecotropicos*, **2**: 1–13.

Bate, G.C., Furniss, P.R. and Pendle, P.G. (1982). Water relations of southern African savannas. In: Huntley, B.J. and Walker, B.H. (eds), *Ecology of tropical savannas*. Ecological Studies 42, Springer-Verlag, Berlin, pp. 336–358.

Bayliss, P. and Yeomans, K.M. (1989). The distribution and abundance of feral livestock in the 'Top End' of the Northern Territory (1985–1986), and their relation to population control. *Australian Wildlife Research*, **16**: 651–676.

Bazzaz, F.A., Garbutt, K. and Williams, W.E. (1985). Effect of increased atmospheric carbon dioxide concentration on plant communities. In: Strain, B.R. and Cure, J.D. (eds), *Direct effects of increasing carbon dioxide on vegetation*. United States Department of Energy, National Technical Information Service, Springfield, pp. 155–170.

Beard, J.S. (1953). The savanna vegetation of northern tropical America. *Ecological Monographs*, **23**: 149–215.

Behling, H. (1998). Late Quaternary vegetational and climatic changes in Brazil. *Review of Palaeobotany and Palynology*, **99**: 143–156.

Behling, H. and Hooghiemstra, H. (1998). Late Quaternary palaeoecology and palaeoclimatology from pollen records of the savannas of the *Llanos* Orientales in Colombia. *Palaeogeography, Palaeoclimatology, Palaeoecology*, **139**: 251–267.

Behnke, R.H. and Abel, N. (1996). Revisited: the overstocking controversy in semiarid Africa. *World Animal Review*, **87**: 4–27.

Behnke, R.H., Scoones, I. and Kerven, C. (eds), (1993). *Range ecology at disequilibrium. New models of natural variability and pastoral adaptation in African savannas*. ODI, London.

Bell, M., Faulkner, R., Hotchkiss, P., Lambert, R., Roberts, N. and Windram, A. (1987). *The use of dambos in rural development, with reference to Zimbabwe*. Loughborough University, Loughborough.

Bell, R.H.V. (1970). The use of the herb layer by grazing ungulates in the Serengeti. In: Watson, A. (ed.), *Animal populations in relation to their food resources*. Blackwell, Oxford, pp. 111–124.

Bell, R.H.V. and Jachmann, H. (1984). Influence of fire on the use of *Brachystegia* woodland by elephants. *African Journal of Ecology*, **22**: 157–163.

Belsky, A.J. (1983). Small-scale pattern in four grassland communities in the Serengeti National Park, Tanzania. *Vegetatio*, **55**: 141–151.

Belsky, A.J. (1984). The role of small browsing mammals in preventing woodland regeneration in the Serengeti National Park, Tanzania. *African Journal of Ecology*, **22**: 271–279.

Belsky, A.J. (1985). Long-term ecological monitoring in the Serengeti National Park, Tanzania. *Journal of Applied Ecology*, **22**: 449–460.

Belsky, A.J. (1986a). Population and community processes in a mosaic grassland in the Serengeti, Tanzania. *Journal of Ecology*, **74**: 841–856.

Belsky, A.J. (1986b). Revegetation of artificial disturbances in grasslands in the Serengeti National Park, Tanzania. I. Colonisation of grazed and ungrazed plots. *Journal of Ecology*, **74**: 419–437.

Belsky, A.J. (1986c). Revegetation of artificial disturbances in grasslands in the Serengeti National Park, Tanzania. II. Five years of successional change. *Journal of Ecology*, **74**: 937–951.

Belsky, A.J. (1988). Regional influences on small-scale vegetational heterogeneity within grasslands in the Serengeti National Park, Tanzania. *Vegetatio*, **74**: 3–10.

Belsky, A.J. (1989). Landscape patterns in a semi-arid ecosystem in East Africa. *Journal of the Arid Environment*, **17**: 265–270.

Belsky, A.J. (1991). Tree/grass ratios in East African savannas: a comparison of existing models. In: Werner, P.A. (ed.), *Savanna ecology and management. Australian perspectives and intercontinental comparisons*. Blackwell, Oxford, pp. 139–145.

Belsky, A.J. (1992a). Effects of trees on nutritional quality of understorey gramineous forage in tropical savannas. *Tropical Grasslands*, **26**(1): 12–20.

Belsky, A.J. (1992b). Effects of grazing, competition, disturbance and fire on species composition and diversity of grassland communities. *Journal of Vegetation Science*, **3**: 187–200.

Belsky, A.J. (1994). Influences of trees on savanna productivity: tests of shade, nutrients and tree–grass competition. *Ecology*, **75**: 922–932.

Belsky, A.J. (1995). Spatial and temporal landscape patterns in arid and semi-arid African savannas. In: Hansson, L., Fahrig, L. and Merriam, G. (eds), *Mosaic landscape and ecological processes*. Chapman and Hall, London, pp. 31–56.

Belsky, A.J. and Canham, C.D. (1994). Gaps, patches, and isolated trees: a comparison of the patch dynamics of forest gaps and savanna trees. *BioScience*, **44**: 77–84.

Belsky, A.J., Amundson, R.G., Duxbury, J.M., Riha, S.J., Ali, A.R. and Mwonga, S.M. (1989). The effects of trees on their physical, chemical, and biological environments in a semi-arid savanna in Kenya. *Journal of Applied Ecology*, **26**: 1005–1024.

Belsky, A.J., Mwonga, S.M., Amundson, R.G., Duxbury, J.M. and Ali, A.R. (1993). Comparative effects of isolated trees on their undercanopy environments in high- and low-rainfall savannas. *Journal of Applied Ecology*, **30**: 143–155.

Benjaminsen, T.A. (1993). Fuelwood and desertification: Sahel orthodoxies discussed on the basis of field data from the Gourma region in Mali. *Geoforum*, **24**(4): 397–409.

Berardi, A. (1994). *Effects of the African grass Melinis minutiflora on the plant community composition and fire characteristics of a central Brazilian savanna*. MSc thesis, University College, University of London, London.

Berrade, F. and Tejos, R. (1984). Productividad primaria aerea neta en diferentes unidades fisiográfica del módulo 'Fernando Corrales', Apure, Venezuela. *Revista UNELLEZ de Ciencia y Tecnología*, **2**: 17–34.

Bilbao, B. and Medina, E. (1991). Nitrogen-use efficiency for growth in a cultivated African grass and a native South American pasture grass. In: Werner, P.A. (ed.), *Savanna ecology and management. Australian perspectives and intercontinental comparisons*. Blackwell, Oxford, pp. 77–81.

Bird, M.I. and Cali, J.A. (1998). A million-year record of fire in sub-Saharan Africa. *Nature*, **394**(20): 767–769.

Blackmore, A.C. (1992). *The functional classification of South African savanna plants based on their ecophysiological characteristics*. MSc Thesis, University of Witswatersrand, Johannesburg.

Blackmore, A.C., Mentis, M.T. and Scholes, R.J. (1991). The origin and extent of nutrient-enriched patches within a nutrient-poor savanna in South Africa. In: Werner, P.A. (ed.), *Savanna ecology and management. Australian perspectives and intercontinental comparisons*. Blackwell, Oxford, pp. 119–126.

Blanchard, C.A. (1987). *Return to country: the Aboriginal Homelands Movement in Australia.* AGPS, Canberra.

Blasco, F. (1983). The transition from open forest to savanna in continental Southeast Asia. In: Bourlière, F. (ed.), *Tropical savannas. Ecosystems of the world 13.* Elsevier Scientific Publishing Company, Amsterdam, pp. 167–181.

Bloch, P. (1958). Thailand forest soils. *Natural History Bulletin of the Siam Society,* **19**: 45–55.

Blydenstein, J. (1963). Cambios en la vegetacion despues de proteccion contra el fuego. Parte I: el aumento annual en materia vegetal en varios sitios quemados y no quemados en la Estacion Biologica. *Boletin Sociedad Venezolana Ciencias Naturales,* **103**: 223–238.

Blydenstein, J. (1967). Tropical savanna vegetation of the *Llanos* of Colombia. *Ecology,* **48**: 1–15.

Boaler, S.B. (1966). *The ecology of Pterocarpus angolensis in Tanzania.* Ministry of Overseas Development, London.

Boaler, S.B. and Sciwale, K.C. (1966). Ecology of a *miombo* site, Lupa North Forest Reserve, Tanzania. III: effects on the vegetation of local cultivation practices. *Journal of Ecology,* **54**: 577–587.

Boast, R. (1990). *Dambos* – a review. *Progress in Physical Geography,* **14**(2): 153–177.

Boberg, J. (1993). Competition in Tanzanian woodfuel markets. *Energy Policy,* **21**: 474–490.

Boland, D.J., Brooker, M.I.H., Chippendale, G.M., Hall, N., Hyland, B.P.M., Johnston, R.D., Kleinig, D.A. and Turner, J.D. (1992). *Forest trees of Australia* (4th edition). CSIRO, Melbourne.

Bonell, M., Coventry, R.J. and Holt, J.A. (1986). Erosion of termite mounds under natural rainfall in semiarid tropical northeastern Australia. *Catena,* **13**: 11–28.

Boonplian, S. (1985). *Effects of fire on soil and plants at Doi Angkhang: the first year results.* MSc Thesis, Faculty of Forestry, Kasetsart University, Bangkok.

Booysen, P. de V. and Tainton, N. (1978). *Pasture management in South Africa.* Shooter and Schuter, Pietermaritzburg.

Botkin, D.B. (1993). *Forest dynamics. An ecological model.* Oxford University Press, Oxford.

Bourlière, F. and Hadley, M. (1983). Present-day savannas: an overview. In: Bourlière, F. (ed.), *Tropical savannas. Ecosystems of the world 13.* Elsevier Scientific Publishing Company, Amsterdam, pp. 1–17.

Boutton, T.W., Archer, S., Midwood, A.J., Zitzer, S.F. and Bol, R. (1998). $^{13}C^{12}C$ values of soil organic carbon and their use in documenting vegetation change in a subtropical savanna ecosystem. *Geoderma,* **85**: 5–41.

Bowman, D.M.J.S. (1986). Stand characteristics, understorey associates and environmental correlates of *Eucalyptus tetrodonta* F. Muell. forests on Gunn Point, northern Australia. *Vegetatio,* **65**: 105–113.

Bowman, D.M.J.S. (1993). Establishment of two dry monsoon forest tree species on a fire-protected monsoon forest–savanna boundary, Cobourg Peninsula, northern Australia. *Australian Journal of Ecology,* **18**: 235–237.

Bowman, D.M.J.S. (1998). Tansley Review No. 101 – The impact of Aboriginal landscape burning on the Australian biota. *New Phytologist,* **140**(3): 385–410.

Bowman, D.M.J.S. and Fensham, R.J. (1991). Response of a monsoon forest–savanna boundary to fire protection, Weipa, northern Australia. *Australian Journal of Ecology,* **16**: 111–118.

Bowman, D.M.J.S. and Kirkpatrick, J.B. (1986). Establishment, suppression and growth of *Eucalyptus delegatensis* R.T. baker in multi-aged forests. III. Intraspecific allelopathy, competition between adult and juvenile for moisture and nutrients, and frost damage to seedlings. *Australian Journal of Botany*, **34**: 81–94.

Bowman, D.M.J.S. and Panton, W. (1991). Sign and habitat impact of banteng (*Bos javanicus*) and pig (*Sus scrofa*), Cobourg Peninsula, Northern Australia. *Australian Journal of Ecology*, **16**: 15–17.

Bowman, D.M.J.S. and Panton, W.J. (1993a). Differences in the stand structure of *Eucalyptus tetrodonta* forests between Elcho Island and Gunn Point, northern Australia. *Australian Journal of Botany*, **41**: 211–215.

Bowman, D.M.J.S. and Panton, W.J. (1993b). Factors that control monsoon-rainforest seedling establishment and growth in northern Australian *Eucalyptus* savanna. *Journal of Ecology*, **81**: 297–304.

Bowman, D.M.J.S. and Panton, W.J. (1994). Fire and cyclone damage to woody vegetation on the north coast of the Northern Territory, Australia. *Australian Geographer*, **25**: 32–35.

Bowman, D.M.J.S. and Panton, W.J. (1995). Munmarlary revisited: response of a north Australian *Eucalyptus tetrodonta* savanna protected from fire for 20 years. *Australian Journal of Ecology*, **20**: 526–531.

Bowman, D.M.J.S., Wilson, B.A. and Hooper, R.J. (1988). Response of *Eucalyptus* forest and woodland to four fire regimes at Munmarlary, Northern Territory, Australia. *Journal of Ecology*, **76**: 215–232.

Bowman, D.M.J.S., Panton, W.J. and McDonough, L. (1990). Dynamics of forest clumps on chenier plains, Cobourg Peninsula, Northern Territory. *Australian Journal of Botany*, **38**: 593–601.

Bowman, D.M.J.S., Wilson, B.A. and Woinarski, J.C.Z. (1991). Floristic and phenological variation in a northern Australian rocky *Eucalyptus* savanna. *Proceedings of the Royal Society of Queensland*, **101**: 79–90.

Bradley, P.N. and McNamara, K. (eds), (1993). *Living with trees: policies for forestry management in Zimbabwe*. World Bank Technical Paper 210, World Bank, Washington, DC.

Bradley, P.N., Chavangi, N. and Van Gelder, A. (1985). Development research and energy planning in Kenya. *Ambio*, **14**(4–5): 228–236.

Braithwaite, R.W. (1987). Effects of fire regimes on lizards in the wet–dry tropics of Australia. *Journal of Tropical Ecology*, **3**: 265–275.

Braithwaite, R.W. (1990). A new savannah fire experiment. *Bulletin of the Ecological Society of Australia*, **20**: 47–48.

Braithwaite, R.W. (1991). Aboriginal fire regimes of monsoonal Australia in the nineteenth century. *Search*, **22**: 247–249.

Braithwaite, R.W. (1996). Biodiversity and fire in the savanna landscape. In: Solbrig, O.T., Medina, E. and Silva, J.F. (eds), *Biodiversity and savanna ecosystem processes: a global perspective*. Springer-Verlag, Berlin, pp. 121–140.

Braithwaite, R.W. and Estbergs, J.A. (1985). Fire pattern and woody vegetation trends in the Alligator Rivers region of northern Australia. In: Tothill, J.C. and Mott, J.J. (eds), *Ecology and management of the world's savannas*. Australian Academy of Science, Canberra, pp. 359–364.

Braithwaite, R.W. and Estbergs, J.A. (1987). Firebirds of the Top End. *Australian Natural History*, **22**: 298–302.

Braithwaite, R.W. and Werner, P.W. (1987). The biological value of Kakadu National Park. *Search*, **18**: 296–301.

Braithwaite, W.R., Dudzinski, M.L., Ridpath, M.G. and Parker, B.S. (1984). The impact of water buffalo on the monsoon forest ecosystem in Kakadu National Park. *Australian Journal of Ecology*, **9**: 309–322.

Brigham, T. (1994). *Trees in the rural cash economy: a case study from Zimbabwe's communal areas*. MA Thesis, Carleton University, Ottawa.

Brigham, T., Chihongo, A. and Chidumayo, E. (1996). Trade in woodland products from the *miombo* region. In: Campbell, B. (ed.), *The miombo in transition: woodlands and welfare in Africa*. Center for International Forestry Research (CIFOR), Bogor, Indonesia, pp. 137–174.

Brockwell, C. (1989). *Archaeological investigations of Kakadu wetlands, Northern Australia*. MA Thesis, ANU, Canberra.

Brockwell, S., Levitus, R., Russell-Smith, J. and Forrest, P. (1995). Aboriginal heritage. In: Press, T., Lea, D., Webb, A. and Graham, A. (eds), *Kakadu. Natural and cultural heritage and management*. Australian Nature Conservation Agency, North Australia Research Unit, The Australian National University, Darwin, pp. 15–63.

Brookman-Amissah, J., Hall, J.B., Swaine, M.D. and Attakorah, J.Y. (1980). A reassessment of a fire protection experiment in north-eastern Ghana savanna. *Journal of Applied Ecology*, **17**: 85–99.

Brown, J.R. and Archer, S. (1987). Woody plant seed dispersal and gap formation in a Northern American subtropical savanna woodland: the role of domestic herbivores. *Vegetatio*, **73**: 73–80.

Brown, J.R. and Archer, S. (1989). Woody plant invasion of grasslands: establishment of honey mesquite (*Prosopis glandulosa* var. *glandulosa*) on sites differing in herbaceous biomass and grazing history. *Oecologia*, **80**: 19–26.

Brown, J.R. and Archer, S. (1990). Water relations of a perennial grass and seedling versus adult woody plants in a subtropical savanna, Texas. *Oikos*, **57**: 366–374.

Bruce, J.W., Jensen, E., Kloeck-Jenson, S., Knox, A., Subramanian, J. and Williams, M. (1998a). Synthesis of trends and issues raised by land tenure country profiles of Southern African countries, 1996. In: Bruce, J.W. (ed.), *Country profiles of land tenure: Africa, 1996*. Research paper No. 130, Land Tenure Centre, University of Wisconsin, Madison, pp. 202–208.

Bruce, J.W., Subramanian, J., Knox, A., Bohrer, K. and Leisz, S. (1998b). Synthesis of trends and issues raised by land tenure country profiles of Greater Horn of Africa countries, 1996. In: Bruce, J.W. (ed.), *Country profiles of land tenure: Africa, 1996*. Research paper No. 130, Land Tenure Centre, University of Wisconsin, Madison, pp. 138–148.

Bryant, J.P., Kuropat, P.J., Cooper, S.M., Frisby, K. and Owen-Smith, N. (1989). Resource availability hypothesis of plant antiherbivore defence tested in a South African savanna ecosystem. *Nature*, **340**: 227–229.

Bryant, J.P., Heitkonig, I., Kuropant, P. and Owen-Smith, N. (1991). Effects of severe defoliation on the long-term resistance to insect attack and on leaf chemistry in six woody species of southern African savanna. *American Naturalist*, **137**: 50–63.

Bucher, E.H. (1982). Chaco and caatinga – South American arid savannas, woodlands and thickets. In: Huntley, B. and Walker, B. (eds), *Ecology of tropical savannas*. Springer-Verlag, Berlin, pp. 48–59.

Bureau of Meteorology (1988). *Climatic averages, Australia*. Australian Government Printing Service, Canberra.

Burnham, P. (1980). Changing agricultural and pastoral ecologies in the West African savanna region. In: Harris, D.R. (ed.), *Human ecology in savanna environments*. Academic Press, London, pp. 147–170.

Bush, M.B. (1994). Amazonian speciation: a necessarily complex model. *Journal of Biogeography*, **21**: 5–17.

Buss, I.O. (1990). *Elephant life: fifteen years of high population density.* Iowa State University Press, Ames.

Cabrera, M. and Baruch, Z., (1983). *Growth responses to watering regimes in native and introduced pasture grasses from the derived savannas of the Coastal Cordillera (Venezuela).* Report of Project S1–1135, CONICIT, Caracas.

Camargo, M.N. and Bennema, J. (1966). Delineamento esquemático dos solos do Brasil. *Pesquisa Agropecúaria Brasileira*, **1**: 47–54.

Campbell, B.M., Frost, P. and Byron, N. (1996). *Miombo* woodlands and their use: overview and key issues. In: Campbell, B. (ed.), *The miombo in transition: woodlands and welfare in Africa.* Center for International Forestry Research (CIFOR), Bogor, Indonesia, pp. 1–10.

Campbell, B.M., Grundy, I. and Matose, F. (1993). Tree and woodland resources – the technical practices of small-scale farmers. In: Bradley, P.N. and McNamara, K. (eds), *Living with trees: policies for forestry management in Zimbabwe.* World Bank Technical Paper No. 210, World Bank, Washington, DC, pp. 29–62.

Campbell, B.M., Clarke, J., Luckert, M., Matose, F., Musvoto, C. and Scoones, I. (1995). *Local-level economic valuation of savanna woodland resources: village cases from Zimbabwe.* Hidden Harvest Project Research Series 3, IIED, London.

Canales, J. and Silva, J. (1987). Efecto de una quema sobre el crecimiento y demografía de vástagog de *Sporobolus cubensis. Acta Oecologica, Oecologia Generalis*, **8**: 391–401.

Canales, J., Trevisan, M.C., Silva, J.F. and Caswell, H. (1994). A demographic study of an annual grass (*Andropogon brevifolius* Schwarz) in burnt and unburnt savanna. *Acta Oecologica*, **15**(3): 261–273.

Carter, J.O. and Cowan, D.C. (1993). Population dynamics of prickly acacia, *Acacia nilotica* subsp. *indica* (Mimosaceae). In: Delfosse, E.S. (ed.), *Pests of pastures, weeds, invertebrate and disease pests.* CSIRO, Melbourne, pp. 128–132.

Castellanos, J., Maass, M. and Kummerow, J. (1991). Root biomass of a dry deciduous tropical forest in Mexico. *Plant and Soil*, **131**: 225–228.

Castro, A.A.J.F. (1994a). *Comparação floristico-geográfica (Brazil) e fitossociológica (Piaui-São Paulo) de amostras de cerrado.* Tese de doctorado, Universidade Estadual de Campinas, São Paulo, Brazil.

Castro, A.A.J.F. (1994b). Comparação floristica de espécies do *cerrado. Silvicultura*, **14**: 16–18.

Castro Neves, B.M. and Miranda, H.S. (1996). Efeitos do fogo no regime térmico do solo de um *campo sujo* de *cerrado.* In Miranda, H.S., Saito, C.H. and Dias, B.F. de S. (eds), *Impactos de queimadas em áreas de cerrado e restinga.* Universidade de Brasilia, Brasilia, pp. 20–30.

Catchpool, V.R. (1984). Cultivating the surface soil to renovate a green panic (*Panicum maximum*) pasture on a brigalow soil in southeast Queensland. *Tropical Grasslands*, **18**: 96–99.

Caughley, G. (1976). The elephant problem – an alternative hypothesis. *East African Wildlife Journal*, **14**: 265–283.

Cavalcanti, L.H. (1978). *Efeito das cinzas resultantes da queimada sobre a productividade da estrato herbáceo subarbustivo do cerrado de Emas.* Tese de doctorado, Universidade de São Paulo, São Paulo.

Cavelier, J., Aide, T.M., Santos, C., Eusse, A.M. and Dupuy, J.M. (1998). The savannization of moist forests in the Sierra Nevada de Santa Marta, Colombia. *Journal of Biogeography*, **25**(5): 901–912.

Cavendish, W.P. (1996). *Environmental resources and rural household welfare*. Mimeo, Centre for the Study of African Economies, University of Oxford.

Central Bureau of Statistics (1996). *Statistical Abstract*. Office of Planning and National Development, Kenya.

Cesar, H.L. (1980). *Efeitos da Queima e Corte sobre a vegetação de um campo sujo na Fazenda Agua Limpa, Distrito Federal*. Tese de mestrado, Universidade de Brasilia, Brasilia, DF.

César, J. (1971). *Etude quantitative de la strate herbacee de la savane de Lamto (Côte d'Ivoire)*. Thesis, Universite de Paris.

César, J. and Menaut, J.C. (1974). Le peuplement végétal. In: *Analyse d'un écosystème tropical humide: la savane de Lamto*. Bull. Liaison Chercheurs Lamto, Numéro Spéc., 2.

Chacón-Moreno, E. and Sarmiento, G. (1995). Dinámica del crecimiento y producción primaria de gramínea forrajera tropical, *Panicum maximum* (tipo común), antes diferentes frecuencias de corte. *Turrialba*, **45**(1–2): 8–18.

Chacón-Moreno, E., Rada, F. and Sarmiento, G. (1995). Intercambio gaseoso, nitrógeno foliar y optimización del manejo de *Panicum maximum* (tipo común) sometido a diferentes frecuencias de corte. *Turrialba*, **45**(1–2): 19–26.

Chaloupka, G. (1981b). The traditional movement of a band of Aborigines in Kakadu. In: Stokes, T. (ed.), *Kakadu National Park: education resources. Appendix 1*. ANPWS and Northern Territory Department of Education, Canberra and Darwin.

Champion, H.G. (1936). A preliminary survey of the forest types of India and Burma. *Indian Forestry Records*, **1**: 1–286.

Chantawong, M. (1992). Infrastructure development policy: legalised deforestation. In: Leungaramsri, P. and Rajesh, N. (eds), *The future of people and forests in Thailand after the logging ban*. Project for Ecological Recovery, Bangkok, pp. 137–147.

Chantawong, M., Katesombun, B., Koohacharoen, O., Leungaramsri, P., Malapetch, P., Paisarnpanichkul, D., Thungsuro, K. and Rajesh, N. (1992). People and forests of Thailand: community forests. In: Leungaramsri, P. and Rajesh, N. (eds). *The future of people and forests in Thailand after the logging ban*. Project for Ecological recovery, Bangkok, pp. 151–196.

Cheater, A. (1990). The ideology of 'communal' land tenure in Zimbabwe: mythogenes is enacted? *Africa*, **60**: 188–206.

Cheney, N.P. (1981). Fire behaviour. In: Gill, A.M., Groves, R.H. and Noble, I.R. (eds), *Fire and the Australian biota*. Australian Academy of Science, Canberra, pp. 151–175.

Cheney, N.P., Raison, R.J. and Khanna, P.K. (1980). Release of carbon to the atmosphere in Australian vegetation fires. In: Pearman, G.I. (ed.), *Carbon dioxide and climate: Australian research*. Australian Academy of Science, Canberra, pp. 153–158.

Chidumayo, E.N. (1987a). A shifting cultivation land use system under population pressure in Zambia. *Agroforestry Systems*, **5**: 15–25.

Chidumayo, E.N. (1987b). Woodland structure, destruction and conservation in the Copp. erbelt area of Zambia. *Biological Conservation*, **40**: 89–100.

Chidumayo, E.N. (1987c). A survey of wood stocks for charcoal production in the *miombo* woodlands of Zambia. *Forest Ecology and Management*, **20**: 105–115.

Chidumayo, E.N. (1988a). A re-assessment of effects of fire on *miombo* regeneration in the Zambian Copperbelt. *Journal of Tropical Ecology*, **4**: 361–372.

Chidumayo, E.N. (1988b). Integration and role of planted trees in a bush-fallow cultivation system in central Zambia. *Agroforestry Systems*, **7**: 63–76.

Chidumayo, E.N. (1988c). Estimating fuelwood production and yield in regrowth dry *miombo* woodland in Zambia. *Forest Ecology and Management,* **24**: 59–66.

Chidumayo, E.N. (1989a). Early post-felling response of *Marquesia* woodland to burning in the Zambian Copperbelt. *Journal of Ecology,* **77**: 430–438.

Chidumayo, E.N. (1989b). Land use, deforestation and reforestation in the Zambian Copperbelt. *Land Degradation and Rehabilitation,* **1**: 209–216.

Chidumayo, E.N. (1990). Above-ground woody biomass structure and productivity in a Zambezian woodland. *Forest Ecology and Management,* **36**: 33–46.

Chidumayo, E.N. (1991). Woody biomass structure and utilisation for charcoal production in a Zambian *miombo* woodland. *Bioresources Technology,* **37**: 43–52.

Chidumayo, E.N. (1992). Seedling ecology of two *miombo* woodland trees. *Vegetatio,* **103**: 51–58.

Chidumayo, E.N. (1993a). *Responses of miombo to harvesting: ecology and management.* Stockholm Environment Institute, Stockholm.

Chidumayo, E.N. (1993b). Zambian charcoal production: *miombo* woodland recovery. *Energy Policy,* **12**: 586–597.

Chidumayo, E.N. (1993c). *Wood use in charcoal production in Zambia.* Interim report for World Wildlife Fund (Biodiversity Support Programme), Washington, DC.

Chidumayo, E.N. (1994a). Phenology and nutrition of *miombo* woodland trees in Zambia. *Trees,* **9**: 67–72.

Chidumayo, E.N. (1994b). Effect of wood carbonisation on soil and seedling productivity in *miombo* woodland. *Forest Ecology and Management,* **70**: 353–357.

Chidumayo, E.N. (1997). *Miombo ecology and management. An introduction.* Stockholm Environment Institute, Stockholm.

Chidumayo, E.N. and Chidumayo, S.B.M. (1984). *The status and impact of woodfuel in urban Zambia.* Department of Natural Resources, Lusaka.

Chidumayo, E.N. and Frost, P. (1996). Population biology of *miombo* trees. In: Campbell, B. (ed.), *The miombo in transition: woodlands and welfare in Africa.* Center for International Forestry Research (CIFOR), Bogor, Indonesia, pp. 59–71.

Chidumayo, E.N., Gambiza, J. and Grundy, I. (1996). Managing *miombo* woodlands. In: Campbell, B. (ed.), *The miombo in transition: woodlands and welfare in Africa.* Center for International Forestry Research (CIFOR), Bogor, Indonesia, pp. 175–193.

Chihongo, A.W. (1993). Pilot country study on non-wood forest products for Tanzania. In: *Non-wood forest products: a regional expert consultation for English-speaking African countries.* Arusha, 17–22 October 1993, Annex IV. FAO and the Commonwealth Science Council, Rome and London.

Childes, S.L. (1989). Phenology of nine common woody species in semi-arid, deciduous Kalahari Sand vegetation. *Vegetatio,* **79**: 151–163.

Choquenot, D. and Bowman, D.M.J.S. (1998). Marsupial megafauna, Aborigines and the overkill hypothesis: application of predator-prey models to the question of Pleistocene extinction in Australia. *Global Ecology and Biogeography Letters,* **7**: 167–180.

Chunkao, K. (1969). *The determination of aggregate stability by waterdrop impact in relation to sediment yields from erosion plots at Mae-Huad Forest, Lampang.* Forest Research Bulletin 4, Kasetsart University, Bangkok.

Chunkao, K., Tangtham, N. and Ungkulpakdikul, S. (1971). *Measurements of rainfall in early wet season under hill- and dry-evergreen, natural teak and dry dipterocarp forests of Thailand.* Kog-Ma Watershed Research Bulletin 10, Kasetsart University, Bangkok.

Clark, D.J. (1975). Stone Age man at the Victoria Falls. In: Phillipson, D.W. (ed.), *Mosi-oa-Tunya: a handbook to the Victoria Falls region.* Longman, London, pp. 28–47.

Clark, J.D. (1980). Early human occupation of African savanna environments. In: Harris, D.R. (ed.), *Human ecology in savanna environments.* Academic Press, London, pp. 41–71.

Clark, R.L. (1983). Pollen and charcoal evidence for the effects of Aboriginal burning on the vegetation of Australia. *Archaeology in Oceania,* **18**: 32–37.

Clarke, A.L. (1986). The impact of Australian practices on Australian soils: cultivation. In: Russell, J.F. and Isbell, R.F. (eds), *Australian soils: the human impact.* The University of Queensland Press, Brisbane, pp. 273–303.

Clarke, J., Cavendish, W. and Coote, C. (1996). Rural households and *miombo* woodlands: use, value and management. In: Campbell, B. (ed.), *The miombo in transition: woodlands and welfare in Africa.* Center for International Forestry Research (CIFOR), Bogor, Indonesia, pp. 101–135.

Clauss, B. (1991). *Bees and beekeeping in the North-Western Province of Zambia.* Mission Press, Ndola, Zambia.

Clements, F.E. (1916). *Plant succession: an analysis of the development of vegetation.* Publication 242, Carnegie Institute, Washington, DC.

Cline-Cole, R.A. (1996). Dryland forestry: manufacturing forests and farming trees in Nigeria. In: Leach, M. and Mearns, R. (eds), *African issues – the lie of the land: challenging received wisdom on the African continent.* International African Institute, London, pp. 122–139.

Cline-Cole, R.A., Main, H.A.C. and Nichol, J.E. (1990). On fuelwood consumption, population dynamics and deforestation in Africa. *World Development,* **18**(4): 513–527.

Clutton-Brock, J. (1995). The spread of domestic animals in Africa. In: Shaw, T., Sinclair, P., Andah, B. and Okpoko, A. (eds), *The archaeology of Africa. Food, metals and towns.* Routledge, London, pp. 61–70.

Cochrane, T.T., Porras, J.A., Henão, M.R. (1988). The relative tendency of the *cerrados* to be affected by *Veranicos*; a provisional assessment. In: *VI Simpósio Sobre o Cerrado. Savannas: Alimento e Energia.* EMBRAPA, Planaltina, DF, pp. 229–239.

Coe M.J., Cumming, D.H.M. and Phillipson, J. (1976). Biomass and production of large African herbivores in relation to rainfall and primary production. *Oecologia,* **22**: 341–354.

Coetzee, B.J., van der Meulen, F., Zwanziger, S., Gonsalves, P. and Weisser, P.J. (1976). A phytosociological classification of the Nylsvley Nature Reserve. *Bothalia,* **12**: 137–160.

Cole, M.M. (1986). *The savannas: biogeography and geobotany.* Academic Press, New York.

Cole, S. (1964). *The prehistory of East Africa.* Weidenfeld and Nicolson, London.

Coley, P.D., Bryant, J.P. and Chapin, F.S. (1985). Resource availability and plant anti-herbivore defence. *Science,* **230**: 895–899.

Collins, G. (1996). Tourism in the tropical savannas. In: Ash, A.J. (ed.), *The future of tropical savannas: an Australian perspective.* CSIRO, Australia, pp. 62–67.

Collins, N.M. (1981). The role of termites in the decomposition of wood and leaf litter in the southern Guinea savanna of Nigeria. *Oecologia,* **51**: 389–399.

Connah, G. (1992). *African civilisations. Precolonial cities and states in tropical Africa: an archaeological perspective.* Cambridge University Press, Cambridge.

Connell, J.H. and Slatyer, R.O. (1977). Mechanisms of succession in natural communities and their role in community stability and organisation. *American Naturalist*, **111**: 1119–1144.

Constantino, R. (1988). *Influência da macrofauna na dinâmica de nutrientes do folhedo em decomposição em cerrado sensu stricto*. Tese de mestrado, Departamento de Biologia Vegetal, Universidade de Brasilia, Brasilia, DF.

Cook, G.D. (1991). Effects of fire regimes on 2 species of epiphytic orchids in tropical savannas of the Northern Territory. *Australian Journal of Ecology*, **16**(4): 537–540.

Cook, G.D. (1992). The effects of fire on nutrient losses from Top End savannas. In: Moffatt, I. and Webb, A. (eds), *Conservation and development issues in northern Australia*. North Australia Research Unit, Darwin, pp. 123–129.

Cook, G.D. (1994). The fate of nutrients during fires in a tropical savanna. *Australian Journal of Ecology*, **19**: 359–365.

Cook, G.D. and Andrew, M.H. (1991). The nutrient capital of indigenous *Sorghum* species and other understorey components of savannas in northwestern Australia. *Australian Journal of Ecology*, **16**: 375–384.

Cook, G.D., Hurst, D. and Griffith, D. (1995). Atmospheric trace gas emissions from tropical Australian savanna fires. *CALMScience Supplement*, **4**: 123–128.

Coombes, H.C., Dargavel, J., Kesteven, J., Ross, H., Smith, D.I. and Young, E. (1990). *The promise of the land: sustainable use by Aboriginal communities*. Centre for Resource and Environmental Studies, Australian National University, Canberra.

Cooper, S.M. (1985). *Factors influencing the utilisation of woody plants and forbs by ungulates*. PhD Thesis, University of Witwatersrand, Johannesburg.

Cooper, S.M. and Owen-Smith, N. (1985). Condensed tannins deter feeding by browsing ungulates in a South African savanna. *Oecologia*, **67**: 142–146.

Cooper, S.M. and Owen-Smith, N. (1986). Effects of plant spinescence on large mammalian herbivores. *Oecologia*, **68**: 446–455.

Correa, M.P. (1984). *Dicionario das Plantas uteis do Brasil*. Ministerio da Agricultura, Instituto Brasileira de Desenvolvimento Florestal, Brasilia, DF.

Coughenour, M.B. and Ellis, J.E. (1993). Landscape and climatic control of woody vegetation in a dry tropical ecosystem: Turkana District, Kenya. *Journal of Biogeography*, **20**: 383–398.

Coutinho, L.M. (1976). *Contribuição ao conhecimento do papel ecológico das queimadas na floração de espècies do cerrado*. Tese de ivre-docente, Universidade de São Paulo, São Paulo.

Coutinho, L.M. (1977). Aspectos ecológicos do fogo no *cerrado*. II – as queimadas e a dispersão de sementes em algumas espècies anemocóricas do estrato herbáceo-subarbustivo. *Boletim Botânica, Univ. S. Paulo*, **5**: 57–64.

Coutinho, L.M. (1978a). O conceito de *cerrado*. *Revista Brasileira de Botânica*, **1**: 17–23.

Coutinho, L.M. (1978b). Aspectos ecológicos do fogo no *cerrado*. I – A temperatura do solo durante as queimadas. *Revista Brasileira de Botânica*, **1**(2): 93–96.

Coutinho, L.M. (1979). Aspectos ecológicos do fogo no *cerrado*. III – A precipitação atmosférica de nutrientes minerais. *Revista Brasileira de Botânica*, **2**(2): 97–101.

Coutinho, L.M. (1982a). Ecological effects of fire in Brazilian *Cerrado*. In: Huntley, B.J. and Walker, B.H. (eds), *Ecology of tropical savannas*. Ecological Studies 42, Springer-Verlag, Berlin, pp. 273–291.

Coutinho, L.M. (1982b). Aspectos ecológicos da saúva no *cerrado* – a saúva, as queimadas a sua possivel relação na ciclagem de nutrientes minerais. *Boletim de Zoologia, Univ. São Paulo*, **8**: 1–9.

Coutinho, L.M. (1990). Fire in the ecology of the Brazilian *Cerrado*. In: Goldammer, J.G. (ed.), *Fire in the tropical biota*. Ecological Studies 84, Springer-Verlag, Berlin, pp. 82–105.

Coutinho, L.M. and Jurkewics, I.R. (1978). Aspectos ecológicos do fogo no *cerrado*. V – O efeito de altas temperaturas ne germinação de uma espécie de *Mimosa*. *Ciência e Cultura*, **30** (resumos): 420.

Coutinho, L.M., Pagano, S.N. and Sartori, A.A. (1978). Sobre o teor de água e nutrientes minerais em xilopódios de algumas espécies de *cerrado*. *Ciência e Cultura*, **30** (resumos): 349–350.

Coventry, R.J., Holt, J.A. and Sinclair, D.F. (1988). Nutrient cycling by mound-building termites in low fertility soils of semi-arid tropical Australia. *Australian Journal of Soil Research*, **26**: 375–390.

Cowling, R.M., Gibbs Russell, G.E., Hoffman, M.T. and Hilton Taylor, C. (1989). Patterns of plant species diversity in southern Africa. In: Huntley, B.J. (ed.), *Biotic diversity in Southern Africa: concepts and conservation*. Oxford University Press, Cape Town, pp. 19–50.

Crawford, D.N. (1979). Effects of grass and fires on birds in the Darwin area. *Emu*, **10**: 150–152.

Cronin, G. (1998). Between-species and temporal variation in *Acacia*–ant–herbivore interactions. *Biotropica*, **30**(1): 135–139.

Cronk, Q.C.B. and Fuller, J.L. (1995). *Plant invaders. The threat to natural ecosystems*. Chapman and Hall, London.

Croze, H. (1974a). The Seronera bull problem. II The trees. *East African Wildlife Journal*, **12**: 29–47.

Croze, H. (1974b). The Seronera bull problem. I The bulls. *East African Wildlife Journal*, **12**: 1–27.

Crutzen, P.J. (1988). Tropospheric ozone: an overview. In: Isaksen, I. (ed.), *Tropospheric ozone: regional and global scale interactions*. Reidel, Norwell, MA, pp. 3–32.

Crutzen, P.J. and Andreae, M.O. (1990). Biomass burning in the tropics: impacts on atmospheric chemistry and biogeochemical cycles. *Science*, **250**: 1669–1678.

Crutzen, P.J., Delany, A.C., Greenberg, J. Haagenson, P., Heidt, L., Lueb, R., Pollock, W., Seiler, W., Wartburg, A. and Zimmerman, P. (1985). Tropospheric chemical composition measurements in Brazil during the dry season. *Journal of Atmospheric Chemistry*, **2**: 233–256.

Cumming, D.H.M., Du Toit, R.F. and Stuart, S.N. (1990). *African elephants and rhinos: status survey and conservation action plan*. IUCN, Gland, Switzerland.

Cunha, A.S, Mueller, C.C., Alves, E.R.A. and da Silva, J.E. (1994). *Uma avaliação da sustentabilidade da agricultural nos cerrados*. Instituto de Pesquisa Economica Aplicada (IPEA) e Programa das Nações Unidas para o Desenvolvimento (PNUD), Brasilia, DF.

Dabadghao, P.M. and Shankarnarayan, K.A. (1973). *The grass cover of India*. ICAR, New Delhi.

Dahlberg, A. (1993). *The degradation debate: is clarification possible?* Overseas Development Institute, Pastoral Development Network Paper 35c, pp. 10–14.

Dall'Aglio, C.G. (1992). *Estabilidade de comunidade de cerrado em relação ao fogo: assimetria de impacto em guildas de aranhas*. Tese de mestrado, Universidade de Brasilia, Brasilia, DF.

Dangerfield, J.M., Perkins, J.S. and Kaunda, S.K. (1996). Shoot characteristics of *Acacia tortilis* (Forsk) in wildlife and rangeland habitats of Botswana. *African Journal of Ecology*, **34**(2): 167–176.

Dansereau, P. (1957). *Biogeography: an ecological perspective.* Ronald Press, New York.

D'Antonio, C.M. and Vitousek, P.M. (1992). Biological invasions by exotic grasses, the grass/fire cycle, and global change. *Annual Review of Ecological Systematics,* **23**: 63–87.

Darkoh, M.B.K. (1998). The nature, causes and consequences of desertification in the drylands of Africa. *Land Degradation and Development,* **9**(1): 1–20.

Darling, F.F. (1960). *An ecological reconnaissance of the Mara Plains in Kenya colony.* Wildlife Monograph No. 5, The Wildlife Society.

Davidson, B. (1990). *Modern Africa. A social and political history.* (2nd edition). Longman, Harlow.

Davies, J.R., Hoare, J.R.L. and Nanninga, P.M. (1986). Developing a fire management expert system for Kakadu National Park, Australia. *Journal of Environmental Management,* **22**: 215–227.

Davies, R.G. (1997). Termite species richness in fire-prone and fire-protected dry deciduous dipterocarp forest in Doi Suthep-Pui National Park, northern Thailand. *Journal of Tropical Ecology,* **13**: 153–160.

Davies, S. (1996). *Adaptable livelihoods. Coping with food insecurity in the Malian Sahel.* Macmillian, London.

Day, T.A. and Detling, J.K. (1990). Grassland patch dynamics and herbivore grazing preference following urine deposition. *Ecology,* **71**: 180–188.

De Almeida, S.P. (1995). Grupos fenológicos da comunidade de gramíneas perenes de um campo *cerrado* no Distrito Federal, Brasil. *Pesquisa Agropecuária Brasileira,* **30**(8): 1067–1073.

De Assis Dansa, C.V. and Duarte Rocha, C.F. (1992). An ant–membracid–plant interaction in a *cerrado* area of Brazil. *Journal of Tropical Ecology,* **8**: 339–348.

De Bie, S. (1991). *Wildlife resources of the West African savanna.* Wageningen Agricultural University Papers No. 91–92, Wageningen Agricultural University, The Netherlands.

De Bie, S., Geerling, C. and Heringa, A. (1987). B-1: Région du Baoulé – utilisation du gibier. In: Geerling, C. and Diakité, M.D. (eds), *Resources sahélo-soudaniennes – rapport final du projet 'Recherche pour l'utilisation rationnelle du gibier au sahel'.* Dir. Nat. des Eaux et Forêts, Bamako/Mali et Dept. de l'Aménagement de la Nature, Univ. Agronomique, Wageningen, Pays Bas.

De Bie, S., Ketner, P., Paasse, M. and Geerling, C. (1998). Woody plant phenology in the West Africa savanna. *Journal of Biogeography,* **25**: 883–900.

De Boer, W.F. and Prins, H.H.T. (1990). Large herbivores that strive mightily but eat and drink like friends. *Oecologia,* **82**: 264–274.

De Castro, E.A. and Kauffman, J.B. (1998). Ecosystem structure in the Brazilian *cerrado*: a vegetation gradient of aboveground biomass, root mass and consumption by fire. *Journal of Tropical Ecology,* **14**: 263–283.

De Leeuw, P.N. (1979). *A review of the ecology and fodder resources of the sub-humid zone.* Paper ILCA Symposium Intensification of livestock production in the sub-humid tropics of West Africa. Kaduma, 23–30 March 1979.

De Queiroz, J.S. (1993). *Range degradation in Botswana: myth or reality?* Overseas Development Institute, Pastoral Development Network Paper, 35b, pp. 1–17.

De Souza, M.A.A. (1994). Relação entre as atividades ocupacionais e a qualidade de água. In: Novaes-Pinto, M. (ed.), *Cerrado, Caracterização, Ocupação e Perspectivas* (2nd edition). Editora Universidade de Brasilia, Brasilia, DF, pp. 189–212.

De Wit, H.A. (1978). *Soils and grassland types of the Serengeti Plains (Tanzania)*. PhD Thesis, Agricultural University, Wageningen.

DeAngelis, D.L. and Waterhouse, J.C. (1987). Equilibrium and non-equilibrium concepts in ecological models. *Ecological Monographs*, **57**: 1–21.

Dedecek, R.A. (1986). *Erosão e práticas conservacionistas nos cerrados*. Circular Técnica da EMBRAPA/CPAC No. 22.

Dedecek, R.A., Resck, D.V.S. and Freitas Jr., E. (1986). Perdas de solo, água e nutrientes por erosão em latossolo vermelho-escuro dos *cerrados* em diferentes cultivos sob chuva natural. *Revista Brasileira de Ciência do Solo*, **10**: 265–272.

Defoliart, G.R. (1995). Edible insects as minilivestock. *Biodiversity and Conservation*, **4**: 306–321.

Del-Claro, K., Berto, V. and Réu, W. (1996). Effect of herbivore deterrence by ants on the fruit set of an extrafloral nectary plant, *Qualea multiflora* (Vochysiaceae). *Journal of Tropical Ecology*, **12**: 887–892.

Delcourt, H.R. and Delcourt, P.A. (1988). Quaternary landscape ecology: relevant scales in space and time. *Landscape Ecology*, **2**: 23–44.

Delmas, R.A., Marenco, A., Tathy, J.P., Cros, B. and Baudet, J.G.R. (1991). Sources and sinks of methane in the African savanna: CH_4 emissions from biomass burning. *Journal of Geophysical Research*, **96**: 7287–7299.

Denbow, J.R. (1979). *Cenchrus ciliaris*: an ecological indicator of Iron Age middens using aerial photography in Eastern Botswana. *South African Journal of Science*, **75**: 405–408.

Denbow, J.R. (1984). Prehistoric herders and foragers of the Kalahari: the evidence for 1500 years of interaction. In: Schrire, C. (ed.), *Past and present in hunter gatherer studies*. Academic Press, London, pp. 175–193.

Department of Energy (1992). *Energy statistics bulletin: 1974–1990*. Ministry of Energy and Water Development, Lusaka.

Desanker, P.V. (1996). Development of a *miombo* woodland dynamics model in Zambezian Africa using Malawi as a case study. *Climatic Change*, **34**: 279–288.

Desanker, P.V. and Prentice, I.C. (1994). MIOMBO – A vegetation dynamics model for the *miombo* woodlands of Zambezian Africa. *Forest Ecology and Management*, **69**: 87–95.

Desanker, P.V., Frost, P.G.H., Frost, C.O., Justice, C.O. and Scholes, R.J. (eds), (1997). *The Miombo Network: framework for a terrestrial transect study of land-use and land-cover change in the miombo ecosystems of Central Africa*. IGBP Report 41, The International Geosphere–Biosphere Programme (IGBP), Stockholm, Sweden.

Deshmukh, I. (1984). A common relationship between precipitation and grassland peak biomass for east and southern Africa. *African Journal of Ecology*, **22**: 181–186.

Dewees, P.A. (1989). The woodfuel crisis reconsidered: observations on the dynamics of abundance and scarcity. *World Development*, **17**(8): 1159–1172.

Dewees, P.A. (1993). *Economic dimensions to the use and management of trees and woodlands in small-holder agriculture in Malawi*. Background paper prepared for the Malawi National Forest Policy Review, World Bank, Washington, DC.

Dewees, P.A. (1994). *Social and economic aspects of miombo woodland management in Southern Africa. Options and opportunities for research*. CIFOR Occasional Paper 2, Bogor, Indonesia.

Dias, B.F. de S. (1994). A conservação da natureza. In: Novaes-Pinto, M. (ed.), *Cerrado. Caracterização, Ocupação e Perspectivas* (2nd edition). Editora Universidade de Brasilia, Brasilia, DF, pp. 607–664.

Dias, B.F. de S. (1997). *Perspectiva histórica do fogo no cerrado*. Unpublished manuscript.

Dias, V.L.B. (1993). *Impactos do fogo sobre cupins, construtores de ninhos epigeos no cerrado.* Tese de mestrado, Universidade de Brasilia, Brasilia, DF.

Dinerstein, E. and Wikramanayake, E.D. (1993). Beyond 'hotspots': how to prioritise investments to conserve biodiversity in the Indo-Pacific region. *Conservation Biology,* 7(1): 53–65.

Dingwall, P.R. (1992). Tourism in protected areas: conflict or saviour. *Proceedings of the Royal Australian Institute of Parks and Recreation,* **28**: 117–122.

Diniz de Araújo Neto, M., Furley, P.A., Haridasan, M. and Johnson, C.E. (1986). The *murundus* of the *cerrado* region of central Brazil. *Journal of Tropical Ecology,* **2**: 17–35.

Dionello, S.B. (1978). *Germinação de sementes e desenvolvimento de plântulas de Kielmeyera coriacea Mart.* Tese de DSc, Universidade de São Paulo, São Paulo.

Dowling, R.K. (1993). Tourist and resident perceptions of the environment-tourism relationship in the Gascoyne Region, Western Australia. *GeoJournal,* **29**: 243–251.

Du Toit, J.T. (1988). *Patterns of resource use within the browsing ruminant guild in the central Kruger National Park.* PhD Thesis, University of Witwatersrand, Johannesburg.

Du Toit, J.T. (1990). Giraffe feeding on *Acacia* flowers: predation or pollination? *African Journal of Ecology,* **28**: 63–68.

Dublin, H.T. (1986). *Decline of the Mara woodlands: the role of fire and elephants.* PhD Thesis, University of British Columbia.

Dublin, H.T. (1991). Dynamics of the Serengeti–Mara woodlands: an historical perspective. *Forest Conservation and History,* **35**: 169–178.

Dublin, H.T. (1995). Vegetation dynamics in the Serengeti–Mara ecosystem: the role of elephants, fire and other factors. In: Sinclair, A.R.E. and Arcese, P. (eds), *Serengeti II. Dynamics, management, and conservation of an ecosystem.* The University of Chicago Press, Chicago, pp. 71–90.

Dublin, H.T. and Douglas-Hamilton, I. (1987). Status and trends of elephants in the Serengeti–Mara ecosystem. *African Journal of Ecology,* **25**: 19–33.

Dublin, H.T., Sinclair, A.R.E. and McGlade, J. (1990a). Elephants and fire as causes of multiple stable states in the Serengeti–Mara woodlands. *Journal of Animal Ecology,* **59**: 1147–1164.

Dublin, H.T., Sinclair, A.R.E., Boutin, S., Anderson, E., Jago, M. and Arcese, P. (1990b). Does competition regulate ungulate populations? Further evidence from Serengeti, Tanzania. *Oecologia,* **82**: 283–288.

Duff, G.A. and Braithwaite, R.W. (1990). Fire and Top End forests – past, present and future research. In: Roberts, B.R. (ed.), *Fire research in rural Queensland.* University of South Queensland, Toowoomba, pp. 84–98.

Duff, G.A., Myers, B.A., Williams, R.J., Eamus, D., Fordyce, I. and O'Grady, A. (1997). Seasonal patterns in canopy cover and microclimate in a tropical savanna near Darwin, northern Australia. *Australian Journal of Botany,* **45**: 211–224.

Duffy, R. (1997). The environmental challenge to the nation-state: superparks and national parks policy in Zimbabwe. *Journal of Southern African Studies,* **23**(3): 441–451.

Dunlop, C.R. and Webb, L.J. (1991). Flora and vegetation. In: Haynes, C.D., Ridpath, M.G. and Williams, M.A.J. (eds), *Monsoonal Australia. Landscape, ecology and man in the northern lowlands.* A.A. Balkema, Rotterdam, pp. 41–59.

Dwyer, E., Gregoire, J.M. and Malingreau, J.P. (1998). A global analysis of vegetation fires using satellite images: spatial and temporal dynamics. *Ambio,* **27**(3): 175–181.

Dyer, C. (1980). *The development of stomata in Ochna pulchra Burch ex DC., Terminalia sericea Hook and Burkea africana Hook.* BSc Honours Report, Botany Department, University of Witwatersrand, Johannesburg.

Eagleson, P.S. and Segarra, R.I. (1985). Water-limited equilibrium of savanna vegetation systems. *Water Resources Research*, **21**: 1483–1493.

Economic Survey (1995). *Kenya Government*. Government Press, Nairobi.

Eden, M.J. (1964). The savanna ecosystem – northern Rupununi, British Guyana. *McGill University Savanna Research Service*, **1**: 1–216.

Eden, M.J. (1974). Palaeoclimatic influences and the development of savanna in southern Venezuela. *Journal of Biogeography*, **1**: 95–109.

Edroma, E.L. (1981). The role of grazing in maintaining high species composition in *Imperata* grassland in Rwenzori National Park, Uganda. *African Journal of Ecology*, **19**: 215–233.

Egler, I. and Haridasan, M. (1987). Alteration of soil properties by *Procornitermes araujoii* (Isoptera, Termitidae) in latosols of the *cerrado* region of Brazil. In: San Jose, J.J. and Montes, R. (eds), *La capacidad bioproductiva de sabanas*. IVIC, Caracas, pp. 280–308.

Ehrlich, D., Lambin, E.F. and Malingreau, J.P. (1997). Biomass burning and broad-scale land-cover changes in Western Africa. *Remote Sensing of Environment*, **61**(2): 201–209.

Eiten, G. (1972). The *cerrado* vegetation of Brazil. *The Botanical Review*, **38**(2): 201–341.

Eiten, G. (1975). The vegetation of the Serra do Roncador. *Biotropica*, **7**: 112–135.

Eiten, G. (1978). Delimitation of the *cerrado* concept. *Vegetatio*, **36**: 169–178.

Eiten, G. (1982). Brazilian 'Savannas'. In: Huntley, B.J. and Walker, B.H. (eds), *Ecology of tropical savannas*. Ecological Studies 42, Springer-Verlag, Berlin, pp. 25–47.

Eiten, G. (1994). Vegetação. In: Novaes-Pinto, M. (ed.), *Cerrado. Caracterização, Ocupação e Perspectivas* (2nd edition). Editora Universidade de Brasilia, Brasilia, DF, pp. 17–73.

Eiten, G. and Goodland, R. (1979). Ecology and management of semiarid ecosystems in Brazil. In: Walker, B.H. (ed.), *Management of semiarid ecosystems*. Elsevier, Amsterdam, pp. 277–300.

Elbow, K., Furth, R., Knox, A., Bohrer, K., Hobbs, M., Leisz, S. and Williams, M. (1998). Synthesis of trends and issues raised by land tenure country profiles of West African countries, 1996. In: Bruce, J.W. (ed.), *Country profiles of land tenure: Africa, 1996.* Research Paper No. 130, Land Tenure Centre, University of Wisconsin, Madison, pp. 2–18.

Ellis, J.E. and Swift, D.M. (1988). Stability of African pastoral ecosystems: alternate paradigms and implications for development. *Journal of Range Management*, **41**: 450–459.

Epstein, T. (1972). *Mineral nutrition of plants: principles and perspectives*. John Wiley, New York.

Ernst, W. (1988). Seed and seedling ecology of *Brachystegia spiciformis*, a predominant tree component in *miombo* woodlands in south central Africa. *Forest Ecology and Management*, **25**: 195–210.

Ewing, G. (1996). Sustainable mining in Australia's tropical savannas. In: Ash, A.J. (ed.), *The future of tropical savannas: an Australian perspective*. CSIRO, Australia, pp. 80–87.

Fanshawe, D.B. (1968). The vegetation of Zambian termitaria. *Kirkia*, **6**: 169–179.

Fanshawe, D.B. (1971). *The vegetation of Zambia*. Government Printer, Lusaka.

FAO (1999). *Agricultural production statistics*, FAO, Rome.

Fariñas, M. and San José, J.J. (1985). Cambios en el estrato herbáceo de una parcela de sabana protegida del fuego y del pastoreo durante 20 años. *Acta Científica Venezolana*, **36**: 199–200.

Fariñas, M. and San José, J.J. (1987). Efectos de la supresión del fuego y el pastoreo sobre la composición de una sabana de *Trachypogon* en los *Llanos* del Orinoco. In: San José, J.J. and Montes, R. (eds), *La capacidad bioproductiva de sabanas*. Centro Internacional de Ecología Tropical, UNESCO-IVIC, Caracas, pp. 513–545.

Farji Brener, A.G. and Silva, J.F. (1995a). Leaf-cutting ants and forest groves in a tropical parkland savanna of Venezuela: facilitated succession? *Journal of Tropical Ecology*, **11**: 651–669.

Farji Brener, A.G. and Silva, J.F. (1995b). Leaf-cutting ant nests and soil fertility in a well-drained savanna in Western Venezuela. *Biotropica*, **27**(2): 250–253.

Felfili, J.M. and Silva Jr, M.C. (1988). Distribuição dos diametros numa faixa de *cerrado* na Fazenda Agua Limpa (FAL) in Brasilia-D.F. *Acta Botânica Brasilica*, **2**(1–2): 85–104.

Felfili, J.M. and Silva Jr, M.C. (1993). A comparative study of *cerrado* (*sensu stricto*) vegetation in Central Brazil. *Journal of Tropical Ecology*, **9**: 277–289.

Fensham, R.J. (1990). Interactive effects of fire frequency and site factors in tropical eucalypt forests. *Australian Journal of Ecology*, **15**: 255–266.

Fensham, R.J. (1994). Phytophagous insect–woody sprout interactions in tropical eucalypt forest. II. Insect community structure. *Australian Journal of Ecology*, **19**: 189–196.

Fensham, R.J. (1997). Aboriginal fire regimes in Queensland, Australia: analysis of the explorers' record. *Journal of Biogeography*, **24**: 11–22.

Fensham, R.J. and Bowman, D.M.J.S. (1992). Stand structure and the influence of overwood on regeneration in tropical eucalypt forest on Melville Island. *Australian Journal of Botany*, **40**: 335–352.

Fensham, R.J. and Cowie, A.D. (1998). Alien plant invasions on the Tiwi Islands. Extent, implications and priorities for control. *Biological Conservation*, **83**(1): 55–68.

Fensham, R.J. and Kirkpatrick, J.B. (1992). Soil characteristics and tree species distribution in the savanna of Melville Island, Northern Territory. *Australian Journal of Botany*, **40**: 311–333.

Ferrar, P. (1982a). Termites of a south African savanna. I. List of species and subhabitat preferences. *Oecologia*, **52**: 125–132.

Ferrar, P. (1982b). *The termites of the Savanna Ecosystem Project study area, Nylsvley*. South African National Scientific Programmes Report 60, CSIR, Pretoria.

Ferrar, P. (1982c). Termites of a South African savanna. IV. Subterranean populations, mass determinations and biomass estimations. *Oecologia*, **52**: 147–151.

Ferrar, P. (1982d). Termites of a South African savanna. III. Comparative attack on toilet roll baits in subhabitats. *Oecologia*, **52**: 139–146.

Ferri, M.G. (1944). Transpiração de plantas permanentes dos 'cerrados'. *Univ. São Paulo, Fac. Filosofia, Ciências Letras*, **41**(4): 155–224.

Ferri, M.G. (1973). Sobre a origem, a manutenção e a transformação dos *cerrados*, tipos de savana do Brasil. *Revta Biológica*, **9**: 1–13.

Ffolliott, P.F., Fisher, J.T., Sachs, M., DeBoer, D.W., Dawson, J.O. and Fulbright, T.E. (1998). Role of demonstration projects in combating desertification. *Journal of Arid Environments*, **39**: 155–163.

Figueiredo, S.V. and Cavalcanti, R.B. (1992). Efeitos do fogo sobre a avifauna do *cerrado*. *Resumes II Congresso Brasiliera Ornitologia, UFMS*, 39. Mato Grosso do Sul, Campo Grande.

Filgueiras, T.S. and Pereira, B.A. da S. (1994). Flora do Distrito Federal. In: Novaes-Pinto, M. (ed.), *Cerrado, Caracterização, Ocupação e Perspectivas* (2nd edition). Editora Universidade de Brasilia, Brasilia, DF, pp. 345–404.

Fischer, F.U. (1993). *Beekeeping in the subsistence economy of the miombo savanna woodlands of south-central Africa*. Rural Development Forestry Network Paper 15c, Overseas Development Institute, London, pp. 1–12.

Fishman, J., Watson, C.E., Larsen, J.C. and Logan, J.A. (1990). Distribution of tropospheric ozone determined from satellite data. *Journal of Geophysical Research*, **95**: 3599–3617.

Fishman, J., Fakhruzzman, K., Cros, B. and Nganga, D. (1991). Identification of widespread pollution in the southern hemisphere deduced from satellite analyses. *Science*, **252**: 1693–1697.

Flenley, J.R. (1979). *The equatorial rain forest: a geological history*. Butterworth, London.

Flenley, J.R. (1982). The evidence for ecological change in the tropics. *The Geographical Journal*, **148**(1): 8–21.

Foldats, E. and Rutkis, E. (1965). Influencia mecánica del suelo sobre la fisonomía de algunas sabanas del *llano* venezolano. *Boletin Sociedad Venezolana Ciencias Naturales*, **108**: 335–392.

Foldats, E. and Rutkis, E. (1975). Ecological studies of chaparro (*Curatella americana* L.) and manteco (*Byrsonima crassifolia* HBK) in Venezuela. *Journal of Biogeography*, **2**: 159–178.

Foley, G. (1987). *The energy question*. Penguin, London.

Ford, J. (1971). *The role of trypanosomiasis in African ecology*. Clarendon Press, Oxford.

Fordyce, B. (1980). *The prehistory of Nylsvley*. Progress report to the National Programme for Environmental Science, CSIR, Pretoria.

Forest Research Institute (1976). *Indian timbers: gurjan*. Information Series 19, Forest Research Institute, Dehra Dun, India.

Foster, T. (1976). *Bushfire: history, prevention, control*. Reed, Sydney.

Franca, H. and Setzer, A.W. (1998). AVHRR temporal analysis of a savanna site in Brazil. *International Journal of Remote Sensing*, **19**(16): 3127–3140.

Freeland, W.J. (1983). Parasites and the co-existence of animal host species. *American Naturalist*, **121**: 223–236.

Freeland, W.J. (1991). Large herbivorous mammals: exotic species in northern Australia. In: Werner, P.A. (ed.), *Savanna ecology and management. Australian perspectives and intercontinental comparisons*. Blackwell, Oxford, pp. 101–105.

Freeman, H.A. and Smith, J. (1996). Intensification of land use and the evolution of agricultural systems in the West African northern Guinea savanna. *Zeitschrift Fur Auslandische Landwirtschaft*, **35**(2): 109–124.

Freire, E. M. da S. (1979). *Influència das propriedades do solo na distribuição de comunidades de vegetação em uma toposeqüência, em área da 2ª Superficie de erosão do planalto central Brasileiro, na Fazenda Agua Limpa – D.F.* Tese de mestrado, Universidade de Brasilia, Brasilia, DF.

Freitas, F.G. and Silveira, C.O. da (1977). Principais solos sob vegetação de *cerrado* e sua aptidão agricola. In: Ferri, M.G. (ed.), *IV Simpósio sobre o Cerrado*. Editora da Universidade de São Paulo, São Paulo, pp. 155–194.

French, D. (1986). Confronting an unsolvable problem: deforestation in Malawi. *World Development*, **14**(4): 531–540.

Freudenberger, M.S., Carney, J.A. and Lebbie, A.R. (1997). Resiliency and change in common property regimes in West Africa: the case of the Tongo in The Gambia, Guinea, and Sierra Leone. *Society and Natural Resources*, **10**: 383–402.

Frost, P.G. (1984). The responses and survival of organisms in fire-prone environments. In: Booysen, P. de V. and Tainton, N.M. (eds), *Ecological effects of fire in south African ecosystems*. Ecological Studies 48, Springer-Verlag, Berlin, pp. 273–309.

Frost, P.G. (1985). Organic matter and nutrient dynamics in a broadleafed African savanna. In: Tothill, J.C. and Mott, J.J. (eds), *Ecology and management of the world's savannas*. Australian Academy of Science, Canberra, pp. 200–206.

Frost, P.G. (1996). The ecology of *miombo* woodlands. In: Campbell, B. (ed.), *The miombo in transition: woodlands and welfare in Africa*. Center for International Forestry Research (CIFOR), Bogor, Indonesia, pp. 11–57.

Frost, P.G. and Robertson, F. (1987). Effects of fire in savannas. In: Walker, B.H. (ed.), *Determinants of tropical savannas*. IRL Press, Oxford, pp. 93–140.

Frost, P.G., Medina, E., Menaut, J.C., Solbrig, O., Swift, M. and Walker, B. (eds), (1986). *Responses of savannas to stress and disturbance*. Biology International Special Issue 10, IUBS, Paris.

Fryxell, J.M., Greever, J. and Sinclair, A.R.E. (1988). Why are migratory ungulates so abundant? *American Naturalist*, **131**: 781–798.

Fuelberg, H.E., Loring Jr., R.O., Watson, M.V., Sinha, M.C., Pickering, K.E., Thompson, A.M. and McNamara, D.P. (1996). Trace A trajectory intercomparison 2, isentropic and kinematic methods. *Journal of Geophysical Research-Atmospheres*, **101**(D19): 23 927–23 939.

Furley, P.A. (1985). *Notes on the soils and plant communities of Fazenda Agua Limpa (Brasilia, D.F., Brazil)*. Department of Geography Occasional Publications, NS, No. 5, University of Edinburgh.

Furley, P.A. and Ratter, J.A. (1988). Soil resources and plant communities of the central Brazilian *cerrado* and their development. *Journal of Biogeography*, **15**: 97–108.

Furstenburg, D. and Vanhoven, W. (1994). Condensed tannin as anti-defoliate agent against browsing by giraffe (*Giraffa camelopardalis*) in the Kruger National Park. *Comparative Biochemistry and Physiology A – Physiology*, **107**(2): 425–431.

Gandar, M.V. (1982a). Description of a fire and its effects in the Nylsvley Nature Reserve: a synthesis report. *South African National Scientific Report Series*, **63**: 1–39.

Gandar, M.V. (1982b). Trophic ecology and plant/herbivore energetics. In: Huntley, B.J. and Walker, B.H. (eds), *Ecology of tropical savannas*. Ecological Studies 42, Springer-Verlag, Berlin, pp. 514–534.

Gandar, M.V. (1982c). The dynamics and trophic ecology of grasshoppers (Acridoidea) in a South African savanna. *Oecologia*, **54**: 370–378.

Ganjanapan, S. (1996). A comparative study of indigenous and scientific concepts in land and forest classification in northern Thailand. In: Hirsch, P. (ed.), *Seeing forests for trees. Environment and environmentalism in Thailand*. National Thai Studies Centre, Australian National University and Asia Research Centre on Social, Political and Economic Change, Murdoch University, Silkworm Books, Chiang Mai, pp. 247–267.

Garcia, E.A.C. (1995). Desenvolvimento econômico sustentável do *cerrado*. *Pesquisa Agropecuária Brasileira*, **30**(6): 759–774.

Gardner, C.J., McIvor, J.G. and Williams, J. (1990). Dry tropical rangelands: solving one problem and creating another. *Proceedings of the Ecological Society of Australia*, **16**: 279–286.

Geerling, C. (1982). *Guide de terrain des ligneux sahéliens et soudano-guinéens*. Meded. Landbouwhogeschool Wageningen No. 82–83.

Geerling, C. (1985). The status of the woody plant species of the Sudan and Sahel zones in West Africa. *Forest Ecology and Management,* **13**: 247–255.

Georgiadis, N.G. and McNaughton, S.J. (1990). Elemental and fibre contents of savanna grasses: variation with grazing, soil type, season and species. *Journal of Applied Ecology,* **27**: 623–634.

Ghazoul, J. (1997). The pollination and breeding system of *Dipterocarpus obtusifolius* (Dipterocarpaceae) in dry deciduous forests of Thailand. *Journal of Natural History,* **31**(6): 901–916.

Giaccaria, B. and Heide, A. (1984). *Xavante: povo autêntico*. Editora Salesiana Dom Bosco, São Paulo.

Gibbs, P.E., Leitão Filho, H.F. and Shepherd, G. (1983). Floristic composition and community structure in an area of *cerrado* in SE Brazil. *Flora,* **173**: 433–449.

Gibbs Russell, G.E. (1987). Preliminary floristic analysis of the major biomes in southern Africa. *Bothalia,* **17**: 213–227.

Gignoux, J., Clobert, J. and Menaut, J.C. (1997). Alternative fire resistance strategies in savanna trees. *Oecologia,* **110**: 576–583.

Gil Beroes, R.A. (1976). Producción y manejo de pastos en las sabanas inundables del Alto Apure. *Boletín de la Sociedad Venezolana de Ciencias Naturales,* **32**: 103–114.

Gill, A.M. and Knight, I.K. (1991). Fire measurement. In: Cheney, N.P. and Gill, A.M. (eds), *Proceedings of the conference on bushfire modelling and fire danger rating systems*. CSIRO, Canberra, pp. 137–146.

Gill, A.M., Hoare, J.R.L. and Cheney, N.P. (1990). Fires and their effects in the wet–dry tropics of Australia. In: Goldammer, J.G. (ed.), *Fire in the tropical biota: ecosystem processes and global challenge*. Springer-Verlag, Berlin, pp. 159–178.

Gill, A.M., Moore, P.H.R. and Williams, R.J. (1996). Fire weather in the wet–dry tropics of the World Heritage Kakadu National Park, Australia. *Australian Journal of Ecology,* **21**: 302–308.

Gill, J. (1993). A simple climatology of severe wind gust-producing thunderstorms in the Northern Territory. *Northern Territory Research Papers,* **2**: 97–100.

Gillard, P. and Winter, W.H. (1984). Animal production from *Stylosanthes* based pastures in Australia. In: Stace, H.M. and Edye, L.A. (eds), *The biology and agronomy of Stylosanthes*. Academic Press, London, pp. 408–430.

Gillard, P., Williams, J. and Moneypenny, R. (1989). Clearing trees from Australia's semi-arid tropics – production, economic and long term hydrological changes. *Agricultural Science,* **2**: 34–39.

Gillison, A.N. (1983). Tropical savannas of Australia and the Southwest Pacific. In: Bourlière, F. (ed.), *Tropical savannas. Ecosystems of the world 13*. Elsevier Scientific Publishing Company, Amsterdam, pp. 183–243.

Gillison, A.N. (1994). Woodlands. In: Groves, R.H. (ed.), *Australian vegetation*. (2nd edition). Cambridge University Press, Cambridge, pp. 227–255.

Gillon, C. (1983). The fire problem in tropical savannas. In: Bourlière, F. (ed.), *Tropical savannas. Ecosystems of the world 13*. Elsevier Scientific Publishing Company, Amsterdam, pp. 617–641.

Glover, P.E. (1963). The elephant problem at Tsavo. *East African Wildlife Journal,* **1**: 30–39.

Glover, P.E., Trump, E.C. and Wateridge, C. (1964). Termitaria and vegetation patterns on the Loita Plains of Kenya. *Journal of Ecology,* **52**: 367–377.

Goedert, W.J. (1983). Management of the *cerrado* soils of Brazil: a review. *Journal of Soil Science*, **34**: 405–428.

Goldammer, J.G. (1987). Wildfires and forest development in tropical and subtropical Asia: outlook for the year 2000. In: *The Proceedings of the Symposium of Wildfire 2000*. General Technical Report PSW-101. Forest Service, US Department of Agriculture, Berkeley, CA, pp. 164–176.

Goldammer, J.G. (1988). Rural land-use and wildfires in the tropics. *Agroforestry Systems*, **6**: 235–252.

Goldammer, J.G. and Price, C. (1998). Potential impacts of climate change on fire regimes in the tropics based on MAGICC and a GISS GCM-derived lightning model. *Climatic Change*, **39**: 273–296.

Goldammer, J.G. and Siebert, B. (1990). The impact of droughts and forest fires on tropical lowland rain forest of East Kalimantan. In: Goldammer, J.G. (ed.), *Fire in the tropical biota, ecosystem processes and global challenges*. Springer-Verlag, New York, pp. 11–31.

Goldstein, G., Sarmiento, G. and Meinzer, F. (1986). Patrones diarios y estacionales en las relaciones hídricas de árboles siempreverdes de la sabana tropical. *Acta Oecologica, Oecolgia Plantarum*, 7: 107–119.

Goldstein, G., Menaut, J.C., Noble, I. and Walker, B.H. (1988). Exploratory research. In: Walker, B.H. and Menaut, J.C. (eds), *Research procedure and experimental design for savanna ecology and management*. RSSD Publication No. 1, IUBS, Paris, pp. 13–20.

Goncalves, C.W.P. (1995). Formação socio-espacial e questal ambiental no Brasil. In: Becker, B.K., Christofoletti, A., Davidovich, F.R. and Geiger, P.P. (eds), *Geografia e meio ambiente no Brasil*. Editora Hucitec, Rio de Janeiro, pp. 309–333.

Gonzalez Jiménez, E. (1979). Tropical grazing land ecosystems of Venezuela. II. Primary and secondary productivity in flooded savannas. In: *Tropical grazing land ecosystems*. UNESCO Natural Resources Research Report No. 16, UNESCO, Paris, pp. 620–625.

Goodland, R.J.A. (1969). *An ecological study of the cerrado vegetation of south-central Brasil*. PhD Thesis, Department of Botany, McGill University, Montreal, Canada.

Goodland, R.J.A. (1971a). A physiognomic analysis of the '*cerrado* 'vegetation of Central Brazil. *Journal of Ecology*, **59**: 411–419.

Goodland, R.J.A. (1971b). Oligotrofismo e aluminio no *cerrado*. In: Ferri, M.G. (ed.), *III Simpósio sobre o cerrado*. Editora da Universidade de São Paulo, São Paulo, pp. 44–60.

Gragson, T.L. (1997). The use of underground plant organs and its relation to habitat selection among the Pumé Indians of Venezuela. *Economic Botany*, **51**(4): 377–384.

Granier, P. and Cabanis, Y. (1976). Les feux courants et l'élevage en savane soudanienne. *Revue d'Elevage de Médecine Vétérinaire des Pays Tropicaux*, **29**: 267–275.

Gray, B. (1996). What lies ahead for the tropical savanna? Industries and management regimes. In: Ash, A.J. (ed.), *The future of tropical savannas: an Australian perspective*. CSIRO, Australia, pp. 149–158.

Gray, R. and Birmingham, D. (1970). *Pre-colonial African trade*. Oxford University Press, Oxford.

Greene, R.S.B., Chartres, C.J. and Hodgkinson, K.C. (1990). The effects of fire on the soil in a degraded semi-arid woodland. I. Cryptogam cover and physical and micromorphological properties. *Australian Journal of Soil Research*, **28**: 755–777.

Greenway, P.J. and Vesey-Fitzgerald, D.F. (1969). The vegetation of Lake Manyara National Park. *Journal of Ecology*, **57**: 127–149.

Grice, A.C. (1996). Seed production, dispersal and germination in *Cryptostegia grandiflora* and *Ziziphus mauritiana*, two invasive shrubs in tropical woodlands of northern Australia. *Australian Journal of Ecology*, **21**: 324–331.

Griffioen, C. and O'Connor, T.G. (1990). The influence of trees and termite mounds on the soils and herbaceous composition of a savanna grassland. *South African Journal of Ecology*, **1**: 18–26.

Griffiths, A.D. (1998). Probe looks at long-term effects of sulphur dioxide. *Savanna Links*, **5**: 3.

Griffiths, A.D. and Christian, K.A. (1996). The effects of fire on the frillneck lizard (*Chlamydosaurus kingii*) in northern Australia. *Australian Journal of Ecology*, **21**(4): 386–398.

Grossman, D. and Gandar, M.V. (1989). Land transformation in South African savanna regions. *South African Geographical Journal*, **71**: 38–45.

Grossman, D., Ferrar, T.A. and du Plessis, P.C. (1992). Socio-economic factors influencing conservation in South Africa. *TRAFFIC Bulletin*, **13**: 29–31.

Grundy, I.M. (1990). *The potential for management of the indigenous woodland in communal farming areas of Zimbabwe with reference to the regeneration of Brachystegia spiciformis and Julbernardia globiflora*. MSc Thesis, University of Zimbabwe, Harare.

Grundy, I.M. (1995a). *Regeneration and management of Brachystegia spiciformis Benth. and Julbernardia globiflora (Benth.) Troupin in miombo woodland, Zimbabwe*. DPhil Thesis, University of Oxford.

Grundy, I.M. (1995b). Wood biomass estimation in dry *miombo* woodland in Zimbabwe. *Forest Ecology and Management*, **72**: 109–117.

Grundy, I.M., Campbell, B.M., Balebereho, S., Cunliffe, R., Tafangenyasha, C., Fergusson, R. and Parry, D. (1993). Availability and use of trees in Mutanda Resettlement Area, Zimbabwe. *Forest Ecology and Management*, **56**: 243–266.

Guedes, D.M. (1993). *Resistencia da árvores do cerrado ao fogo: papel de casca como isolante termico*. Tese de mestrado, University of Brasilia, Brasilia, DF.

Guidon, N. and Delibrias, G. (1986). Carbon-14 dates point to man in the Americas 32,000 years ago. *Nature*, **321**: 769–771.

Gusmão, R.P (1988). A expansão da agricultura e suas consequências no meio ambiente. In: Silva, S.T. (ed.), *Brasil uma visão geográfica dos anos 80*. Fundação Instituto Brasileiro de Geografia e Estatistica, Rio de Janeiro, pp. 323–332.

Guy, G.L. (1970). *Adansonia digitata* and its rate of growth in relation to rainfall in south central Africa. *Proceedings of the Transactions of the Rhodesian Scientific Association*, **54**: 68–84.

Guy, P.R. (1976). The feeding behaviour of elephant (*Loxodonta africana*) in the Sengwa area, Rhodesia. *South African Journal of Wildlife Research*, **6**: 55–63.

Guy, P.R. (1981). Changes in the biomass and productivity of woodlands in the Sengwa Wildlife Research Area, Zimbabwe. *Journal of Applied Ecology*, **18**: 507–519.

Guy, P.R. (1989). The influence of elephants and fire on a *Brachystegia–Julbernardia* woodland in Zimbabwe. *Journal of Tropical Ecology*, **5**: 215–226.

Haffer, J. (1969). Speciation in Amazonian forest birds. *Science*, **165**: 131–137.

Hall, J., McLeod, R.A. and Mitchell, V. (1987). *Pequeno Dicionário Xavante/Português e Português/Xavante*. Summer Institute of Linguistics, Brasilia, DF.

Hall, M. (1984). Man's historical and traditional use of fire in southern Africa. In: Booysen, P. de V. and Tainton, N.M. (eds), *Ecological effects of fire in south African ecosystems*. Ecological Studies 48, Springer-Verlag, Berlin, pp. 39–52.

Hall, M. (1987). *The changing past: farmers, kings and traders in southern Africa*. David Philip, Cape Town.

Hallam, S.J. (1975). *Fire and Hearth*. Australian Institute of Aboriginal Studies, Canberra.

Hallam, S.J. (1985). The history of Aborignial firing. In: Ford, J. (ed.), *Fire ecology and management in Western Australia*. Environmental Studies Group Bulletin 14, Western Australia Institute of Technology, Bentley, Western Australia, pp. 7–20.

Hall-Martin, A.J. (1992). Distribution and status of the African elephant *Loxodonta africana* in South Africa, 1652–1992. *Koedoe*, **35**: 65–88.

Hannah, L.D., Lohse, D., Hutchinson, C., Carr, J.L. and Lankerani, A. (1994). A preliminary inventory of human disturbance of world ecosystems. *Ambio*, **23**: 246–250.

Hao, W.M., Liu, M.H. and Crutzen, P.J. (1990). Estimates of annual and regional releases of CO_2 and other trace gases to the atmosphere from fires in the tropics, based on the FAO statistics for the period 1975–1980. In: Goldammer, J.G. (ed.), *Fire in the Tropical Biota*. Springer-Verlag, Berlin, pp. 440–462.

Hardin, G. (1968). The tragedy of the commons. *Science*, **162**: 1243–1248.

Haridasan, M. (1982). Aluminium accumulation by some *cerrado* native species of central Brazil. *Plant and Soil*, **65**(2): 265–273.

Haridasan, M. (1992). Estresse nutricional. In: Dias, B.F. de S. (ed.), *Alternativas de desenvolvimento dos cerrados: manejo e conservação dos recursos naturais renováveis*. Fundação pró-natureza (FUNATURA), Brasilia, DF, pp. 27–30.

Haridasan, M. (1994). Solos. In: Novaes-Pinto, M. (ed.), *Cerrado. Caracterização, Ocupação e Perspectivas* (2nd edition). Editora Universidade de Brasilia, Brasilia, DF, pp. 321–344.

Haridasan, M. and Araújo, G.M. (1988). Aluminium-accumulating species in two forest communities in the *cerrado* region of central Brazil. *Forest Ecology and Management*, **24**: 15–26.

Haridasan, M., Hill, P.G. and Russel, D.G. (1987). Semi-quantitative estimates of Al and other cations in the leaf tissues of some Al-accumulating species using electron probe microanalysis. *Plant and Soil*, **104**: 99–102.

Harris, P. (1992). The economy of northern Australia. In: *Research and technology in tropical Australia – symposia*. ASTEC Occasional paper No. 23, AGPS, Canberra.

Harrison, T.D. (1978). *Report on maximum temperature measurements during the Nylsvley veld fire of 5 September 1978 and on postfire micrometeorological measurements*. Unpublished report to the National Programme for Environmental Sciences, CSIR, Pretoria.

Hasse, R. (1990). Community composition and soil properties in northern Bolivian savanna vegetation. *Journal of Vegetation Science*, **1**: 345–352.

Hatton, J.C. and Smart, N.O.E. (1984). The effect of long-term exclusion of large herbivores on soil nutrient status in Murchison Falls National Park, Uganda. *African Journal of Ecology*, **22**: 23–30.

Haynes, C.D. (1978). Land, trees and man (Gunret, gundulk, djabining). *Commonwealth Forestry Review*, **57**: 99–106.

Haynes, C.D. (1985). The pattern and ecology of munwag: traditional Aboriginal fire regimes in north-central Arnhemland. *Proceedings of the Ecological Society of Australia*, **13**: 203–214.

Haynes, C.D. (1991). Use and impact of fire. In: Haynes, C.D., Ridpath, M.G. and Williams, M.A.J. (eds), *Monsoonal Australia: landscape, ecology and man in the northern lowlands*. A.A. Balkema, Rotterdam, pp. 61–71.

Head, L. (1994). Landscapes socialised by fire: post-contact changes in Aboriginal fire use in northern Australia, and implications for prehistory. *Archaeology in Oceania,* **29**: 172–181.

Hees, D.R., de Sá, M.E.P.C. and Aguiar, T.C. (1987). A evolução da agricultura na região Centro-Oeste na década de 70. *Revista Brasileira de Geografia,* **49**: 197–257.

Henderson, L. and Wells, M.J. (1986). Alien plant invasions in grassland and savanna biomes. In: Macdonald, I.A.W., Kruger, F.J. and Ferrar, A.A. (eds), *The ecology and management of biological invasions in Southern Africa.* Oxford University Press, Cape Town, pp. 109–118.

Henriques, R.P.B. (1993). *Organização e estrutura das comunidades vegetais de cerrado em un gradiente topográfico no Brasil Central.* Tese de Doctorado, Universidade de Campinas, Campinas, São Paulo.

Heringer, E.P. (1971). Propogação e sucessao de especies arboreas do *cerrado* em função de fogo, do capim, da capina e de aldrin. In: Ferri, M.G. (ed.), *III Simpósio sobre o Cerrado.* Editora da Universidade de São Paulo, São Paulo, pp. 167–179.

Herlocker, D. (1976). *Woody vegetation of the Serengeti National Park.* Texas A and M University Press, College Station, Texas.

Hernandez A., Montilla, G. and Baruch Z. (1983*). Fenologia y reparticion de biomasa en gramineas nativas e introducidas en las Cordillera de la Costa.* Report of Project S1–1135 CONICIT, Caracas.

Hiatt, L. (1962). Local organisation among the Australian Aborigines. *Oceania,* **32**: 267–286.

Hibajene, S.H. (1994). *Woodfuel transportation and distribution in Zambia.* Energy, Environment and Development Series 28, Stockholm Environment Institute, Stockholm.

Hiscock, P. and Kershaw, A.P. (1992). Palaeoenvironment and prehistory of Australia's tropical top end. In: Dodson, J. (ed.), *The naive lands: prehistory and environmental change in Australia and the south-west Pacific.* Longman Cheshire, Melbourne, pp. 43–75.

Hoare, J.R.L., Hooper, R.J., Cheney, N.P. and Jacobsen, K.L.S. (1980). *A report on the effect of fire in tall open forest and woodland with particular reference to fire management in Kakadu National Park in the Northern Territory.* Australian National Parks and Wildlife Service, Canberra.

Hobane, P.A. (1994). *Amacimbi: the gathering, processing, consumption and trade of edible caterpillars in Bulilimamangwe District.* Centre for Applied Social Sciences, University of Zimbabwe.

Hobbs, R.J. and Hopkins, A.J.M. (1990). From frontier to fragments: European impact on Australia's vegetation. *Proceedings of the Ecological Society of Australia,* **16**: 93–114.

Hobbs, R.J. and Mooney, H.A. (1986). Community changes following shrub invasion of grassland. *Oecologia,* **70**: 508–513.

Hochberg, M.E., Menaut, J.C. and Gignoux, J. (1994). The influences of tree biology and fire in the spatial structure of the West African savanna. *Journal of Ecology,* **82**: 217–226.

Hofer, H., Campbell, K.L.I., East, M.L. and Huish, S.A. (1996). The impact of game meat hunting on target and non-target species in the Serengeti. In: Taylor, V.J. and Dunstone, N. (eds), *The exploitation of mammal populations.* Chapman and Hall, London, pp. 117–146.

Hoffman, M.T. (1997). Human impacts on vegetation. In: Cowling, R.M., Richardson, D.M. and Pierce, S.M. (eds), *Vegetation of Southern Africa.* Cambridge University Press, Cambridge, pp. 507–534.

Högberg, P. (1982). Mycorrhizal associations in some woodland and forest trees and shrubs in Tanzania. *New Phytologist,* **92**: 407–415.

Högberg, P. (1989). Root symbioses of trees in savannas. In: Proctor, J. (ed.), *Mineral nutrients in tropical forest and savanna ecosystems.* Blackwell, Oxford, pp. 121–136.

Högberg, P. (1992). Root symbioses of trees in African dry tropical forests. *Journal of Vegetation Science,* **3**: 393–400.

Högberg, P. and Piearce, G.D. (1986). Mycorrhizas in Zambian trees in relation to host taxonomy, vegetation type and successional pattern. *Journal of Ecology,* **74**: 775–785.

Holden, S. (1991). Edible caterpillars – a potential agroforestry resource? *The Food Insects Newsletter,* **4**: 3–4.

Holdridge, L.R. (1947). Determination of world plant formations from simple climatic data. *Science,* **105**: 367–368.

Holland, G.J. (1984). Tropical cyclones in the Australian/southwest Pacific region. *Australian Metereological Magazine,* **32**: 33–46.

Holling, C.S. (1973). Resilience and stability of ecological systems. *Annual Review of Ecology and Systematics,* **4**: 1–23.

Holmes, J.H. (1990). Ricardo revisited: submarginal land and non-viable cattle enterprises in the Northern Territory Gulf District. *Journal of Rural Studies,* **6**: 45–65.

Holmes, J.H. (1996). Changing resource values in Australia's tropical savanna: priorities in institutional reform. In: Ash, A.J. (ed.), *The future of tropical savannas: an Australian perspective.* CSIRO, Australia, pp. 28–43.

Holmes, J.H. and Mott, J.J. (1993). Towards the diversified use of Australia's savannas. In: Young, M.D. and Solbrig, O.T. (eds), *The world's savannas. Economic driving forces, ecological constraints and policy options for sustainable land use.* Man and the Biosphere Series, Volume 12, UNESCO, Paris, pp. 283–317.

Holmgren, P., Masakha, E.J. and Sjöholm, H. (1994). Not all African land is being degraded: a recent survey of trees on farms in Kenya reveals rapidly increasing forest resources. *Ambio,* **23**: 390–395.

Holt, J.A. (1987). Carbon mineralisation in semi-arid northeastern Australia: the role of termites. *Journal of Tropical Ecology,* **3**: 255–263.

Holt, J.A. (1988). Carbon mineralisation in semi-arid tropical Australia: the role of mound-building termites. PhD Thesis. University of Queensland, Brisbane, Queensland.

Holt, J.A. and Coventry, R.J. (1991). Nutrient cycling in Australian savannas. In: Werner, P.A. (ed.). *Savanna ecology and management. Australian perspectives and intercontinental comparisons.* Blackwell, Oxford, pp. 83–88.

Holt, J.A. and Easey, J.F. (1984). Biomass of mound-building termites in a red and yellow earth landscape, north Queensland. In: *Proceedings of the Natural Soils Conference.* Australian Society of Soil Science, Brisbane, Australia, p. 363.

Hoogesteijn, R. and Chapman, C.A. (1997). Large ranches as conservation tools in the Venezuelan *llanos. Oryx,* **31**(4): 274–284.

Hoogesteijn, R. and Mondolfi, E. (1992). *The jaguar.* Armitano Editores, Caracas.

Hope, J. (1984). The Australian Quaternary. In: Archer, M. and Clayton, G. (eds). *Vertebrate zoogeography and evolution in Australasia.* Hesperian Press, Victoria Park, pp. 69–81.

Hopkins, B. (1963). The role of fire in promoting the sprouting of some savanna species. *Journal of West African Science Association,* **7**: 154–162.

Hopkins, B. (1965). Observations on savanna burning in the Olokemeji Forest Reserve, Nigeria. *Journal of Applied Ecology,* **2**: 367–381.

Hopkins, M.S., Ash, J., Graham, A.W., Head, J. and Hewett, R.K. (1993). Charcoal evidence of the spatial extent of the *Eucalyptus* woodland expansions and rainforest contractions in North Queenland during the late Pleistocene. *Journal of Biogeography*, **20**: 357–372.

Horton, D.R. (1982). The burning question: Aborigines, fire and Australian ecosystems. *Mankind*, **13**: 237–251.

Hughes, F., Vitousek P.M. and Tunison T. (1991). Alien grass invasion and fire in the seasonal submontane zone of Hawaii. *Ecology*, **72**: 743–746.

Hulme, M. (1998). *Decades of drought: looking back on what they mean to tomorrow's Sahel.* ID21 Report, IDS, University of Sussex.

Hulme, M. and Viner, D. (1998). A climate change scenario for the tropics. *Climatic Change*, **39**: 145–176.

Humphries, S.E., Groves, R.H. and Mitchell, D.S. (1991). Plant invasions of Australian ecosystems: a status review and management directions. *Kowari*, **2**: 1–116.

Huntley, B.J. (1982). Southern African savannas. In: Huntley, B.J. and Walker, B.H. (eds), *Ecology of tropical savannas.* Ecological Studies 42, Springer-Verlag, Berlin, pp. 101–119.

Huntley, B.J. and Morris, J.W. (1978). *Savanna ecosystem project: Phase I summary and Phase II progress.* South African National Scientific Programmes Report 29, CSIR, Pretoria.

Huntley, B.J. and Morris, J.W. (1982). Structure of the Nylsvley savanna. In: Huntley, B.J. and Walker, B.H. (eds), *Ecology of tropical savannas.* Ecological Studies 42, Springer-Verlag, Berlin, pp. 433–455.

Hurst, D.F., Griffith, D.W.T. and Cook, G.D. (1994). Trace gas emissions from biomass burning in tropical Australian savannas. *Journal of Geophysical Research*, **99**: 441–456.

IBGE (1960–1987). *Anuário estatistico do Brasil.* Fundação Instituto Brasileiro de Geografia e Estatistica, Rio de Janeiro.

Igboanugo, A.B.I., Omijeh, J.E. and Adegbehin, J.O. (1990). Pasture floristic composition in different *Eucalyptus* species plantations in some parts of northern Guinea savanna zone of Nigeria. *Agroforestry Systems*, **12**: 257–268.

Iliffe, J. (1979). *A modern history of Tanganyika.* Cambridge University Press, Cambridge.

Iliffe, J. (1995). *Africans. The history of a continent.* Cambridge University Press, Cambridge.

Inglis, J.M. (1976). Wet season movements of individual wildebeests of the Serengeti migratory herd. *East African Wildlife Journal*, **14**: 17–34.

IPCC (1999). *Intergovernmental Panel on Climate Change.* http://www.ipcc.ch

Isbell, R.F. (1986). The tropical and subtropical north and northeast. In: Russell, J.F. and Isbell, R.F. (eds), *Australian soils: the human impact.* The University of Queensland Press, Brisbane, pp. 3–35.

Isichei, A.O. and Muoghalu, J.I. (1992). The effects of tree canopy cover on soil fertility in a Nigerian savanna. *Journal of Tropical Ecology*, **8**: 329–338.

IUCN (1999). *Endangered species database.* http://www.iucn.org.

Jachmann, H. (1989). Food selection by elephants in the 'miombo' biome, in relation to leaf chemistry. *Biochemical Systematics and Ecology*, **17**: 15–24.

Jachmann, H. and Bell, R.H.V. (1985). Utilisation by elephants of the *Brachystegia* woodlands of the Kasungu National Park, Malawi. *African Journal of Ecology*, **23**: 245–258.

Jachmann, H. and Croes, T. (1991). Effects of browsing by elephants on the *Combretum/Terminalia* woodland at the Nazinga Game Ranch, Burkino Faso, West Africa. *Biological Conservation*, **57**: 13–24.

Jackson, R.B., Canadell, J., Ehleringer, J.R., Mooney, H.A., Sala, O.E. and Schulze, E.D. (1996). A global analysis of root distributions for terrestrial biomes. *Oecologia*, **108**: 389–411.

Jaeger, F. (1945). Zur gliederung und benenming des tropischen graslandgürtels. *Verhandlunger der Naturforschenden Gesellschaft in Basel*, **56**: 509–520.

Jager, T. (1982). *Soils of the Serengeti Woodlands, Tanzania.* Pudoc, Wageningen.

Jansma, R. (1994). *Ecology of some northern Suriname savannas.* Koeltz Scientific Books, Koenigstein, Germany.

Jarman, P.J. and Sinclair, A.R.E. (1979). Feeding strategy and the pattern of resource partitioning in ungulates. In: Sinclair, A.R.E. and Norton-Griffiths, M. (eds), *Serengeti: dynamics of an ecosystem.* University of Chicago Press, Chicago, pp. 130–163.

Jarmillo, V.J. and Detling, J.K. (1988). Grazing history, defoliation and competition: effects on shortgrass production and nitrogen accumulation. *Ecology*, **69**: 1599–1608.

Jarvis, P.G. and McNaughton, K.G. (1986). Stomatal control of transpiration: scaling up from leaf to region. *Advances in Ecological Research*, **15**: 1–45.

Jeltsch, F., Milton, S.J., Dean, W.R.J. and van Rooyen, N. (1996). Tree spacing and coexistence in semiarid savannas. *Journal of Ecology*, **84**: 583–595.

Jeltsch, F., Milton, S.J., Dean, W.R.J. and van Rooyen, N. (1997a). Simulated pattern formation around artificial waterholes in the semi-arid Kalahari. *Journal of Vegetation Science*, **8**(2): 177–188.

Jeltsch, F., Milton, S.J., Dean, W.R.J. and van Rooyen, N. (1997b). Analysing shrub encroachment in the southern Kalahari: a grid-based modelling approach. *Journal of Applied Ecology*, **34**(6): 1497–1508.

Jeltsch, F., Milton, S.J., Dean, W.R.J., van Rooyen, N. and Moloney, K.A. (1998). Modelling the impact of small-scale heterogeneities on tree-grass coexistence in semi-arid savannas. *Journal of Ecology*, **86**: 780–793.

Jensen, C.L. and Belsky, A.J. (1989). Grassland homogeneity in Tsavo National Park (West), Kenya. *African Journal of Ecology*, **27**: 35–44.

Jensen, E. (1998). South Africa country profile. In: Bruce, J.W. (ed.), *Country profiles of land tenure: Africa, 1996.* Research Paper No. 130, Land Tenure Centre, University of Wisconsin, Madison, pp. 253–260.

Jin, F.F. (1996). Tropical ocean–atmosphere interaction, the Pacific cold tongue, and the El Niño Southern Oscillation. *Science*, **274**(5284): 76–78.

Johns, T., Mhoro, E.B. and Uiso, F.C. (1996). Edible plants of Mara Region, Tanzania. *Ecology of Food and Nutrition*, **35**: 71–80.

Johnston, J. (1985). Factors affecting financial viability of remote area beef properties. In: *Brucellosis and Tubercolosis Eradication Campaign Workshop.* Department of Primary Industries, Brisbane.

Johnston, M.C. (1983). Past and present grasslands of southern Texas and north-eastern Mexico. *Ecology*, **44**: 456–466.

Joly, C.A. and Crawford, R.M.M. (1982). Variation in tolerance and metabolic responses to flooding in some tropical trees. *Journal of Experimental Botany*, **33**: 799–809.

Jones, J.A. (1989). Environmental influences on soil chemistry in central semiarid Tanzania. *Soil Science Society of America Journal*, **53**: 1748–1758.

Jones, J.A. (1990). Termites, soil fertility and carbon cycling in dry tropical Africa: a hypothesis. *Journal of Tropical Ecology*, **6**: 291–305.

Jones, R. (1969). Fire-stick farming. *Australian Natural History*, **16**: 224–228.

Jones, R. (1980). Hunters in the Australian coastal savanna. In: Harris, D.R. (ed.), *Human ecology in savanna environments*. Academic Press, New York, pp. 107–146.

Jones, R. and Bowler, J. (1980). Struggle for the savanna. In: Jones, R. (ed.), *Northern Australia: options and implications*. Research School of Pacific Studies, ANU, Canberra, pp. 3–31.

Jones, R.J. (1997). Steer gains, pasture yield and pasture composition on native pasture and native pasture oversewn with Indian couch (*Bothriochloa pertusa*) at three stocking rates. *Australian Journal of Experimental Agriculture*, **37**(7): 755–765.

Jonkman, J.C.M. (1976). Biology and ecology of the leaf-cutting ant *Atta collenweideri*. *Zeitschrift für Angewandte Entomologie*, **81**: 140–148.

Jonkman, J.C.M. (1978). Nests of the leaf-cutting ant *Atta collenweideri* as accelerators of succession in pastures. *Zeitschrift für Angewandte Entomologie*, **86**: 25–34.

Josens, G. (1974). Les termites de la savana de Lamto. In: Analyse d'un écosystème tropical humide: la savane de Lamto (Côte d'Ivoire). V. Les organismes endogés. *Bulletin de Liaison des Chercheurs de Lamto*, no. spécial **5**: 91–131.

Josens, G. (1983). The soil fauna of tropical savannas. II. The termites. In: Bourlière, F. (ed.), *Tropical savannas. Ecosystems of the world 13*. Elsevier Scientific Publishing Company, Amsterdam, pp. 505–524.

Justice, C., Scholes, B. and Frost, P. (1994). *African savannas and the global atmosphere: research agenda*. IGBP Report 31, International Geosphere–Biosphere Programme, Stockholm, Sweden.

Kanjanavanit, S. (1992). *Aspects of the temporal pattern of dry season fires in the dry dipterocarp forests of Thailand*. PhD Thesis, Department of Geography, School of Oriental and African Studies, London.

Kasusya, P. (1998). Combating desertification in northern Kenya (Samburu) through community action: a community case experience. *Journal of Arid Environments*, **39**: 325–329.

Kauffman, J.B., Cummings, D.L. and Ward, D.E. (1994). Relationships of fire, biomass and nutrient dynamics along a vegetation gradient in the Brazilian *cerrado*. *Journal of Ecology*, **82**: 519–531.

Kaufman, Y.J., Setzer, A., Ward, D., Tanre, D., Holben, B.N., Menzel, P., Pereira, M.C. and Rasmussen, R. (1992). Biomass burning airborne and spaceborne experiment in the Amazonas (Base-A). *Journal of Geophysical Research-Atmopsheres*, **97**(D13): 14 581–14 599.

Keen, I. (1980). *The Alligator Rivers Stage II Land Claim*. Report to the Northern Land Council, Darwin.

Keith, J.O. and Plowes, D.C.H. (1997). *Considerations of wildlife resources and land use in Chad*. Sustainable Development Technical Paper No. 45, USAID, Washington, DC.

Kellman, K., Miyanishi, K. and Hieburt, P. (1985). Nutrient retention by savanna ecosystems: 2. retention after fire. *Journal of Ecology*, **73**: 953–962.

Kellman, M. (1979). Soil enrichment by neotropical savanna trees. *Journal of Ecology*, **67**: 565–577.

Kemp, E.M. (1981). Pre-Quaternary fire in Australia. In: Gill, A.M., Groves, R.H. and Noble, I.R. (eds), *Fire and the Australian biota*. Australian Academy of Science, Canberra, pp. 3–21.

Kerr, W.E. and Posey, D.A. (1984). Informações adicionais sobre a agricultura dos Kayapó. *Interciencia*, **9**(6): 392–400.

Kershaw, A.P. (1985). An extended late Quaternary vegetation record from north-eastern Queensland and its implications for the seasonal tropics of Australia. *Proceedings of the Ecological Society of Australia*, **13**: 179–189.

Kershaw, A.P. (1989). Was there a 'Great Australian Arid Period'? *Search*, **20**: 89–92.

Kessler, J.J. and Bremen, H. (1991). The potential of agroforestry to increase primary production in the Sahelian and Sudanian zones of West Africa. *Agroforestry Systems*, **13**: 41–62.

Khemnark, C., Wacharakitti, S., Aksornkoae, S. and Kaewlaiad, T. (1972*). Forest production and soil fertility at Nikhom Doi Chiangdao, Chiangmai Province*. Forest Research Bulletin 22, Faculty of Forestry, Kasetsart University, Bangkok.

King, D.A. (1997). The functional significance of leaf angle in *Eucalyptus*. *Australian Journal of Botany*, **45**: 619–639.

King, J.A. and Campbell, B.M. (1994). Soil organic matter relations in five land cover types in the *miombo* region (Zimbabwe). *Forest Ecology and Management*, **67**: 225–239.

King, N.K. and Vines, R.G. (1969). *Variation in the flammability of the leaves of some Australian forest species*. CSIRO Mimeograph Report, Division of Applied Chemistry, Melbourne.

Kirchoff, V.W.J.H. and Alvala, P.C. (1996). Overview of an aircraft expedition into the Brazilian *cerrado* for the observation of atmospheric trace gases. *Journal of Geophysical Research – Atmospheres*, **101**(D19): 23 973–23 981.

Kirchoff, V.W.J.H. and Marinho, E.V.A. (1994). Layer enhancements of tropospheric ozone in regions of biomass burning. *Atmospheric Environment Part A – General Topics*, **28**(1): 69–74.

Kirchoff, V.W.J.H. and Nobre, C.M. (1986). Atmospheric chemistry research in Brazil: ozone measurements at Natal, Manaus and Cuiabá. *Geofisica*, **24**: 95–108.

Klink, C.A., (1992). *A Comparative Study of the Ecology of Native and Introduced African Grasses of the Brazilian Savannas*. PhD Thesis, Harvard University, Cambridge, MA.

Klink, C.A., Moreira, A.G. and Solbrig, O.T. (1993). Ecological impact of agricultural development in the Brazilian *cerrados*. In: Young, M.D. and Solbrig, O.T. (eds), *The world's savannas. Economic driving forces, ecological constraints and policy options for sustainable land use. Man and the biosphere Vol. 12*. UNESCO, Paris, pp. 259–282.

Kock, R.A. (1995). Wildlife utilisation: use it or lose it – a Kenyan perspective. *Biodiversity and Conservation*, **4**: 241–256.

Komarek, E.V. (1971). Lightning and fire ecology in Africa. *Proceedings of the Tall Timbers Fire Ecology Conference*, **11**: 473–511.

Komkris, T., Naraballobh, V., Chunkao, K., Ngampongsai, C. and Tangtham, N. (1969). *Effect of fire on soil and water losses at Mae-Huad Forest, Amphur Ngao, Lampang Province*. Forest Research Bulletin 6, Kasetsart University, Bangkok.

Koohacharoen, O. (1992). Commercial reforestation policy. In: Leungaramsri, P. and Rajesh, N. (eds), *The future of people and forests in Thailand after the logging ban*. Project for Ecological Recovery, Bangkok, pp. 56–78.

Koohacharoen, O. and Paisarnpanichkul, D. (1992a). Promotion of commercial crop cultivation. In: Leungaramsri, P. and Rajesh, N. (eds), *The future of people and forests in Thailand after the logging ban*. Project for Ecological Recovery, Bangkok, pp. 79–102.

Koohacharoen, O. and Paisarnpanichkul, D. (1992b). Tourism. In: Leungaramsri, P. and Rajesh, N. (eds), *The future of people and forests in Thailand after the logging ban*. Project for Ecological Recovery, Bangkok, pp. 123–136.

Koponen, J. (1994). *Development for exploitation: German colonial policies in mainland Tanzania, 1884–1914*. Finnish Historical Society, Studia Historica 49, Helsinki/Hamburg.

Köppen, W. (1884). Die wärmezonen der erde, nach der dauer der heissen, gemässigten und kalten zeit und nach der wirkung der wärme auf die organische. *Welt betrachtet. Meteorologische Zeitschrift*, **1**: 215–226.

Köppen, W. (1900). Versuch einer klassifikation der klimate, vorzugsweise nach ihren beziehungen zur pflanzenwelt. *Geographische Zeitschrift*, **6**: 593–611.

Köppen, W. (1931). *Grundriss der Klimakunde* (2nd edition). W. de Gruyter, Berlin.

Korn, H. (1987). Densities and biomasses of non-fossorial southern African savanna rodents during the dry season. *Oecologia*, **72**: 410–413.

Körner, C. and Bazzaz, F.A. (eds) (1996). *Carbon dioxide, population and communities.* Physiological Ecology Series of Academic Press, London.

Kotliar, N.B. and Wiens, J.A. (1990). Multiple scales of patchiness and patch structure: a hierarchical framework for the study of heterogeneity. *Oikos*, **59**: 253–260.

Kotze, K. (1993). *Study on the export potential of the Zambian handicraft industry.* Report prepared for the European Development Fund, Lusaka.

Kreulen, D.A. (1975). Wildebeest habitat selection in the Serengeti plains, Tanzania, in relation to calcium and lactation: a preliminary report. *East African Wildlife Journal*, **13**: 297–304.

Küchler, A.W. (1964). *The potential natural vegetation of the conterminous United States.* American Geographical Society, New York.

Kurz, S. (1875). *Preliminary report on the forest and other vegetation of Pegu.* Office of the Superintendent of Government Printing, Calcutta.

Kurz, S. (1877). *Forest flora of British Burma. Volumes I and II.* Office of the Superintendent of Government Printing, Calcutta.

Kutintara, U. (1975). *Structure of dry dipterocarp forest.* PhD Thesis, Colorado State University, Fort Collins.

Lacaux, J.P., Delmas, R., Jambert, C. and Kuhlbusch, T.A.J. (1996). NO_x emissions from African savanna fires. *Journal of Geophysical Research-Atmospheres*, **101**(D19): 23 585–23 595.

Lacey, C.J. (1979). Forestry in the Top End of the Northern Territory: part of the Northern Myth. *Search*, **10**: 174–180.

Lacey, C.J. and Whelan, P.I. (1976). Observations on the ecological significance of vegetative reproduction in the Katherine–Darwin region of the Northern Territory. *Australian Forester*, **39**: 131–139.

Lacey, C.J., Walker, J. and Noble, I.R. (1982). Fire in Australian tropical savannas. In: Huntley, B.J. and Walker, B.H. (eds), *Ecology of tropical savannas.* Springer-Verlag, Berlin, pp. 246–272.

Lamont, B.B. and Bergl, S.M. (1991). Water relations, shoot and root architecture, and phenology of three co-occurring *Banksia* species: no evidence for niche differentiation in the pattern of water use. *Oikos*, **60**: 291–298.

Lamprey, H.F. (1983). Pastoralism yesterday and today: the overgrazing problem. In: Bourlière, F. (ed.), *Tropical savannas. Ecosystems of the world 13.* Elsevier Scientific Publishing Company, Amsterdam, pp. 643–666.

Lamprey, H.F., Gover, P.E., Turner, M.I.M. and Bell, R.H.V. (1967). Invasion of the Serengeti National Park by elephants. *East African Wildlife Journal*, **5**: 151–166.

Lane, C. and Scoones, I. (1993). Barabaig natural resource management. In: Young, M.D. and Solbrig, O.T. (eds), *The world's savannas. Economic driving forces, ecological constraints and policy options for sustainable land use. Man and the biosphere Vol. 12.* UNESCO, Paris, pp. 93–120.

Langkamp, P.J., Swinden, L.B. and Dalling, M.H. (1979). Nitrogen fixation (acetylene reduction) by *Acacia pellita* on areas restored after mining at Groote Eylandt, Northern territory. *Australian Journal of Botany*, **27**: 353–361.

Langworthy, H.W. (1971). Pre-colonial kingdoms and tribal migrations, AD 1500–1900. In: Davies, D.H. (ed.), *Zambia in maps*. University of London Press, London, pp. 32–33.

Lauer, W. (1952). *Humide und aride jahreszeiten in afrika und Sudamerika und ihre beziehungen zu der vegetationsgürteln*. Bonner Geographische Abhandlungen, 9.

Laut, P. and Nanninga, P.M. (1985). *Landscape data for herd disease eradication in northern Australia*. CSIRO Division of Land and Water Resources, Canberra.

Laws, R.M. (1969). The Tsavo research project. *Journal of Reproductive Fertility Supplement*, **6**: 495–531.

Laws, R.M. (1970). Elephants as agents of habitat and landscape change in East Africa. *Oikos*, **21**: 1–15.

Laws, R.M., Parker, I.S.C. and Johnstone, R.C.B. (1970). Elephants and habitats in North Bunyoro, Uganda. *East African Wildlife Journal*, **8**: 163–180.

Lawton, R.M. (1978). A study of the dynamic ecology of Zambian vegetation. *Journal of Ecology*, **66**: 175–198.

Lawton, R.M. (1980). Browse in *miombo* woodland. In: Le Houerou, H.N. (ed.), *Browse in Africa: the current state of knowledge*. ILCA, Addis Ababa, pp. 25–31.

Leach, M. and Mearns, R. (1988). *Beyond the woodfuel crisis*. Earthscan Publications, London.

Leach, M. and Mearns, R. (eds), (1996). *African issues – the lie of the land: challenging received wisdom on the African continent*. International African Institute, London.

Ledru, M.P. (1993). Late Quaternary environmental and climatic changes in central Brazil. *Quaternary Research*, **39**: 90–98.

Ledru, M.P., Salgado-Labouriau, M.L. and Lorscheitter, M.L. (1998). Vegetation dynamics in southern and central Brazil during the last 10,000 yr B.P. *Review of Palaeobotany and Palynology*, **99**: 131–142.

Lee, K.E. and Wood, T.G. (1971) *Termites and Soils*. Academic Press, London.

Lees, H.M.N. (1962). *Working plan for the forests supplying the Copperbelt, Western Province*. Government Printer, Lusaka.

LeHouérou, H.N. (1980). Chemical composition and nutritive value of browse in tropical West Africa. In: LeHouérou, H.N. (ed.), *Browse in Africa – the current state of knowledge*. ILCA, Addis Ababa, pp. 83–100.

LeHouérou, H.N. (1989). *The grazing land ecosystems of the African Sahel*. Ecological Studies 75, Springer-Verlag, Berlin.

LeHouérou, H.N. (1996). Climate change, drought and desertification. *Journal of Arid Environments*, **34**: 133–185.

LeHouérou, H.N. (1997). Climate, flora and fauna changes in the Sahara over the past 500 million years. *Journal of Arid Environments*, **37**(4): 619–647.

Lepage, M. (1972). Recherches écologiques sur une savane sahélienne du Ferlo septentrional, Sénégal: données préliminaires sur l'écologie des termites. *La Terra et la Vie*, **26**: 383–409.

Lepage, M., Abbadie, L. and Mariotti, A. (1993). Food habits of sympatric termite species (Isoptera, Macrotermitinae) as determined by stable carbon isotope analysis in a Guinean savanna (Lamto, Côte d'Ivoire). *Journal of Tropical Ecology*, **9**: 303–311.

LeRoux, X., Bariac, T. and Mariotti, A. (1995). Spatial partitioning of the soil water resource between grass and shrub components in a West African humid savanna. *Oecologia*, **104**: 147–155.

LeRoux, X., Gauthier, H., Begue, A. and Sinoquet, H. (1997). Radiation absorption and use by humid savanna grassland: assessment using remote sensing and modelling. *Agricultural and Forest Meteorology*, **85**(1–2): 117–132.

Leungaramsri, P. and Malapetch, P. (1992). Illegal logging. In: Leungaramsri, P. and Rajesh, N. (eds), *The future of people and forests in Thailand after the logging ban*. Project for Ecological Recovery, Bangkok, pp. 29–55.

Leungaramsri, P. and Rajesh, N. (eds) (1992). *The future of people and forests in Thailand after the logging ban*. Project for Ecological Recovery, Bangkok.

Leuthold, W. (1977). Changes in tree populations of Tsavo East National Park, Kenya. *East African Wildlife Journal*, **15**: 61–69.

Levine, J.S., Winstead, E.L., Parsons, D.A.B., Scholes, M.C., Scholes, R.J., Cofer, W.R., Cahoon, D.R. and Sebacher, D.I. (1996). Biogenic soil emissions of nitric oxide (NO) and nitrous oxide (N_2O) from savannas in South Africa: the impact of wetting and burning. *Journal of Geophysical Research – Atmospheres*, **101**(D19): 23 689–23 697.

Lewis, D. (1988). *The rock paintings of Arnhem Land, Australia. Social, ecological and material culture change in the post-glacial period*. BAR Internation Series, Oxford.

Lewis, H.T. (1989). Ecological and technological knowledge of fire: Aborigines versus Park Rangers in Northern Australia. *American Anthropologist*, **91**: 940–961.

Lindesay, J.A., Andreae, M.O., Goldammer, J.G., Harris, G., Annegarn, H.J., Garstang, M., Scholes, R.J. and Van Wilgen, B.W. (1996). International Geosphere–Biosphere Programme International Global Atmospheric Chemistry SAFARI-92 field experiment: background and overview. *Journal of Geophysical Research-Atmospheres*, **101**(D19): 23 521–23 530.

Little, P.D. (1996). Pastoralism, biodiversity, and the shaping of savanna landscapes in East Africa. *Africa*, **66**: 37–51.

Little, P.D. and Brokensha, D.W. (1987). Local institutions, tenure and resource management in East Africa. In: Anderson, D. and Grove, R. (eds), *Conservation in Africa: people, policies and practice*. Cambridge University Press, Cambridge, pp. 193–209.

Lock, J.M. (1972). The effects of hippopotamus grazing on grasslands. *Journal of Ecology*, **60**: 445–468.

Lock, J.M. (1977). Preliminary results from fire and elephant exclusion plots in Kabalega National Park, Uganda. *East African Wildlife Journal*, **15**: 229–232.

Lohmann, L. (1990). Commercial tree plantations in Thailand: deforestation by any other name. *The Ecologist*, **20**(1), Jan/Feb.

Lohmann, L. (1991). Peasants, plantations and pulp: the politics of Eucalyptus in Thailand. *Bulletin of Concerned Asian Scholars*, **24**(4), Oct–Dec.

Lonsdale, M. and Braithwaite, R. (1988). The shrub that conquered the bush. *New Scientist*, **120**(1634): 52–55.

Lonsdale, W.M. and Braithwaite, R.W. (1991). Assessing the effects of fire on vegetation in tropical savannas. *Australian Journal of Ecology*, **16**: 363–374.

Lonsdale W.M., Braithwaite, R.W., Lane, A.M. and Farmer, J. (1998). Modelling the recovery of an annual savanna grass following a fire-induced crash. *Australian Journal of Ecology*, **23**: 509–513.

Lonsdale, W.M. and Lane, A.M. (1994). Tourist vehicles as vectors of weed seeds in Kakadu National Park, northern Australia. *Biological Conservation*, **69**(3): 277–283.

Lopes, A.S. and Cox, F.R. (1977). *Cerrado* vegetation in Brazil: an edaphic gradient. *Agronomy Journal*, **69**: 828–831.

Lopes-Naranjo, H.J. (1975). *Estrutura morfológica de Anacardium pumilum St. Hil. Anacardiaceae*. Tese de mestrado, Universidade de São Paulo, São Paulo.

López-Hernández, D., García, M. and Niño, M. (1994). Input and output of nutrients in a diked flooded savanna. *Journal of Applied Ecology*, **31**: 303–312.

López-Hernández, D., Niño, M., García, M. and Carrion, N. (1983). Annual budgets of some elements in a flooded savanna (Módulo Experimental, Mantecal, Venezuela). *Ecological Bulletin*, **35**: 541–545.

Loth, P.E. and Prins, H.H.T. (1986). Spatial patterns of the landscape and vegetation of Lake Manyara National Park, Tanzania. *ITC Journal*, **1986**(2): 115–130.

Lowore, J.D., Abbot, P.G. and Werren, M. (1994). Stackwood volume estimations for *miombo* woodlands in Malawi. *Commonwealth Forestry Review*, **73**: 193–197.

Lowore, J.D., Coote, H.C., Abbot, P.G., Chapola, G.B. and Malembo, L.N. (1995). *Community use and management of indigenous trees and forest products in Malawi: the case of four villages close to Chimaliro Forest Reserve*. Forestry Research Institute of Malawi Report No. 93008, Zomba.

Ludlow, A.E. (1987). *A development study of the anatomy and fine structure of the leaves of Ochna pulchra Hook*. PhD Thesis, University of Witwatersrand, Johannesburg.

Ludwig, J. (1991). SHRUBKILL: a decision support system for management burns in Australian savannas. In: Werner, P.A. (ed.), *Savanna ecology and management. Australian perspectives and intercontinental comparisons*. Blackwell, Oxford, pp. 203–206.

Luke, R.H. and McArthur, A.G. (1978). *Bushfires in Australia*. Australian Government Publishing Service, Canberra.

Lukesch, A. (1969). *Mito e vida dos indios Caiapós*. Editora da Universidade de São Paulo, São Paulo.

Lungu, O.I. and Chinene, V.R.N. (1993). *Cropping and soil management systems and their effect on soil productivity in Zambia*. Agricultural University of Norway (Ecology and Development Programme), Norway.

Luy, A. (1992). Consideraciones sobre la factibilidad de establecer un programa de manejo del jaguar mediante la caceria deportiva. In: Clement, R. (ed.), *Felinos de Venezuela Biologia, Ecologia y Conservacion*. Memorias del Symposium organizado por Fudeci, Valencia, Venezuela, pp. 161–173.

Lyaruu, H.V.M. (1995). *Seed bank dynamics of the formerly overgrazed Kondoa Irangi Hills, Central Tanzania*. A preliminary report, EDSU Working Paper No. 29, Stockholm University.

Lykke, A.M. and Sambou, B. (1998). Structure, floristic composition, and vegetation forming factors of three vegetation types in Senegal. *Nordic Journal of Botany*, **18**(2): 129–139.

McCosker, T.H. and Emerson, C.A. (1982). Failure of legume pastures to improve animal production in the monsoonal dry tropics of Australia: a management review. *Animal Production in Australia*, **14**: 337–340.

McDonald, N.S. and McAlpine, J. (1991). Floods and droughts: the northern climate. In: Haynes, C.D., Ridpath, M.G. and Williams, M.A.J. (eds), *Monsoonal Australia. Landscape, ecology and man in the northern lowlands*. A.A. Balkema, Rotterdam, pp. 19–29.

Macedo, J. (1995). *Prospectives for the rational use of the Brazilian cerrados for food production*. Brazilian Agriculture Research Enterprise-CPAC, Planaltina, Brasilia, DF.

McGregor, J. (1995). Gathered produce in Zimbabwe's communal areas: changing resource availability and use. *Ecology of Food and Nutrition*, **33**: 163–193.

McIvor, J.G. (1981). Seasonal changes in the growth, dry matter distribution and herbage quality of three native grasses in northern Queensland. *Australian Journal of Experimental Agriculture and Animal Husbandry*, **21**: 600–609.

McKeon, G.M., Day, K.A., Howden, M., Mott, J.J., Orr, D.M., Scattini, W.J. and Weston, E.J. (1991). Northern Australian savannas: management for pastoral production. In: Werner, P.A. (ed.), *Savanna ecology and management. Australian perspectives and intercontinental comparisons.* Blackwell, Oxford, pp. 11–28.

Macleod, J.C. (1971). *Forest fire control in Thailand.* Final Report to RFD, Bangkok.

MacLeod, N.D. and Ludwig, J.A. (1991). Using BurnEcon to evaluate the economics of fire management in semi-arid woodlands, Eastern Australia. *Journal of Environmental Management,* **33**: 65–77.

McLeod, R. and Mitchell, V. (1980). *Aspectos da Lingua Xavante.* Summer Institute of Linguistics, Brasilia, DF.

McNaughton, S.J. (1979). Grassland–herbivore dynamics. In: Sinclair, A.R.E. and Norton-Griffiths, M. (eds), *Serengeti: dynamics of an ecosystem.* University of Chicago Press, Chicago, pp. 46–81.

McNaughton, S.J. (1983). Serengeti grassland ecology: the role of composite environmental factors and contingency in community organisation. *Ecological Monographs,* **53**: 291–320.

McNaughton, S.J. (1984). Grazing lawns: animals in herds, plant form and coevolution. *American Naturalist,* **124**: 863–886.

McNaughton, S.J. (1985). Ecology of a grazing ecosystem: the Serengeti. *Ecological Monographs,* **55**: 259–294.

McNaughton, S.J. (1988). Mineral nutrition and spatial concentrations of African ungulates. *Nature,* **334**: 343–345.

McNaughton, S.J. (1989). Interactions of plants of the field layer with large herbivores. In: Jewell, P.A. and Maloiy, G.M.O. (eds), *The biology of large African mammals in their environment.* Zoological Society of London Symposium No. 61, Clarendon Press, Oxford, pp. 15–29.

McNaughton, S.J. (1990). Mineral nutrition and seasonal movements of African migratory ungulates. *Nature,* **345**: 613–615.

McNaughton, S.J. (1991). Evolutionary ecology of large tropical herbivores. In: Price, P.W., Lewinsohn, T.M., Fernandes, G.W. and Benson, W.W. (eds), *Plant–animal interactions: evolutionary ecology in tropical and temperate regions.* Wiley, New York, pp. 509–522.

McNaughton, S.J. and Banyikwa, F.F. (1995). Plant communities and herbivory. In: Sinclair, A.R.E. and Arcese, P. (eds), *Serengeti II. Dynamics, management, and conservation of an ecosystem.* The University of Chicago Press, Chicago, pp. 49–70.

Madden, D. and Young, T.P. (1992). Symbiotic ants as an alternative defence against giraffe herbivory in spinescent *Acacia drepanolobium. Oecologia,* **91**(2): 235–238.

Maddock, L. (1979). The 'migration' and grazing succession. In: Sinclair, A.R.E. and Norton-Griffiths, M. (eds), *Serengeti: dynamics of an ecosystem.* University of Chicago Press, Chicago, pp. 104–129.

Maenhaut, W., Salma, I., Cafmeyer, J., Annegarn, H.J. and Andreae, M.O. (1996). Regional atmospheric aerosol composition and sources in the eastern Transvaal, South Africa, and impact of biomass burning. *Journal of Geophysical Research – Atmospheres,* **101**(D19): 23 631–23 650.

Maggs, T. (1984). The Iron Age south of the Zambezi. In: Klein, R.G. (ed.), *Southern African prehistory and palaeoenvironments.* A.A. Balkema, Rotterdam, pp. 329–360.

Mainguet, M. and de Silva, G.G. (1998). Desertification and drylands development: what can be done? *Land Degradation and Development,* **9**: 375–382.

Maitelli, G.T. (1987). *Balanço de energia e evapotranspiração de um cerrado (sens. strict.) no Distrito Federal.* Tese de mestrado, Universidade de Brasilia, Brasilia, DF.

Malaisse, F. (1973). *Contribution à l'étude de l'écosystème forêt claire (miombo). Note 8 – Le projet Miombo.* Annales Université Abidjan, Série E (Ecologie), Tome VI, pp. 227–250.

Malaisse, F. (1974). Phenology of the Zambezian woodland area with emphasis on the *miombo* ecosystem. In: Leith, H. (ed.), *Phenology and seasonality modelling.* Ecological Studies 8, Springer-Verlag, Berlin, pp. 269–286.

Malaisse, F. (1978a). The *miombo* ecosystem. *Natural Resources Research* (Unesco, Paris), **14**: 589–606.

Malaisse, F. (1978b). High termitaria. In: Werger, M.J.A. (ed.), *Biogeography and ecology of southern Africa.* Junk, The Hague, pp. 1279–1300.

Malaisse, F., Freson, R., Goffinet, G. and Malaisse-Mousset, M. (1975). Litter fall and litter breakage in *miombo.* In: Golley, F.B. and Medina, E. (eds), *Tropical ecological systems: trends in terrestrial and aquatic research.* Springer-Verlag, New York, pp. 137–152.

Malaisse-Mousset, M., Malaisse, F. and Watula, C. (1970). *Contribution à l'étude de l'écosystème forêt claire (miombo). Note 2 – Le cycle biologique d'Elaphrodes lactea (Gaede) (Notodontidae) et son influence sur l'écosystème 'miombo'.* Lubumbashi, Congo.

Malimbwi, R.E., Solberg, B. and Luoga, E. (1994). Estimation of biomass and volume in *miombo* woodland at Kitulangalo Forest Reserve, Tanzania. *Journal of Tropical Forest Science,* **7**: 230–242.

Mares, M.A., Ernest, K.A. and Gettinger, D. (1986). Small mammals community structure and composition in the *cerrado* province of central Brazil. *Journal of Tropical Ecology,* **2**: 301–325.

Marini, M.A. and Cavalcanti, R.B. (1996). Influência do fogo na avifauna do sub-bosque de uma mata de galeria do Brasil central. *Revista Brasileria de Biologia,* **56**(4): 749–754.

MARNR, (1995). *Inventario preliminar de emisiones de gases de efecto invernadero en Venezuela.* Proyecto PNUMA, GF/4102–92–40, Ministerio Ambiente Recursos Naturales Renovables y Ministerio Energía Minas, Caracas, Venezuela.

Martin, M.A. (1971). *Introduction á l'ethnobotanique du Cambodge.* Centre National de la Recherche Scientifique, Paris.

Martin, R.B. (1974). *Structure, biomass and utilisation of vegetation in the mopane and miombo woodlands of the Sengwa Wildlife Research Area.* Certificate in Field Ecology Thesis, University of Rhodesia, Salisbury.

Mather, A.D. (1978a). *Fire weather data.* Technical Note 15, FAO/Royal Forest Department, Chiangmai, Thailand.

Mather, A.D. (1978b). *Prescribed burning.* Technical Note 17, FAO/Royal Forest Department, Chiangmai, Thailand.

Matose, F. and Wily, L. (1996). Institutional arrangements governing the use and management of *miombo* woodlands. In: Campbell, B. (ed.), *The miombo in transition: woodlands and welfare in Africa.* Center for International Forestry Research (CIFOR), Bogor, Indonesia, pp. 195–219.

Matose, F., Mudhara, M. and Mushove, P. (1996). *Woodcraft industry study report.* Forestry Commission of Zimbabwe.

Maybury-Lewis, D. (1984). *A sociedade Xavante.* Francisco Alves, Rio de Janeiro.

Mearns, R. (1995). Institutions and natural resource management: access to and control over woodfuel in East Africa. In: Binns, T. (ed.), *People and environment in Africa.* John Wiley, Chichester, pp. 103–114.

Medeiros, J.X. (1995). Aspectos economico-ecologicos da produção e utilização de carvão vegetal na siderurgia brasileira. In: May, P.H. (ed.), *Economica ecologia.* Editora Campus, Rio de Janeiro, pp. 83–114.

Medellin, R.A. and Redford, K.H. (1992). The role of mammals in neotropical forest-savanna boundaries. In: Furley, P.A., Proctor, J. and Ratter, J.A. (eds), *Nature and dynamics of forest–savanna boundaries*, Chapman and Hall, London, pp. 519–548.

Medina, E. (1987). Requirements, conservation and cycles of nutrients in the herbaceous layer. In: Walker, B.H. (ed.), *Determinants of tropical savannas*. IRL Press, Oxford, pp. 39–65.

Medina, E. (1993). Mineral nutrition: tropical savannas. *Progress in Botany*, **54**: 237–253.

Medina, E. and Huber, O. (1994). The role of biodiversity in the functioning of savanna ecosystems. In: Solbrig, O.T., van Emden, H.M. and van Oordt, P.G.W.J. (eds), *Biodiversity and global change*. CAB International, Wallingford, pp. 141–160.

Medina, E. and Silva, J.F. (1991). Savannas of northern South America: a steady state regulated by water–fire interactions on a background of low nutrient availability. In: Werner, P.A. (ed.), *Savanna ecology and management. Australian perspectives and intercontinental comparisons*. Blackwell, Oxford, pp. 59–69.

Medwecka-Kornas, A. (1980). *Gardenia subaculus* Stapf and Hutch.: a pyrophytic suffrutex of the African savanna. *Acta Botanica Academiae Scientiarum Hungaricae*, **26**: 131–137.

Medwecka-Kornas, A. and Kornas, J. (1985). Fire-resistant sedges (Cyparaceae) in Zambia. *Flora*, **176**: 61–71.

Meinzer, F., Seymour, V. and Goldstein, G. (1983). Water balance in developing leaves of four tropical savanna woody species. *Oecologia* (Berlin), **60**: 237–243.

Menaut, J.C. (1977). Evolution of plots protected from fire since 13 years in a Guinean savanna of Ivory Coast. In: *Actas del IV Symposium Internacional de Ecologia Tropical*, Panama City, Panama.

Menaut, J.C. and César, J. (1979). Structure and primary productivity of Lamto savannas. *Ecology*, **60**: 1197–1210.

Menaut, J.C. and César, J. (1982). The structure and dynamics of a west African savanna. In: Huntley, B.J. and Walker, B.H. (eds), *Ecology of tropical savannas*. Ecological Studies 42, Springer-Verlag, Berlin, pp. 80–100.

Menaut, J.C., Barbault, R., Lavalle, P. and Lepage, M. (1985). African savannas: biological systems of humification and mineralisation. In: Tothill, J.C. and Mott, J.J. (eds). *Ecology and management of the world's savannas*. Australian Academy of Science, Canberra, pp. 13–44.

Menaut, J.C., Gignoux, J., Prado, C. and Clobert, J. (1991). Tree community dynamics in a humid savanna of the Côte d'Ivoire: modelling the effects of fire and competition with grass and neighbours. In: Werner, P.A. (ed.), *Savanna ecology and management. Australian perspectives and intercontinental comparisons*. Blackwell, Oxford, pp. 127–137.

Mentis, M.T. and Bailey, A.W. (1990). Changing perceptions of fire management in savanna parks. *Journal of the Grassland Society of Southern Africa*, **7**: 81–85.

Mentis, M.T. and Seijas, N. (1993). Rangeland bioeconomics in revolutionary South Africa. In: Young, M.D. and Solbrig, O.T. (eds), *The world's savannas. Economic driving forces, ecological constraints and policy options for sustainable land use. Man and the biosphere Vol. 12*. UNESCO, Paris, pp. 179–204.

Mesquita, O.V. (1989). Agricultura. In: Duarte, A.C. (ed.), *Geografia do Brasil, região Centro-Oeste* (vol. 1). Fundação Instituto Brasileiro de Geografia e Estatistica, Rio de Janeiro, pp. 149–170.

Millington, A.C., Critchley, R.W., Douglas, T.D. and Ryan, P. (1994). *Estimating woody biomass in Sub-Saharan Africa*. The World Bank, Washington, DC.

Miranda, A.C. and Miranda, H.S. (1993). *Efeitos de diferentes regimes de queima na estrutura e dinâmica de comunidades de cerrado.* Relatório final, Dept. de Ecologia, Universidade de Brasilia, DF.

Miranda, A.C., Miranda, H.S., Dias, I.O. and Dias, B.F. (1993). Soil and air temperatures during prescribed *cerrado* fires in Central Brazil. *Journal of Tropical Ecology*, **9**(3): 313–320.

Misana, S., Mung'ong'o, C. and Mukamuri, B. (1996). *Miombo* woodlands in the wider context: macro-economic and inter-sectoral influences. In: Campbell, B. (ed.), *The miombo in transition: woodlands and welfare in Africa.* Center for International Forestry Research (CIFOR), Bogor, Indonesia, pp. 73–99.

Misra, R. (1983). Indian savannas. In: Bourlière, F. (ed.), *Tropical savannas. Ecosystems of the world 13.* Elsevier Scientific Publishing Company, Amsterdam, pp. 151–166.

Mistry, J. (1996). *Corticolous lichens as potential bioindicators of fire history: a study in the cerrado of the Distrito Federal, central Brazil.* PhD Thesis, Department of Geography, School of Oriental and African Studies, University of London, London.

Mistry, J. (1998a). Fire in the *cerrado* (savannas) of Brazil: an ecological review. *Progress in Physical Geography*, **22**: 425–448.

Mistry, J. (1998b). Decision-making for fire use among farmers in the savannas of central Brazil. *Journal of Environmental Management*, **54**: 321–334.

Mistry, J. (1998c). Corticolous lichens as potential bioindicators of fire history: a study in the *cerrado* of the Distrito Federal, central Brazil. *Journal of Biogeography*, **25**: 409–441.

Mistry, J. (1998d). A preliminary Lichen Fire History (LFH) Key for the *cerrado* of the Distrito Federal, central Brazil. *Journal of Biogeography*, **25**: 443–452.

Mistry, J. and Stott, P. (1993). The savanna forests of Manipur State, India: an historical overview. *Global Ecology and Biogeography Letters*, **3**: 10–17.

Mittermeier, R.A., Werner, T., Ayres, J.M. and Fonseca, G.A.B. (1992). O pais da megadiversidade. *Ciencia Hoje*, **14**(81): 20–27.

Monasterio, M. (1970). Ecología de las Sabanas de América Tropical. II. Caracterización ecológica del clima en los *llanos* de Calabozo, Venezuela. *Revista Geográfica* (ULA Mérida), **9**(21): 5–38.

Monasterio, M. and Sarmiento, G. (1976). Phenological strategies of plant species in the tropical savanna and the semi-deciduous forest of the Venezuelan *Llanos*. *Journal of Biogeography*, **3**: 325–356.

Monela, G.C., O'Ktin'ati, A. and Kiwele, P.M. (1993). Socio-economic aspects of charcoal consumption and environmental consequences along the Dar Es Salaam–Morogoro highway, Tanzania. *Forest Ecology and Management*, **58**: 249–258.

Monnier, Y. (1981). *La poussière et la cendre.* Agence de Coopération Culturelle et Technique, Paris.

Montes, R. and Medina, E. (1977). Seasonal changes in nutrient content of leaves of savanna trees with different ecological behaviour. *Geó-Eco-Tropical*, **1**: 295–307.

Montes, R. and San José, J.J. (1992). Changes in bulk precipitation reactivity throughout the vegetation soil continuum in a *Trachypogon* savanna (Venezuela). *Communications in Soil Science and Plant Analysis*, **23**(15–16): 1753–1766.

Montes, R. and San José, J.J. (1995). Vegetation and soil analysis of topo-sequences in the Orinoco *llanos*. *Flora*, **190**: 1–33.

Montgomery, R.F. and Askew, G.P. (1983). Soils of tropical savannas. In: Bourlière, F. (ed.), *Tropical savannas. Ecosystems of the world 13.* Elsevier Scientific Publishing Company, Amsterdam, pp. 63–78.

Moore, P.D., Chaloner, B. and Stott, P. (1996). *Global environmental change.* Blackwell Science, Oxford.

Moore, P.H.R., Gill, A.M. and Kohnert, R. (1995). Quantitating bushfires for ecology using two electronic devices and biological indicators. *CALMScience Supplement,* **4**: 83–88.

Mordelet, P. and Menaut, J.C. (1995). Influence of trees on above-ground production dynamics of grasses in a humid savanna. *Journal of Vegetation Science,* **6**: 223–228.

Mordelet, P., Abbadie, L. and Menaut, J.C. (1993). Effects of tree clumps on soil characteristics in a humid savanna of West Africa (Lamto, Côte d'Ivoire). *Plant Soil,* **153**: 103–111.

Mordelet, P., Barot, S. and Abbadie, L. (1996). Root foraging strategies and soil patchiness in a humid savanna. *Plant Soil,* **182**: 171–176.

Mordelet, P., Menaut, J.C. and Mariotti, A. (1997). Tree and grass rooting patterns in an African humid savanna. *Journal of Vegetation Science,* **8**: 65–70.

Moreira, A.G. (1992). *Fire protection and vegetation dynamics in the Brazilian cerrado.* PhD Thesis, Harvard University, Cambridge, MA.

Moreira, J.R. and Macdonald, D.W. (1996). Capybara use and conservation in South America. In: Taylor, V.J. and Dunstone, N. (eds), *The exploitation of mammal populations.* Chapman and Hall, London, pp. 88–101.

Morison, C.G.T., Hoyle, A.C. and Hope-Simpson, J.F. (1948). Tropical soil vegetation catenas and mosaics. *Journal of Ecology,* **36**: 1–84.

Mortimore, M. (1998). *Roots in the African dust. Sustaining the Sub-Saharan dryland.* Cambridge University Press, Cambridge.

Mott, J.J. (1987). Planned invasions of Australian tropical savannas. In: Groves, R.H. and Bourdon, J.J. (eds), *Ecology of biological invasions: an Australian perspective.* Australian Academy of Science, Canberra, pp. 97–105.

Mott, J.J. and Andrew, M.H. (1985). The effect of fire on the population dynamics of native grasses in tropical savannas of north-west Australia. *Proceedings of the Ecological Society of Australia,* **13**: 231–239.

Mott, J.J., Bridge, B.J. and Arndt, W. (1979). Soil seals in tropical tallgrass pastures of northern Australia. *Australian Journal of Soil Research,* **30**: 483–494.

Mott, J.J., Williams, J., Andrew, M.H. and Gillison, A. (1985). Australian savanna ecosystems. In: Tothill, J.C. and Mott, J.J. (eds), *Ecology and management of the world's savannas.* Australian Academy of Science, Canberra, pp. 56–82.

Moura, L., Pagani, M. and Campos, M. (1989). Avaliação de predação em algumas espécies vegetais de *cerrado* (Itirapina, SP). *Congresso Nacional de Botânica,* **40**: 477.

Moyo, S., O'Keefe, P and Sill, M. (1993). *The Southern African environment: profiles of the SADC Countries.* Earthscan, London.

Mtambanengwe, F. and Kirchmann, H. (1995). Litter from a tropical savanna woodland (*miombo*): chemical composition and C and N mineralisation. *Soil Biology and Biochemistry,* **27**: 1639–1651.

Mucunguzi, P. and Oryem-Origa, H. (1996). Effects of heat and fire on the germination of *Acacia sieberiana* D.C. and *Acacia gerrardii* Benth in Uganda. *Journal of Tropical Ecology,* **12**: 1–10.

Mueller, B., Alston, L., Libecap, G. and Schneider, R. (1994). Land, property rights and privatisation in Brazil. *The Quarterly Review of Economics and Finance,* **34**: 261–280.

Mueller, C.C. (1995). *A sustentabilidade da expansão agricola nos cerrados.* Documento de Trabalho No. 36, Instituto Sociedade, População e Natureza, Brasilia, DF.

Murphree, M.W. and Cumming, D.H.M. (1993). Savanna land use: policy and practice in Zimbabwe. In: Young, M.D. and Solbrig, O.T. (eds), *The world's savannas. Economic driving forces, ecological constraints and policy options for sustainable land use. Man and the biosphere Vol. 12.* UNESCO, Paris, pp. 139–178.

Musonda, F.B. (1986). Plant food in the diet of the prehistoric inhabitants of the Lunsenfwa drainage basin, Zambia, during the last 20,000 years. *Zambia Geographical Journal,* **36**: 17–27.

Mworia-Maitima, J. (1997). Prehistoric fires and land-cover change in western Kenya: evidences from pollen, charcoal, grass cuticles and grass phytoliths. *The Holocene,* 7(4): 409–417.

Myers, B.A., Duff, G.A., Eamus, D., Williams, R.J., Fordyce, I. and O'Grady, A. (1997). Seasonal variation in water relations of trees in a tropical savanna near Darwin, northern Australia. *Australian Journal of Botany,* **45**: 225–240.

Nakhasathien, S. and Stewart-Cox, B. (1990). *Nomination of the Thung Yai-Huai Kha Khaeng Wildlife Sanctuary to be a UNESCO World Heritage Site.* Wildlife Conservation Division, Royal Forest Department, Bangkok.

Nalamphun, A., Santisuk, T. and Smitinand, T. (1969). *The defoliation of teng (Shorea obtusa Wall.) and rang (Pentacme suavis A.DC.) at ASRCT Sakaerat Experiment Station, Amphoe Pak Thong Chai, Changwat Nakhon Ratchasima.* Report 27/8, 1, ASRCT, Bangkok.

Nascimento, M.T. (1987). Notas preliminares sobre a herbivoria foliar em *Vochysia divergens* Phol. *Ciência e Cultura,* **39**(7): 588–589.

Nascimento, M.T. and Lewinsohn, T.M. (1992). Impacto dos herbivoros. In: Dias, B.F. de S. (ed.), *Alternativas de desenvolvimento dos cerrados: manejo e conservação dos recursos naturais renováveis.* Fundação pró-natureza (FUNATURA), Brasilia, DF, pp. 38–42.

Nascimento, M.T., Villela, D. and Lacerda, L. (1990). Foliar growth, longevity and herbivory in two *cerrado* species near Cuiabá, MT, Brazil. *Revista Brasileira Botanica,* **13**: 27–32.

Naves, M.A. (1996). Efeito do fogo na população de formigas (Hymenoptera – Formicidae) em *cerrado* do Distrito Federal. In Miranda, H.S., Saito, C.H. and Dias, B.F. de S. (eds), *Impactos de queimadas em áreas de cerrado e restinga.* Universidade de Brasilia, Brasilia, DF, pp. 170–177.

Neumann, R.P. (1995). Local challenges to global agendas: conservation, economic liberalisation and pastoralists' rights movement in Tanzania. *Antipode,* **27**(4): 363–382.

Nicholls, N., Lavery, B., Frederiksen, C., Drosdowsky, W. and Torok, S. (1996). Recent apparent changes in relationships between the El Nino – Southern oscillation and Australian rainfall and temperature. *Geophysical Research Letters,* **23**: 3357–3360.

Nicholson, P.H. (1981). Fire and the Australian Aborigine: an enigma. In: Gill, A.M., Groves, R.H. and Noble, I.R. (eds), *Fire and the Australian biota.* Australian Academy of Science, Canberra.

Nicholson, S.E. (1994). Recent rainfall fluctuations in Africa and their relationship to past conditions over the continent. *The Holocene,* **4**: 121–131.

Nicholson, S.E. and Flohn, H. (1980). African environmental and climatic changes and the general atmospheric circulation in late Pleistocene and Holocene. *Climatic Change,* **2**: 313–348.

Nimuendajú, C. (1983). *Os Apinayé.* Museu Paraense Emilio Goeldi, Tradução da edição inglesa de 1939, Belém.

Nix, H.A. (1983). Climate of tropical savannas. In: Bourlière, F. (ed.), *Tropical savannas. Ecosystems of the world 13*. Elsevier Scientific Publishing Company, Amsterdam, pp. 37–62.

Njovu, F.C. (1993). Non-wood forest products: Zambia. In: *Non-wood forest products: a regional expert consultation for English-speaking African countries*, Arusha, 17–22 October 1993, Annex IV. FAO and the Commonwealth Science Council, Rome and London.

Noller, B.N., Currey, N.A., Cusbert, P.J., Tuor, M., Bradley, P. and Harrison, A. (1985). Temporal variability in atmospheric nutrient flux to the Magela and Nourlangie Creek system, Northern territory. *Proceedings of the Ecological Society of Australia*, **13**: 21–31.

Norman, M.J.T. (1963). The pattern of dry matter and nutrient content changes in native pastures at Katherine, NT. *Australian Journal of Experimental Agriculture and Animal Husbandry*, **3**: 119–124.

Norman, M.J.T. and Wetselaar, R. (1960). Losses of nitrogen on burning native pasture at Katherine, NT. *Journal of Australian Institute of Agricultural Science*, **26**: 272–273.

Norton, G.A. and Walker, B.H. (1985). A decision analysis approach to savanna management. *Journal of Environmental Management*, **21**, 15–31.

Norton-Griffiths, M. (1979). The influence of grazing, browsing and fire on vegetation dynamics. In: Sinclair, A.R.E. and Norton-Griffiths, M. (eds), *Serengeti: dynamics of an ecosystem*. University of Chicago Press, Chicago, pp. 310–352.

Nott, J. and Price, D. (1994). Plunge pools and paleoprecipitation. *Geology*, **22**(11): 1047–1050.

Novaes-Pinto, M. (1994). Caracterização geomorfológica. In: Novaes-Pinto, M. (ed.), *Cerrado. Caracterização, Ocupação e Perspectivas* (2nd edition). Editora Universidade de Brasilia, Brasilia, DF, pp. 285–320.

Nyamapfene, K., Hussein, J. and Asumadu, K. (eds), (1988). *The red soils of East and Southern Africa*. International Development Research Centre, Ottawa, Canada.

Nyathi, P. and Campbell, B.M. (1993). Leaf quality of *Sesbania sesban*, *Leucaena leucocephala* and *Brachystegia spiciformis*: potential agroforestry species. *Forest Ecology and Management*, **64**: 259–264.

Oba, G. (1998). Effects of excluding goat herbivory on *Acacia tortilis* woodland around pastoralist settlements in northwest Kenya. *Acta Oecologica*, **19**(4): 395–404.

O'Connor, T.G. (1977) *Habitat selection of impala Aepyceros melampus in the Nylsvley Nature Reserve*. BSc Honours Thesis, University of Witwatersrand, Johannesburg.

O'Connor, T.G. (1985). *A synthesis of field experiments concerning the grass layer in the savanna regions of southern Africa*. South African National Scientific Programmes Report 114, CSIR, Pretoria.

O'Connor, T.G. and Pickett, G.A. (1992). The influence of grazing on seed production and seed banks of some African savanna grasslands. *Journal of Applied Ecology*, **29**: 247–260.

Ohiagu, C.E. and Wood, T.G. (1979). Grass production and decomposition in Southern Guinea savanna, Nigeria. *Oecologia* (Berlin), **40**: 155–165.

Ojasti, J. (1978). *The relation between population and production of the Capybara*. PhD Thesis, University of Georgia, Athens.

Ojasti, J. (1983). Ungulates and large rodents of South America. In: Bourlière, F. (ed.), *Tropical savannas. Ecosystems of the world 13*. Elsevier Scientific Publishing Company, Amsterdam, pp. 427–439.

Ojasti, J. (1991). Human exploitation of capybara. In: Robinson, J.G. and Redford,

K.H. (eds), *Neotropical wildlife use and conservation*. The University of Chicago Press, Chicago, pp. 236–252.

Oliveira, P.S. (1997). The ecological function of extrafloral nectaries: herbivore deterrence by visiting ants and reproductive output in *Caryocar brasiliense* (Caryocaraceae). *Functional Ecology*, **11**: 323–330.

Oliveira-Filho, A.T. (1992). Floodplain 'murundus' of Central Brazil: evidence for the termite-origin hypothesis. *Journal of Tropical Ecology*, **8**: 1–19.

Oliveira-Filho, A.T. and Martins, F.R. (1986). Distribução, caracterização e composição floritica das formações vegetais da régião da Salgadeira, na Chapada dos Guimarães (MT). *Revista Brasileira de Botânica*, **9**: 207–223.

Oliveira-Filho, A.T. and Martins, F.R. (1991). A comparative study of five *cerrado* areas in southern Mato Grosso, Brazil. *Edinburgh Journal of Botany*, **48**: 307–332.

Oliveira-Filho, A.T., Shepherd, G.H., Martins, F.R. and Stubblebine, W.H. (1989). Environmental factors affecting physiognomic and floristic variation in an area of *cerrado* in central Brazil. *Journal of Tropical Ecology*, **5**(4): 413–431.

O'Neill, A.L., Head, L.M. and Marthick, J.K. (1993). Integrating remote sensing and spatial analysis techniques to compare Aboriginal and pastoral fire patterns in the East Kimberley, Australia. *Applied Geography*, **13**: 67–85.

O'Rourke, P.L., Winks, L. and Kelly, A.M. (1992). *North Australia beef producer survey 1990*. Queensland Department of Primary Industries and Meat Research Corporation, Brisbane.

Orta, C.S. (1974). Los obstaculos al crecimiento autosostenido de la agricultura venezolana. In: Maza Zavala, D.F. (ed.), *Venezuela, crecimiento sin desarrollo*. Ediciones Nuestro Tiempo, Mexico, DF, pp. 201–237.

Owen-Smith, R.N. (1988). *Megaherbivores: the influence of very large body size on ecology*. Cambridge University Press, Cambridge.

Owen-Smith, R.N. (1993). Woody plants, browsers and tannins in southern African savannas. *South African Journal of Science*, **89**: 505–510.

Owen-Smith, R.N. and Cooper, S.M. (1987). Palatability of woody plants to browsing ungulates in a South African savanna. *Ecology*, **68**: 319–331.

Owen-Smith, R.N. and Cumming, D.H.M. (1994). Comparative foraging strategies of grazing ungulates in African savanna grasslands. *Proceedings of the XVII International Grassland Congress*, **1993**: 691–698.

Owen-Smith, R.N. and Danckwerts, J.E. (1997). Herbivory. In: Cowling, R.M., Richardson, D.M. and Pierce, S.M. (eds), *Vegetation of Southern Africa*. Cambridge University Press, Cambridge, pp. 397–420.

Palmer, A.R. and van Rooyen, A.F. (1998). Detecting vegetation change in the southern Kalahari using Landsat TM data. *Journal of Arid Environments*, **39**: 143–153.

Palo, R.T., Gowda, J. and Högberg, P. (1993). Species height and root symbiosis: two factors influencing antiherbivore defense of woody plants in East African savanna. *Oecologia*, **93**: 322–326.

Paovongsar, S. (1976). *Litterfall and mineral nutrient content of litter in dry dipterocarp forest*. MSc Thesis, Faculty of Forestry, Kasetsart University, Bangkok.

Parada, J.M. and Andrade, S.M. (1977). *Cerrados*: recursos minerais. In: Ferri, M.G. (ed.), *IV Simpósio sobre o Cerrado: bases para utilização agropecuária*. Editora da Universidade de São Paulo, São Paulo, pp. 195–209.

Parsons, D.A.B., Scholes, M.C., Scholes, R.J. and Levine, J.S. (1996). Biogenic NO emissions from savanna soils as a function of fire regime, soil type, soil nitrogen, and water status. *Journal of Geophysical Research-Atmospheres*, **101**(D19): 23 683–23 688.

Parsons, J.J. (1972). Spread of African pasture grasses to the American tropics. *Journal of Range Management*, **25**: 12–17.

Parsons, J.J. (1980). Europeanisation of the savanna lands of northern South America. In: Harris, D.R. (ed.), *Human ecology in savanna environments*. Academic Press, London, pp. 267–289.

Pearce, D., Jackson, A. and Braithwaite, R. (1996). Aboriginal people of the tropical savannas: resource utilisation and conflict resolution. In: Ash, A.J. (ed.), *The future of tropical savannas: an Australian perspective*. CSIRO, Australia, pp. 88–103.

Pearce, D.W. (1993). Developing Botswana's savannas. In: Young, M.D. and Solbrig, O.T. (eds), *The world's savannas. Economic driving forces, ecological constraints and policy options for sustainable land use. Man and the biosphere Vol. 12*. UNESCO, Paris, pp. 205–220.

Pell, A.S. and Tidemann, C.R. (1997). The impact of two exotic hollow-nesting birds on two native parrots in savannah and woodland in eastern Australia. *Biological Conservation*, **79**: 145–153.

Pellew, R.A.P. (1981). *The giraffe (Giraffa camelopardalis tippelskirchi Matschie) and its Acacia food resource in the Serengeti National park*. PhD Thesis, Cambridge University.

Pellew, R.A.P. (1983). The impacts of elephants, giraffe and fire upon the *Acacia tortilis* woodlands of the Serengeti. *African Journal of Ecology*, **21**: 41–74.

Pendle, B.G. (1982). *The estimation of water use by the dominant species in the Burkea africana savanna*. MSc Thesis, University of Witwatersrand, Johannesburg.

Perkins, J.S. and Thomas, D.S.G. (1993). Spreading deserts or spatially confined environmental impacts? Land degradation and cattle ranching in the Kalahari desert of Botswana. *Land Degradation and Rehabilitation*, **4**: 179–194.

Peters, D.V. (1950). *Land usage in Serenje District*. Rhodes-Livingstone Paper No. 19, Manchester University Press for the Institute for African Studies, University of Zambia.

Phillips, J. (1974). Effects of fire in forest and savnna ecosystems of sub-Saharan Africa. In: Kozlowski, T.T. and Ahlgren, C.E. (eds), *Fire and ecosystems*. Academic Press, New York, pp. 435–481.

Phillipson, D.W. (1971). Early man. In: Davies, D.H. (ed.), *Zambia in maps*. University of London Press, London, pp. 28–31.

Piearce, G.D. (1993). Natural regeneration of indigenous trees: the key to their successful management. In: Piearce, G.D. and Gumbo, D.J. (eds), *The ecology and management of indigenous forests in southern Africa*. Zimbabwe Forestry Commission and SAREC, Harare, pp. 103–123.

Pimbert, M.P. and Pretty, J.N. (1997). *Diversity and sustainability in community based conservation*. IIED Paper, IIED, London.

Pittock, A.B. (1975). Climatic change and the patterns of variation in Australian rainfall. *Search*, **6**: 498–504

Pivello, V.R. (1992). *An expert system for the use of prescribed fires in the management of Brazilian savannas*. PhD Thesis, Imperial College, University of London.

Pivello, V.R. and Coutinho, L.M. (1992). Transfer of macro-nutrients to the atmosphere during experimental burnings in an open *cerrado* (Brazilian savanna). *Journal of Tropical Ecology*, **8**(4): 487–497.

Pivello, V.R. and Coutinho, L.M. (1996). A qualitative successional model to assist in the management of Brazilian *cerrados*. *Forest Ecology and Management*, **82**: 127–138.

Pivello, V.R. and Norton, G.A. (1996). FIRETOOL: an expert system for the use of prescribed fires in Brazilian savannas. *Journal of Applied Ecology*, **33**: 348–356.

Pivello-Pompêia, V.R. (1985). *Exportação de macronutrientes para a atmosfera durante queimadas realizadas no campo cerrado de Emas (Pirassununga, S.P.).* Tese de mestrado, Universidade de São Paulo, São Paulo.

Place, F. and Otsuka, K. (1997). *Population density, land tenure and resource management in Uganda.* ICRAF and IFPRI, Nairobi, Kenya.

Poffenberger, M. (1990). The evolution of forest management systems in Southeast Asia. In: Poffenberger, M. (ed.), *Keepers of the forest. Land management alternatives in Southeast Asia.* Kumarian Press, Connecticut, pp. 7–26.

Polley, H.W., Johnson, H.B. and Mayeux, H.S. (1992). Determination of root biomasses of three species grown in a mixture using stable isotopes of carbon and nitrogen. *Plant and Soil,* **142**: 97–106.

Ponce, V.T. and Da Cunha, C.N. (1993). Vegetated earthmounds in tropical savannas of central Brazil: a synthesis. *Journal of Biogeography,* **20**: 219–225.

Poorter, H. (1993). Interspecific variation in the growth response of plants to an elevated ambient CO_2 concentration. *Vegetatio,* **104/105**: 77–97.

Poth, M., Anderson, I.C., Miranda, H.S., Miranda, A.C. and Riggan, P.J. (1995). The magnitude and persistence of soil NO, N_2O, CH_4 and CO fluxes from burned tropical savanna in Brazil. *Global Biogeochemical Cycles,* **9**(4): 503–513.

Prada, M., Marini, O.J. and Price, P.W. (1995). Insects in flower heads of *Aspilia foliacea* (*Asteraceae*) after a fire in a central Brazilian savanna – evidence for the plant vigour hypothesis. *Biotropica,* **27**(4): 513–518.

Prado, J. (1989). *Herbivoria por insetos em um gradiente de cerrado com diferentes regimes de fogo.* Tese de mestrado, Universidade de Brasilia, Brasilia, DF.

Pragtong, K. and Thomas, D.E. (1990). Evolving management systems in Thailand. In: Poffenberger, M. (ed.), *Keepers of the forest. Land management alternatives in Southeast Asia.* Kumarian Press, Connecticut, pp. 167–186.

Prance, G.T. (1973). Phytogeographic support for the theory of Pleistocene forest refuges in the Amazon basin, based on evidence from distribution patterns in Caryocaraceae, Dichapetalaceae and Lecythidaceae. *Acta Amazônica,* **3**: 5–28.

Prance, G.T. (1982). A review of the phytogeographic evidences for Pleistocene climate changes in the Neotropics. *Annales Missouri Botanical Garden,* **69**: 594–624.

Prance, G.T. and Schaller, G.B. (1982). Preliminary study on some vegetation types of the Pantanal, Mato Grosso, Brazil. *Brittonia,* **34**: 228–251.

Prayurasiddhi, T., Petchkong, T. and Laohawat, O. (1988). *Impact of forest fire upon wild fauna in Huai Kha Khaeng Wildlife Sanctuary.* Paper No. 3, Khao Nang Rum Wildlife Research Station, Wildlife Conservation Division, Royal Forest Department, Bangkok.

Press, A.J. (1988). Comparisons of the extent of fire in different land management systems in the Top End of the Northern Territory. *Proceedings of the Ecological Society of Australia,* **15**: 167–175.

Press, T., Brock, J. and Andersen, A. (1995). Fauna. In: Press, T., Lea, D., Webb, A. and Graham, A. (eds), *Kakadu. Natural and cultural heritage and management.* Australian Nature Conservation Agency, North Australia Research Unit, The Australian National University, Darwin, pp. 167–216.

PREVFOGO (1995). *Como fazer uma queimada controlada.* PREVFOGO, Brasilia, DF.

PREVFOGO/IBAMA/SUPES-MT (1996). *Conversando com o homen do campo.* PREVFOGO, Brasilia, DF.

Prince, S.D., Brown de Colstoun, E. and Kravitz, L.L. (1998). Evidence from rain-use efficiencies does not indicate extensive Sahelian desertification. *Global Change Biology,* **4**: 359–374.

Prins, E.M. and Menzel, W.P. (1994). Trends in South American biomass burning detected with the GOES visible infrared spin scan radiometer atmospheric sounder from 1983 to 1991. *Journal of Geophysical Research – Atmospheres*, **99**(D8): 16 719–16 735.

Prins, H.H.T. (1988). Plant phenology patterns in Lake Manyara National Park. *Journal of Biogeography*, **15**: 465–480.

Prins, H.H.T. (1989). Condition changes and choice of social environment in African buffalo bulls. *Behaviour*, **108**: 297–324.

Prins, H.H.T. (1996). *Ecology and behaviour of the African buffalo: social inequality and decision making*. Chapman and Hall, London.

Prins, H.H.T. and Beekman, J.H. (1987). A balanced diet as a goal of grazing: the food of the Manyara buffalo. In: Prins, H.H.T. (ed.), *The buffalo of Manyara*. Krips Repro, Meppel, The Netherlands, pp. 69–98.

Prins, H.H.T. and Iason, G.R. (1988). Dangerous lions and nonchalant buffalo. *Behaviour*, **108**: 262–297.

Prins, H.H.T. and Loth, P.E. (1988). Rainfall patterns as background to plant phenology in northern Tanzania. *Journal of Biogeography*, **15**: 451–463.

Prins, H.H.T. and Olff, H. (1998). Species richness of African grazer assemblages: towards a functional explanation. In: Newbery, D.M., Prins, H.H.T. and Brown, N.D. (eds), *Dynamics of tropical communities*. British Ecological Symposium Volume No. 37, Blackwell Science, Oxford, pp. 449–490.

Prins, H.H.T. and Van der Jeugd, H.P. (1993). Herbivore population crashes and woodland structure in East Africa. *Journal of Ecology*, **81**: 305–314.

Puigdefábregas, J. (1998). Ecological impacts of global change on drylands, and their implications for desertification. *Land Degradation and Development*, **9**: 393–406.

Puri, G.S. (1960). *Indian forest ecology. Volumes I and II*. Oxford Book and Stationery Co., New Delhi.

Puzo, B. (1978). Patterns of man–land relations. In: Werger, M.J.A. (ed.), *Biogeography and ecology of southern Africa*. Junk, The Hague, pp. 1049–1113.

Queiróz Neto, de J.P. (1982). Solos da região dos *cerrados* e suas interpretações. *Revista Brasileira de Ciência do Solo*, **6**(1): 1–12.

Rabinowitz, A. (1990). Fire, dry dipterocarp forest, and the carnivore community in Huai Kha Khaeng Wildlife Sanctuary, Thailand. *Natural History Bulletin of the Siam Society*, **38**: 99–115.

Rachid, M. (1947). Transpiraçâo e sistemas subterrâneos da vegetação de verão dos *campos cerrados* de Emas. *Boletim de Faculdade de Filosofia, Ciências e Letras da USP, Botânica*, **13**: 37–69.

Rachid-Edwards, M. (1956). Alguns dispositivos para proteção de plantas contra a sêca e o fogo. *Boletim de Botânica, USP*, **13**: 35–68.

Radford, I. (1998). Cattle and spread of exotic weeds. *Savanna Links*, **5**: 9.

Ramage, C.S. (1986). El Niño. *Scientific American*, **254**: 76–83.

Ramanathan, V., Cicerone, R.J., Singh, H.B. and Kiehl, J.T. (1985). Trace gas trends and their potential role in climate change. *Journal of Geophysical Research*, **90**: 5547–5566.

Ranzani, G. (1971). Solos do *cerrado* do Brasil. In: Ferri, M.G. (ed.), *III Simpósio sobre o Cerrado*. Editora da Universidade de São Paulo, São Paulo, pp. 26–43.

Ratter, J.A. (1971). Some notes on two types of *cerradão* occurring in north eastern Mato Grosso. In: Ferri, M.G. (ed.), *III Simpósio sobre o Cerrado*. Editora da Universidade de São Paulo, São Paulo, pp. 26–43.

Ratter, J.A. (1986). *Notas sobre a vegetação da Fazenda Agua Limpa (Brasilia, D.F., Brazil)*. Editora UnB, Textos Universitários No. 003, Brasilia, DF.

Ratter, J.A. (1987). Notes on the vegetation of the Parque Nacional do Araguaia (Brazil). *Notes of the Royal Botanic Garden of Edinburgh*, **44**: 311–342.

Ratter, J.A. (1991). *Notas sobre a vegetação da Fazenda Agua Limpa (Brasilia, D.F.)*. Royal Botanic Garden, Edinburgh.

Ratter, J.A. (1995). Profile of the *cerrado* (tree savanna biome) of central Brazil: modern agriculture and conservation of biodiversity. *Tropical Agriculture Association Newsletter*, **15**(4): 33–34.

Ratter, J.A. and Dargie, T.C.D. (1992). An analysis of the floristic composition of 26 *cerrado* areas in Brazil. *Edinburgh Journal of Botany*, **49**(2): 235–250.

Ratter, J.A. and Ribeiro, J.F. (1996). Biodiversity of the flora of the *cerrado*. In: Pereira, R.C. and Nasser, L.C.B. (eds), *Anais de VIII Simpósio sobre o cerrado*. EMBRAPA/CPAC, Planaltina, DF, Brazil, pp. 3–6.

Ratter, J.A., Richards, P.W., Argent, G. and Gifford, D.R. (1973). Observations on the vegetation of north-eastern Mato Grosso. 1. The woody vegetation types of the Xavantina–Cachimbo expedition area. *Philosophical Transactions of the Royal Society of London (B)*, **226**: 449–492.

Ratter, J.A., Askew, G.P., Montgomery, R.F. and Gifford, D.R. (1977). Observações adicionais sobre o *cerradão* de solos mesotróficos no Brasil Central. In: Ferri, M.G. (ed.), *IV Simpósio sobre o Cerrado*. Editora da Universidade de São Paulo, São Paulo, pp. 303–316.

Ratter, J.A., Askew, G.P., Montgomery, R.F. and Gifford, D.R. (1978). Observations on the vegetation of north-eastern Mato Grosso. II. Forests and soils of the Rio Suiá-Missu area. *Proceedings of the Royal Society of London (Series B)*, **203**: 191–208.

Ratter, J.A., Leitão Filho, H.F., Argent, G., Gibbs, P.E., Semir, J., Shepherd, G. and Tamashiro, J. (1988). Floristic composition and community structure of a southern *cerrado* area in Brazil. *Notes of the Royal Botanic Garden of Edinburgh*, **45**(1): 137–151.

Ratter, J.A., Bridgewater, S., Atkinson, R. and Ribeiro, J.F. (1996). Analysis of the floristic composition of the Brazilian *cerrado* vegetation II: comparison of the woody vegetation of 98 areas. *Edinburgh Journal of Botany*, **53**: 153–180.

Ratter, J.A., Ribeiro, J.F. and Bridgewater, S. (1997). The Brazilian *cerrado* vegetation and threats to its biodiversity. *Annals of Botany*, **80**: 223–230.

Raventós, J. and Silva, J.F. (1988). Architecture, seasonal growth and interference in three grass species with different flowering phenologies in a tropical savanna. *Vegetatio*, **75**: 115–123.

Raventós, J. and Silva, J.F. (1995). Competition effects and responses to variable numbers of neighbours in two tropical savanna grasses in Venezuela. *Journal of Tropical Ecology*, **11**(1): 39–52.

Rawitscher, F. (1942). Problemas de fitoecologia com considerações especies sobre o Brasil meridional. *Univ. São Paulo, Fac. Filosofia, Ciências Letras*, **28**(3): 3–111.

Rawitscher, F. (1948). The water economy of the vegetation of the 'campos cerrados' in southern Brazil. *Journal of Ecology*, **36**: 237–268.

Rawitscher, F. and Rachid, M. (1946). Troncos subterrâneos de plantas brasileiras. *Anais da Academia Brasileira de Ciências*, **17**: 261–280.

Rawitscher, F., Ferri, M.G. and Rachid, M. (1943). Profundidade dos solos e vegetação em *campos cerrados* do Brasil meridional. *Anais da Academia Brasileira de Ciências*, **15**(4): 267–294.

Rebelo, A.G. (1997). Conservation. In: Cowling, R.M., Richardson, D.M. and Pierce, S.M. (eds), *Vegetation of Southern Africa*. Cambridge University Press, Cambridge, pp. 571–590.

Redmond, E.M. and Spencer, C.S. (1994). Pre-Columbian Chiefdoms. *Research and Exploration*, **10**(4): 422–439.

Ribeiro, J.F. (1983). *Comparação da concentração de nutrientes na vegetação arbórea e nos solos de um cerrado e um cerradão no Distrito Federal, Brasil.* Tese de mestrado, Universidade de Brasilia, Brasilia, DF.

Richards, A.I. (1939). *Land, labour and diet in Northern Rhodesia: an economic study of the Bemba tribe.* International African Institute, Oxford University Press, London.

Richards, P.W. (1952). *The tropical rain forest: an ecological study.* Cambridge University Press, Cambridge.

Richardson, D.M., Macdonald, I.A.W., Hoffman, J.H. and Henderson, L. (1997). Alien plant invasions. In: Cowling, R.M., Richardson, D.M. and Pierce, S.M. (eds), *Vegetation of Southern Africa*. Cambridge University Press, Cambridge, pp. 535–570.

Ridley, W.F. and Gardner, A. (1961). Fires and rain forest. *Australian Journal of Science*, **23**: 227–228.

Righetti, S.M. (1992). *O papel do fogo na interação inseto-planta: danos foliares e regimes de queima.* Tese de mestrado, Universidade de Brasilia, Brasilia, DF.

Ringrose, S., Vanderpost, C. and Matheson, W. (1996). The use of integrated remotely sensed and GIS data to determine causes of vegetation cover change in southern Botswana. *Applied Geography*, **16**(3): 225–242.

Rivero Blanco, C. (1990). *Hacia Donde Debe ir el Programa Baba? III Taller Sobre Conservacion y Manejo de la Babab (Caiman crocodilus).* Relatoria Final, Fundafauna, Caracas.

Rizzini, C.T. (1963). A flora do *cerrado*. Analise floristica das savanas centrais. In: Ferri, M.G. (ed.), *III Simpósio sobre o Cerrado*. Editora da Universidade de São Paulo, São Paulo, pp. 125–177.

Rizzini, C.T. and Heringer, E.P. (1962). *Preliminares acerca das formaçoes vegetais e do reflorestamento do Brasil Central.* Serviço de Informaçao, M. da Agricultura, Rio de Janeiro.

Robbins, G.B., Bushell, J.J. and Butler, K.L. (1987). Decline in plant and animal production from ageing pastures of green panic (*Panicum maximum* var. *trichoglume*). *Journal of Agricultural Science*, **108**: 407–417.

Roberts, R.G. and Jones, R. (1994). Luminescence dating of sediments: new light on the human colonisation of Australia. *Australian Aboriginal Studies*, **2**: 2–17.

Roberts, R.G., Jones, R. and Smith, M.A. (1990a). Thermoluminescence dating of a 50,000-year-old human occupation site in northern Australia. *Nature*, **345**: 153–156.

Roberts, R.G., Jones, R. and Smith, M.A. (1990b). Early dates at Malakunanja II, a reply to Bowdler. *Australian Archaeology*, **31**: 94–97.

Roberts, R.G., Jones, R. and Smith, M.A. (1990c). Stratigraphy and statistics at Malakunanja II, a reply to Hiscock. *Archaeology in Oceania*, **25**(3): 125–129.

Roberts, R.G., Jones, R. and Smith, M.A. (1993). Optical dating at Deaf Adder Gorge, Northern Territory, indicates human occupation between 53,000 and 60,000 years ago. *Australian Archaeology*, **37**: 58–59.

Roberts, R.G., Jones, R. and Smith, M.A. (1994). Beyond the radiocarbon barrier in Australian prehistory. *Antiquity*, **68**: 611–616.

Robinson, E.T., Gibbs Russel, G.E., Trollope, W.S.W. and Downing, B.H. (1979).

Assessment of short-term burning treatments on the herb layer of False Thornveld of the eastern Cape. *Proceedings of the Grassland Society of Southern Africa*, **14**: 79–84.

Rocha, D.M.S. (1988). *Levantamento floristico da area que devera ser inundada pela construção da futura usina hidreletrica São Domingos II, Municipio de São Domingos, Goiás*. Geabrasil, Brasilia.

Rodgers, A. and Salehe, J. (1996). Community resource use in the *miombo* woodlands of Tanzania. In: Campbell, B. (ed.), *The miombo in transition: woodlands and welfare in Africa*. Center for International Forestry Research (CIFOR), Bogor, Indonesia, pp. 212

Rodgers, A., Salehe, J. and Howard, G. (1996). The biodiversity of *miombo* woodlands. In: Campbell, B. (ed.), *The miombo in transition: woodlands and welfare in Africa*. Center for International Forestry Research (CIFOR), Bogor, Indonesia, p. 12.

Rodgers, W.A. (1979). The implications of woodland burning for wildlife management. In: Ajayi, S.S. and Halstead, L.B. (eds), *Wildlife management in savannah woodland*. Taylor Francis, London, pp. 151–159.

Rodrigues, F.H.G. (1996). Influência do fogo e da seca na disponibilidade de alimento para herbivoros do *cerrado*. In Miranda, H.S., Saito, C.H. and Dias, B.F. de S. (eds), *Impactos de queimadas em áreas de cerrado e restinga*. Universidade de Brasilia, Brasilia, DF, pp. 76–83.

Rodriguez Mirabal, A.G. (1987). *La formación del latifundio ganadero en los llanos de Apure: 1750–1800*. Academia Nacional de la Historia, Caracas.

Rolim, W.J.O. and Correia, S.D. (1995). *Cerrado: o que vôce precisa saber para preserva-lo*. Federal University of Goiás, Goiana.

Rollet, B. (1953). Note sur les forêts claires de Sud de l'Indochine. *Bois et Forêts des Tropiques*, **31**: 3–13.

Rose, D.B. (ed.) (1995). *Country in flames*. Biodiversity Series, Paper No. 3, DEST Biodiversity Unit, Canberra.

Rowell, M.N. and Cheney, N.P. (1979). Firebreak preparation in tropical areas by rolling and burning. *Australian Forestry*, **42**: 8–12.

Royal Forest Department (1988). *Forest fire control in Thailand*. Royal Forest Department. Forest Fire Control Sub-Division, Bangkok.

Ruangpanit, N. and Pongumphai, S. (1983). *Accumulation of energy and nutrient content of Arundinaria pusilla in the dry dipterocarp forest*. Faculty of Forestry, Kasetsart University, Bangkok.

Rull, V. (1991). Contribución a la paleoecologia de Pantepui la Gran Sabana (Guayana Venezolana): clima, biogeografia, y ecologia. *Ciencia Guaianae*, **2**: 1–333.

Rull, V. (1992). Successional patterns of the Gran Sabana (south-eastern Venezuela) vegetation during the last 5000 years, and its responses to climatic fluctuations and fire. *Journal of Biogeography*, **19**: 329–338.

Russell, J.S. (1988). The effects of climatic change on the productivity of Australian agroecosystems. In: Pearman, G.I. (ed.). *Greenhouse: planning for climate change*. CSIRO, Melbourne, pp. 491–505.

Russell-Smith, J. (1995). Fire management. In: Press, T., Lea, D., Webb, A. and Graham, A. (eds), *Kakadu. Natural and cultural heritage and management*. Australian Nature Conservation Agency, North Australia Research Unit, The Australian National University, Darwin, pp. 217–237.

Russell-Smith, J. and Dunlop, C.R. (1987). The status of monsoon vine forests in the Northern Territory: a perspective. In: Werren, G.L and Kershaw, A.P. (eds), *The rainforest legacy: Australian national rainforests study, Vol. 1*. Australian Government Publishing Service, Canberra, pp. 227–228.

Russell-Smith, J., Lucas, D., Gapindi, M., Gunbunuka, B., Kapirigi, N., Namingum, G., Lucas, K., Giuliani, P. and Chaloupka, G. (1997). Aboriginal resource utilisation and fire management practice in western Arnhem Land, monsoonal northern Australia: notes for prehistory, lessons for the future. *Human Ecology*, **25**(2): 159–195.

Rutherford, M.C. (1978). Primary production ecology in southern Africa. In: Werger, M.J.A. (ed.), *Biogeography and ecology of southern Africa*. Junk, The Hague, pp. 621–659.

Rutherford, M.C. (1979). *Aboveground biomass subdivisions in woody species of the savanna ecosystem project study area*. South African National Scientific Programmes Report 36, CSIR, Pretoria.

Rutherford, M.C. (1981). Survival, regeneration and leaf biomass changes in woody plants following spring burns in *Burkea africana – Ochna pulchra* savanna. *Bothalia*, **13**: 531–552.

Rutherford, M.C. (1983). Growth rates, biomass and distribution of selected woody plant roots in *Burkea africana– Ochna pulchra* savanna. *Vegetatio*, **52**: 45–63.

Rutherford, M.C. (1997). Categorisation of biomes. In: Cowling, R.M., Richardson, D.M. and Pierce, S.M. (eds), *Vegetation of Southern Africa*. Cambridge University Press, Cambridge, pp. 91–98.

Saint-Hilaire, A. (1824). *Histoire des plantes les plus remarcables du Brésil et du Paraguay, I*. Belin Imprimeur-Libraire, Paris.

Salgado-Labouriau, M.L., Casseti, V., Ferraz-Vicentini, K.R., Martin, L., Soubiès, F., Suguio, K. and Turcq, B. (1997). Late Quaternary vegetational and climatic changes in *cerrado* and palm swamp from Central Brazil. *Palaeogeography, Palaeoclimatology, Palaeoecology*, **128**: 215–226.

Salgado-Labouriau, M.L., Barberi, M., Ferraz-Vicentini, K.R. and Parizzi, M.G. (1998). A dry climatic event during the late Quaternary of tropical Brazil. *Review of Palaeobotany and Palynology*, **99**: 115–129.

San José, J.J. and Fariñas, M. (1983). Changes in tree density and species composition in a protected *Trachypogon* savanna in Venezuela. *Ecology*, **64**: 447–458.

San José, J.J. and Fariñas, M. (1991). Temporal changes in the structure of a *Trachypogon* savanna protected for 25 years. *Acta Oecologica*, **12**(2): 237–247.

San José, J.J. and Medina, E. (1975). Effects of fire on organic matter production and water balance in a tropical savanna. In: Medina, E. and Golley, F.B. (eds), *Tropical Ecological Systems*. Ecological Studies 11, Springer-Verlag, Berlin, pp. 251–264.

San José, J.J. and Montes, R. (1991). Regional interpretation of environmental gradients which influence *Trachypogon* savannas in the Orinoco *llanos*. *Vegetatio*, **95**(1): 21–32.

San José, J.J., Montes, R., Garcia-Miragaya, J. and Orihuela, B. (1985). Bioproduction of *Trachypogon* savannas in a latitudinal cross-section of the Orinoco *Llanos*, Venezuela. *Oecologia Generalis*, **6**: 25–43.

San José, J.J., Fariñas, M.R. and Rosales, J. (1991a). Spatial patterns of trees and structuring factors in a *Trachypogon* savanna of the Orinoco *Llanos*. *Biotropica*, **23**(2): 114–123.

San José, J.J., Montes, R. and Nikonovacrespo, N. (1991b). Carbon dioxide and ammonia exchange in the *Trachypogon* savannas of the Orinoco *llanos*. *Annuals of Botany*, **68**(4): 321–328.

San José, J.J., Montes, R.A. and Fariñas, M.R. (1998). Carbon stocks and fluxes in a temporal scaling from a savanna to a semi-deciduous forest. *Forest Ecology and Management*, **105**: 251–262.

Sanchez, P.A., Buresh, R.J. and Leakey, R.R.B. (1997). Trees, soils, and food security. *Philosophical Transactions of the Royal Society of London B*, **352**: 949–961.

Santisuk, T. (1988). An account of the vegetation of northern Thailand. In: Schweinfurth, U. (ed.), *Geoecological Research 5*. Verlag Wiesbaden GMBH, Stuttgart.

Sangtongpraow, S. and Dhamanonda, P. (1973*). Ecological study of stands in dry dipterocarp forest of Thailand*. Unpublished Research Note, Kasetsart University, Bangkok.

Sangtongpraow, S. and Sukwong, S. (1981). *Economic assessment of forest resources in the SERS, Amphur Pakthongchai, Nakhonratchasima Province*. Technical Paper No. 13. Faculty of Forestry, Kasetsart University, Bangkok.

Santongpraow, S. (1982). *Responses of Pentacme suavis A.DC., Dipterocarpus intricatus Dyer and Pterocarpus parvifolius Pierre seedlings to drought*. Technical Paper No. 16, Department of Forest Biology, Faculty of Forestry, Kasetsart University, Bangkok.

Santongpraow, S. (1985). *Mechanisms of forest fire resistance of Shorea obtusa Wall. and Shorea siamensis Miq*. Kasetsart University Conference, Bangkok.

Santongpraow, S. (1986). *Forest fire smoke in the north of Thailand*. Thai Journal of Forestry, Bangkok.

SarDesai, D.R. (1997). *Southeast Asia. Past and present* (4th edition). Westview Press, Colorado.

Sarmiento, G. (1983a). The savannas of tropical America. In: Bourlière, F. (ed.), *Tropical savannas. Ecosystems of the World 13*. Elsevier, Amsterdam, pp. 245–288.

Sarmiento, G. (1983b). Patterns of specific and phenological diversity in the grass community of the Venezuelan tropical savannas. *Journal of Biogeography*, **10**: 373–391.

Sarmiento, G. (1984). *The ecology of neotropical savannas*. Harvard University Press, Cambridge, MA.

Sarmiento, G. (1990). Ecología comarada de ecosistemas de sabana en América del Sul. In: Sarmiento, G. (ed.), *Las sabanas Americanas. Aspectos de su biogeografía, ecología y utilización*. Fondo Editorial Acta Científica Venezolana, Caracas, pp. 15–56.

Sarmiento, G. (1992). Adaptive strategies of perennial grasses in South American savannas. *Journal of Vegetation Science*, **3**: 325–336.

Sarmiento, G. and Acevedo, D. (1991). Dinámica del agua en el suelo, evaporación y transpiración en una pastura y un cultivo de maíz sobre un alfisol en los *llanos* occidentales de Venezuela. *Ecotrópicos*, **4**: 27–42.

Sarmiento, G. and Monasterio, M. (1983). Life forms and phenology. In: Bourlière, F. (ed.), *Tropical savannas. Ecosystems of the World 13*. Elsevier, Amsterdam, pp. 79–108.

Sarmiento, G. and Silva, J.F. (1997). Un modelo de estados y transiciones de la sabana estacional de los *llanos* venezolanos. *Ecotrópicos*, **10**(2): 51–64.

Sarmiento, G. and Vera, M. (1977). La marcha annual del agua en el suelo en sabanas y bosques tropicales de los *llanos* de Venezuela. *Agronomia Tropical*, **27**: 629–649.

Sarmiento, G. and Vera, M. (1979). Composición, estructura, biomana y producción de diferentes sabanas en los Llanos de Venezuela. *Boletin Sociedad Venezolana Ciencias Naturales*, **136**: 5–41.

Sarmiento, G., Goldstein, G. and Meinzer, F. (1985). Adaptive strategies of woody species in neotropical savannas. *Biological Reviews*, **60**: 315–355.

Saunders, I. and Young, A. (1983). Rates of surface processes on slopes, slope retreat and denudation. *Earth Surface Processes and Landforms*, **8**: 473–501.

Savory, B.M. (1962). Rooting habits of important *miombo* species. *Forest (Department) Research Bulletin*, **6**: 1–120.

Scanlon, J.C. and Burrows, W.H. (1990). Woody overstorey impact on herbaceous understorey in *Eucalyptus* spp. communities in central Queensland. *Australian Journal of Ecology*, **15**: 191–197.

Schiavini, I. (1984). Ciclagem de nutrientes no *cerrado*. I – Agua da chuva. *Ciência Cultura, São Paulo*, **36** (Suppl.).

Schimper, A.F.W. (1903). *Plant geography upon a physiological basis*. Clarendon Press, Oxford.

Schmidt, W. (1975). Plant communities in permanent plots of the Serengeti Plains. *Vegetatio*, **30**: 133–145.

Schoffeleers, J. (1971). The religious significance of bushfires in Malawi. *Cahiers des Religions Africaines*, **10**: 271–287.

Scholes, R.J. (1990). The influence of soil fertility on the ecology of southern African savannas. In: Werner, P.A. (ed.), *Savanna ecology and management. Australian perspectives and intercontinental comparisons*. Blackwell, Oxford, pp. 71–75.

Scholes, R.J. (1993). *Nutrient cycling in semi-arid grasslands and savannas: its influence on pattern, productivity and stability*. Proceedings of the 17th International Grasslands Congress, International Grasslands Society, Palmerston North.

Scholes, R.J. (1997). Savanna. In: Cowling, R.M., Richardson, D.M. and Pierce, S.M. (eds), *Vegetation of Southern Africa*. Cambridge University Press, Cambridge, pp. 158–277.

Scholes, R.J. and Archer, S.R. (1997). Tree–grass interactions in savannas. *Annual Review of Ecology and Systematics*, **28**: 517–544.

Scholes, R.J. and Walker, B.H. (1993). *An African savanna. Synthesis of the Nylsvley study*. Cambridge University Press, Cambridge.

Scholes, R.J., Kendall, J. and Justice, C.O. (1996a). The quantity of biomass burned in southern Africa. *Journal of Geophysical Research – Atmospheres*, **101**(D19): 23 667–23 676.

Scholes, R.J., Ward, D. and Justice, C.O. (1996b). Emissions of trace gases and aerosol particles due to vegetation burning in southern hemisphere Africa. *Journal of Geophysical Research – Atmospheres*, **101** (D19): 23 677–23 682.

Scholtz, C.H. (1982). *Trophic ecology of Lepidoptera larvae associated with woody vegetation in a savanna ecosystem*. South African National Scientific Programmes Report 55, CSIR, Pretoria.

Schüle, W. (1990). Landscapes and climate in prehistory: interactions of wildlife, man and fire. In: Goldammer, J.G. (ed.), *Fire in the tropical biota: ecosystem processes and global challenges*. Ecological Studies 84, Springer-Verlag, Berlin, pp. 273–318.

Schulze, B.R. (1965). *Climate of South Africa. Part 8. General Survey*. SA Weather Bureau Report 28, Government Printer, Pretoria.

Schulze, R.E. (1997). Climate. In: Cowling, R.M., Richardson, D.M. and Pierce, S.M. (eds), *Vegetation of Southern Africa*. Cambridge University Press, Cambridge, pp. 21–42.

Scoones, I. (1989). Patch use by cattle in a dryland environment: farmer knowledge and ecological theory. In: Cousins, B. (ed.), *People, land and livestock*. Centre for Applied Social Studies, University of Zimbabwe, Harare, pp. 277–309.

Scoones, I. (ed.) (1995). *Living with uncertainty: new directions in pastoral development in Africa*. Intermediate Technology Publications, Southampton.

Scoones, I., Reij, C. and Toulmin, C. (1996). Sustaining the soil. Indigenous soil and water conservation in Africa. In: Reij, C., Scoones, I. and Toulmin, C. (eds), *Sustaining the soil. Indigenous soil and water conservation in Africa*. Earthscan, London, pp. 1–27.

Scott, L. (1984). Palynological evidence for Quaternary palaeoenvironments in southern

Africa. In: Klein, R.G. (ed.), *Southern African prehistory and palaeoenvironments*. A.A. Balkema, Rotterdam, pp. 65–80.

Scott, L. (1995). Pollen evidence for vegetation and climatic change in Southern Africa during the Neogene and Quaternary. In: Vrba, E.S., Denton, G.H., Partridge, T.C. and Burckle, L.H. (eds), *Palaeoclimate and evolution with emphasis on human origins*. Yale University Press, New Haven, pp. 56–76.

Scott, L. (1996). Palynology of hydrax middens: 2000 years of palaeoenvironmental history in Namibia. *Quaternary International*, **33**: 73–79.

Scott, L., Steenkamp, M. and Beaumont, P.B. (1995). Palaeoenvironments in South Africa at the Pleistocene–Holocene transition. *Quaternary Science Reviews*, **14**: 937–947.

Seastedt, T.R. (1984). The role of microarthropods in decomposition and mineralisation processes. *Annual Review of Entomology*, **29**: 25–46.

Seely, M.K. (1998). Can science and community action connect to combat desertification? *Journal of Arid Environments*, **39**: 267–277.

Sefe, F., Ringrose, S. and Matheson, W. (1996). Desertification in north-central Botswana: causes, processes and impacts. *Journal of Soil and Water Conservation*, **51**(3): 241–248.

Seghieri, J., Floret, C. and Pontanier, R. (1995). Plant phenology in relation to water availability: herbaceous and woody species in the savannas of northern Cameroon. *Journal of Tropical Ecology*, **11**: 237–254.

Seiler, W., Conrad, R. and Scharffe, D. (1984). Field studies of methane emission from termite nests into the atmosphere and measurements of methane uptake by tropical soils. *Journal of Atmospheric Chemistry*, **1**(2): 171–186.

Senft, R.L., Coughenour, M.B., Bailey, D.W., Rittenhouse, L.R., Sala, O.E. and Swift, D.W. (1987). Large herbivore foraging and ecological hierarchies. *BioScience*, **37**: 789–799.

Serca, D., Delmas, R., LeRoux, X., Parsons, D.A.B., Scholes, M.C., Abbadie, L., Lensi, R., Ronce, O. and Labroue, L. (1998). Comparison of nitrogen monoxide emissions from several African tropical ecosystems and influence of season and fire. *Global Biogeochemical Cycles*, **12**(4): 637–651.

Serenje, W., Chidumayo, E.N., Chipuwa, J.H., Egneus, H. and Ellegård, A. (1994). *Environmental impact assessment of the charcoal production and utilisation system in central Zambia*. Energy, Environment and Development Series No. 32, Stockholm Environment Institute, Stockholm.

Seri, P. (1993). *Indigenous knowledge and rural development*. Phumpanya Foundation and Village Foundation, Amarin Printing Group, Bangkok (in Thai).

Seshadri, B. (1969). *The twilight of India's wildlife*. John Baker Publishers, London.

Settarak, A., Songporn, N. and Kanjanakunchorn, S. (1986). *Public attitude and opinion towards forest fire control (a case study: Bhuping Forest Fire Control Project)*. Forest Fire Control Unit, Royal Forest Department.

Setterfield, S.A. (1997). The impact of experimental fire regimes on seed production in two tropical eucalypt species in northern Australia. *Australian Journal of Ecology*, **22**: 279–287.

Setterfield, S.A. and Williams, R.J. (1996). Patterns of flowering and seed production in *Eucalyptus miniata* and *E. tetrodonta* in a tropical savanna woodland, northern Australia. *Australian Journal of Botany*, **44**: 107–122.

Seyffarth, J.A.S., Calouro, A.M. and Price, P.W. (1996). Leaf rollers in *Ouratea hexasperma* (Ochnaceae): fire effect and the plant vigor hypothesis. *Revista Brasileira de Biologia*, **56**: 135–137.

Seyler, J.R. (1993). *A systems analysis of the status and potential of Acacia albida in the peanut basin of Senegal.* PhD Thesis, Department of Agricultural Economics, University of Michigan, Michigan.

Shackley, M. (1998). Designating a protected area at Karanambu Ranch, Rupununi Savannah, Guyana: resource management and indigenous communities. *Ambio*, **27**(3): 207–210.

Shalardchai, R., Anan, G. and Ganjanapan, S. (1993). *Community forestry in Thailand. Directions of development. Volume 2. Community forestry in the north.* Local Development Institute, Bangkok (in Thai).

Shawcross, W. (1975). Thirty thousand years and more. *Hemisphere*, **19**: 26–31.

Shepherd, G. (1992). *Managing Africa's tropical dry forests: a review of indigenous methods.* ODI Agricultural Occasional Paper 14, ODI, London.

Shipton, P. (1994). Land and culture in tropical Africa: soils, symbols and metaphysics of the Mundane. *Annual Review of Anthropology*, **23**: 347–377.

Sick, H. (1965). A fauna do *cerrado. Arquivos de Zoologia, Univ. São Paulo*, **12**: 71–93.

Siddle, D. and Swindell, K. (1990). *Rural change in tropical Africa. From colonies to nation-states.* The Institute of British Geographers Special Publication No. 23, Basil Blackwell, Oxford.

Siebert, B.D., Playne, M.J. and Edye, L.A. (1976). The effects of climate and nutrient supplementation on the fertility of heifers in north Queensland. *Proceedings of the Australian Society for Animal Production*, **11**: 249–252.

Silva, J. and Ataroff, M. (1985). Phenology, seed crop and germination of coexisting grass species from a tropical savanna in Western Venezuela. *Acta Oecologica*, **6**(1): 41–51.

Silva, J., Raventos, J. and Caswell, H. (1990). Fire exclusion effects on the growth and survival of two savanna grasses. *Acta Oecologica*, **11**(6): 783–800.

Silva, J., Raventos, J., Caswell, H. and Trevisan, C. (1991). Population responses to fire in a tropical savanna grass *Andropogon semiberbis*: a matrix model approach. *Journal of Ecology*, **79**: 345–356.

Silva, J.F. and Castro, F. (1989). Fire, growth and survivorship in a Neotropical savanna grass *Andropogon semiberbis* in Venezuela. *Journal of Tropical Ecology*, **5**: 387–400.

Silva, J.F. and Moreno, A. (1993). Land use in Venezuela. In: Young, M.D. and Solbrig, O.T. (eds), *The world's savannas: economic driving forces, ecological constraints and policy options for sustainable land use* (MAB 12). UNESCO and Parthenon Press, Paris, pp. 239–257.

Silva, J.F. and Sarmiento, G. (1997). Densidad de leñosas de la sabana estacional y frecuencia de quemas: la hipotesis del equilibrio fluctuante. *Ecotropicos*, **10**(2): 65–78.

Silviconsult Ltd. (1991). *Northern Nigeria household energy study.* Consultancy report to the Federal Forestry Management Evaluation and Co-ordinating Unit, Ibadan, Nigeria. Silviconsult Ltd, Bjared, Sweden.

Simoes, M. and Baruch, Z. (1991). Responses to simulated herbivory and water stress in two tropical C_4 grasses. *Oecologia*, **88**: 173–180.

Sinclair, A.R.E. (1975). The resource limitation of trophic levels in tropical grassland ecosystems. *Journal of Animal Ecology*, **44**: 497–520.

Sinclair, A.R.E. (1977). *The African Buffalo: a study of resource limitation of populations.* University of Chicago Press, Chicago.

Sinclair, A.R.E. (1979). Dynamics of the Serengeti ecosystem: process and pattern. In: Sinclair, A.R.E. and Norton-Griffiths, M. (eds), *Serengeti: dynamics of an ecosystem.* University of Chicago Press, Chicago, pp. 1–30.

Sinclair, A.R.E. (1995). Equilibria in plant–herbivore interactions. In: Sinclair, A.R.E.

and Arcese, P. (eds), *Serengeti II. Dynamics, management, and conservation of an eco-system*. The University of Chicago Press, Chicago, pp. 91–113.

Sinclair, A.R.E. and Arcese, P. (eds), (1995). *Serengeti II. Dynamics, management, and conservation of an ecosystem*. The University of Chicago Press, Chicago.

Sinclair, A.R.E. and Fryxell, J.M. (1985). The Sahel of Africa: ecology of a disaster. *Canadian Journal of Zoology*, **63**: 987–994.

Sinclair, A.R.E. and Norton-Griffiths, M. (eds) (1979). *Serengeti: dynamics of an eco-system*. University of Chicago Press, Chicago.

Sindiga, I. (1995). Wildlife-based tourism in Kenya: land use conflicts and govern-ment compensation policies over protected areas. *Journal of Tourism Studies*, **6**(2): 45–55.

Singh, G. and Geissler, E.A. (1985). Late Cenozoic history of vegetation, fire, lake levels and climate at Lake George, New South Wales, Australia. *Philosophical Transactions of the Royal Society of London, Series B*, **311**: 379–447.

Skarpe, C. (1992). Dynamics of savanna ecosystems. *Journal of Vegetation Science*, **3**: 293–300.

Skeat, A. (1986). Kakadu National Park weed control. *Australian Ranger Bulletin*, **3**: 16.

Skinner, J.D., Monro, R.H. and Zimmerman, I. (1984). Comparative food intake and growth of cattle and impala on mixed tree savanna. *South African Journal of Wildlife Research*, **14**: 1–9.

Slatyer, R.O. (1954). A note on the available moisture range of some northern soils. *Journal of the Australian Institute of Agricultural Science*, **20**: 46–47.

Smart, N.O.E., Hatton, J.C. and Spence, D.H.N. (1985). The effect of long-term exclusion of large herbivores on vegetation in Murchison Falls National Park, Uganda. *Biological Conservation*, **33**: 229–245.

Smith, C.W. and Tunison, T. (1992). Fire and alien plants in Hawaii: research and management implications for native ecosystems. In: Stone, C.P., Smith, C.W. and Tunison, J.T. (eds), *Alien plant invasion in Hawaii: management and research in native ecosystems*. Honolulu University Press, Hawaii, pp. 394–408.

Smith, J., Barau, A.D., Goldman, A. and Mareck, J.H. (1994). The role of technology in agricultural intensification: the evolution of maize production in the northern Guinea savanna of Nigeria. *Economic Development and Cultural Change*, **42**(3): 537–554.

Smith, W., Meredith, T.C. and Johns, T. (1996). Use and conservation of woody veg-etation by the Batemi of Ngorongoro District, Tanzania. *Economic Botany*, **50**(3): 290–299.

Smitinand, T. (1962). *Types of forests of Thailand*. Royal Forest Department, Ministry of Agriculture, Bangkok.

Smitinand, T. (1968) Identification keys to the genera and species of the Dipterocarpaceae of Thailand. *Natural History Bulletin of the Siam Society*, **19**: 57–83.

Smitinand, T. (1977). *Vegetation and ground cover of Thailand*. Department of Forest Biology, Faculty of Forestry, Kasetsart University, Bangkok.

Solbrig, O.T. (1991). Savanna modelling for global change. *Biology International* Special Issue No. 24, IUBS, Paris.

Solbrig, O.T. (1996). The diversity of the savanna ecosystem. In: Solbrig, O.T., Medina, E. and Silva, J.F. (eds), *Biodiversity and savanna ecosystem processes: a global perspective*. Ecological Studies 121, Springer-Verlag, Berlin, pp. 1–27.

Soper, R.C. (1969). Radio carbon dating of 'Dimpledbase ware' in west Kenya. *Azania*, **4**: 148–152.

Spinage, C.A. (1973). A review of ivory exploitation and elephant population trends in Africa. *East African Wildlife Journal*, **11**: 281–289.

Spinage, C.A. and Guinness, F.E. (1971). Tree survival in the absence of elephants in the Akagera National Park, Rwanda. *Journal of Applied Ecology*, **8**: 723–728.

Sricharatchanya, P. (1988). Thailand: politics of power. *Far Eastern Economic Review*, **31**: 24.

Srikosamatara, S. and Suteethorn, V. (1995). Populations of gaur and banteng and their management in Thailand. *National History Bulletin of the Siam Society*, **43**: 55–83.

Stafford Smith, D.M. and Foran, B.D. (1991). RANGEPACK: the philosophy underlying the development of a microcomputer-based decision support system for pastoral land management. In: Werner, P.A. (ed.), *Savanna ecology and management. Australian perspectives and intercontinental comparisons*. Blackwell, Oxford, pp. 197–202.

Stager, J.C., Cumming, B. and Meeker, L. (1997). A high-resolution 11,400-yr diatom record from Lake Victoria, East Africa. *Quaternary Research*, **47**: 81–89.

Stamp, L.D. (1925). *The vegetation of Burma from an ecological standpoint*. Calcutta.

Stander, P.E., Nott, T.B. and Mentis, M.T. (1993). Proposed burning strategy for a semi-arid African savanna. *African Journal of Ecology*, **31**: 282–289.

Stanner, W. (1965). Aboriginal territorial organisation: estate, range, domain and regime. *Oceania*, **36**: 1–26.

Starfield, A.M., Cumming, D.H.M., Taylor, R.D. and Quadling, M.S. (1993). A frame-based paradigm for dynamic ecosystem models. *AI Applications*, 7(2/3): 1–13.

Stevenson, P.M. (1985). Traditional Aboriginal resource management in the wet–dry tropics: Tiwi case study. *Proceedings of the Ecological Society of Australia*, **13**: 309–315.

Stewart, J.W. (1996). Savanna users and their perspectives: grazing industry. In: Ash, A.J. (ed.), *The future of tropical savannas: an Australian perspective*. CSIRO, Australia, pp. 47–53.

Stocker, G. (1976). *Report on cyclone damage to natural vegetation in the Darwin area after Cyclone Tracey, 25 December 1974*. Forestry and Timber Bureau, Leaflet No. 17, Australian Government Publishing Service, Canberra.

Stoneman, G.L., Dell, B. and Turner, N.C. (1994). Mortality of *Eucalyptus marginata* (jarrah) seedlings in Mediterranean-climate forest in response to overstorey, site, seedbed, fertiliser application and grazing. *Australian Journal of Ecology*, **19**: 103–109.

Storrs, A.E.G. (1979). *Know your trees. Some of the common trees found in Zambia*. Forest Department, Ndola.

Storrs, A.E.G. (1982). *More about trees: a sequel to 'Know your trees'*. Forest Department, Ndola.

Story, R. (1969). Vegetation of the Adelaide-Alligator Area. *Land Research Series*, **25**: 114–130.

Story, R. (1976). Vegetation of the Alligator Rivers area. *Land Research Series*, **38**: 89–111.

Stott, P. (1976). Recent trends in the classification and mapping of dry deciduous dipterocarp forest in Thailand. In: Ashton, P. and Ashton, M. (eds), *The classification and mapping of Southeast Asian ecosystems*. Transaction of the Aberdeen–Hull Symposium on Malesian ecology, IV. University of Hull, pp. 22–56.

Stott, P. (1984). The savanna forests of mainland South East Asia: an ecological survey. *Progress in Physical Geography*, **8**: 315⊦335.

Στοττ, Π. (1986). Τηε σπατιαλ παττερν οφ δρψ σεασον φιρεσ ιν τηε σαϖαν Τηαιλανδ. ϑουρναλ οφ Βιογεογραπηψ, *13*: 345

Stott, P. (1988a). Savanna forest and seasonal fires in South East Asia. *Plants Today*, **Nov–Dec**: 196–200.

Stott, P. (1988b). The forest as Phoenix: towards a biogeography of fire in mainland South East Asia. *The Geographical Journal*, **154**(3): 337–350.

Stott, P. (1991a). Recent trends in the ecology and management of the world's savanna formations. *Progress in Physical Geography*, **15**(1): 18–28.

Stott, P. (1991b). Stability and stress in the savanna forests of mainland Southeast Asia. In: Werner, P.A. (ed.), *Savanna ecology and management. Australian perspectives and intercontinental comparisons.* Blackwell, Oxford, pp. 29–39.

Strang, R.M. (1965). *Bush encroachment and secondary succession in the southern Rhodesian highveld with special reference to soil moisture relations.* PhD Thesis, University of London, London.

Strang, R.M. (1974). Some man-made changes in successional trends on the Rhodesian highveld. *Journal of Applied Ecology*, **111**: 249–263.

Stromgaard, P. (1984). The immediate effect of burning and ash fertilisation. *Plant and Soil*, **80**: 307–320.

Stromgaard, P. (1985). A subsistence society under pressure: the Bemba of northern Zambia. *Africa*, **55**: 40–59.

Stromgaard, P. (1986). Early secondary succession on abandoned shifting cultivator's plots in the *miombo* of South Central Africa. *Biotropica*, **18**: 97–106.

Stromgaard, P. (1989). *Crop potential and adaptive strategies in Zambian agriculture.* Institute of Geography, University of Copenhagen.

Strugnell, R.G. and Pigott, R.G. (1978). Biomass, shoot-production and grazing of two grasslands in the Rwenzori National Park, Uganda. *Journal of Ecology*, **66**: 73–96.

Stuth, J.W. and Stafford Smith, M. (1993). Decision support for grazing lands: an overview. In: Stuth, J.W. and Lyons, B.G. (eds), *Decision support systems for the management of grazing lands. Emerging issues.* Man and the Biosphere Series, Volume 11. UNESCO, Paris, pp. 1–35.

Subhadira, S., Apichatvullop, Y., Kunarat, P. and Hafner, J.A. (1987). *Case studies of human–forest interactions in Northeast Thailand.* KU/KKU/Ford Foundation, 850-0391, Final Report 2, Northeast Thailand Upland Social Forestry Project, Khon Kaen.

Sukwong, S. (1982). Analysis of dry dipterocarp forest vegetation in Thailand. *Journal of the National Council of Thailand*, **14**: 55–65.

Sukwong, S. and Dhamanitayakul, P. (1977). Fire ecology investigations in dry dipterocarp forest. In: *The Proceedings of National Forestry Conference.* Royal Forest Department, Bangkok, pp. 41–56.

Sukwong, S., Dhamanitayakul, P. and Pongumphai, S. (1975). Phenology and seasonal growth of dry dipterocarp forest tree species. *The Kasetsart Journal*, **9**: 105–113.

Sullivan, S. (1996). Towards a non-equilibrium ecology: perspectives from an arid land. *Journal of Biogeography*, **23**:1–5.

Suthivanit, S. (1989). *The effects of fire frequency on vegetation in the dry dipterocarp forest at Sakaerat, Nakorn Ratchasrima province.* MSc Thesis, Faculty of Forestry, Kasetsart University, Bangkok.

Talbot, L.M. and Talbot, H.M. (1963). *The wildebeest in western Masaailand, East Africa.* Wildlife Monograph No. 12, The Wildlife Society, Bethesda, MD, USA.

Tarboton, W.R. (1980). Avian populations in a Transvaal savanna. *Proceedings of the 4th Pan African Ornithological Conference*, pp. 113–124.

Taylor, D.M. (1990). Late Quaternary pollen records from two Ugandan mires:

evidence for environmental change in the Rukiga Highlands of Southwest Uganda. *Palaeogeography, Palaeoclimatology, Palaeoecology,* **80**: 283–300.

Taylor, J.A. and Braithwaite, R.W. (1996). Interactions between land uses in Australia's savannas – it's largely in the mind. In: Ash, A.J. (ed.), *The future of tropical savannas: an Australian perspective.* CSIRO, Australia, pp. 107–118.

Taylor, J.A. and Dunlop, C.R. (1985). Plant communities of the wet–dry tropics of Australia: the Alligator Rivers Region. *Proceedings of the Ecological Society of Australia,* **13**: 83–127.

Taylor, J.A. and Tulloch, D. (1985). Rainfall in the wet–dry tropics: extreme events at Darwin and similarities between years in the period 1870–1983. *Australian Journal of Ecology,* **10**: 281–295.

Taylor, J.L. (1993). Social activism and resistance on the Thai frontier: the case of Phra Prajak Khuttajitto. *Bulletin of Concerned Asian Scholars,* **25**(2): 3–17.

Taylor, J.L. (1996). 'Thamma-chaat': activist monks and competing discourses of nature and nation in Northeastern Thailand. In: Hirsch, P. (ed.), *Seeing forests for trees. Environment and environmentalism in Thailand.* National Thai Studies Centre, Australian National University and Asia Research Centre on Social, Political and Economic Change, Murdoch University. Silkworm Books, Chiang Mai, pp. 37–52.

Taylor, R.D. (1992). *Wildlife management and utilisation in a Zimbabwean communal land: a preliminary evaluation in Nyaminyami District, Kariba.* Proceedings of the 3rd International Wildlife Ranching Symposium, Pretoria, Republic of South Africa.

Tejos, R. (1987). Producción y valor nutritivo de pastos nativos de sabanas inundables de Apure, Venezuela. In: San José, J.J. and Montes, R. (eds), *La capacidad bioproductiva de sabanas.* Centro Internacional de Ecología Tropical, Caracas, pp. 369–450.

Tejos, R., Schargel, R. and Berrade, F. (1990). Características y perspectivas de utilización de sabanas inundables en Venezuela. In: Sarmiento, G. (ed.), *Las sabanas Americanas. Aspectos de su biogeografía, ecología y utilización.* Fondo Editorial Acta Científica Venezolana, Caracas, pp. 163–190.

Thackeray, A.I., Deacon, J., Hall, S., Humphreys, A.J.B., Morris, A.G., Malherbe, V.C. and Catchpole, R.M. (1990). *The early history of Southern Africa to AD1500.* The South African Archaeological Society, Cape Town.

Thailand Development Research Institute (1988). *Thailand natural resource profile,* Thailand Development Research Institute, Bangkok.

Thebaud, B. (1998). *Sahelian Shepherds still struggling 25 years after the big drought.* ID21 Report, IDS, University of Sussex.

Thomas, D.S.G. (1997). Science and the desertification debate. *Journal of Arid Environments,* **37**: 599–608.

Thomas, D.S.G. and Middleton, N.J. (1994). *Desertification: exploding the myth.* John Wiley, Chichester.

Thomson, P.J. (1975). The role of elephants, fire and other agents in the decline of a *Brachystegia boehmii* woodland. *Journal of South African Wildlife Management Assessment,* **5**: 11–18.

Thorbjarnarson, J. (1991). An analysis of the spectacled caiman (*Caiman crocodilus*) harvest program in Venezuela. In: Robinson, J.G. and Redford, K.H. (eds), *Neotropical wildlife use and conservation.* The University of Chicago Press, Chicago, pp. 216–235.

Thurgate, N. (1998). Reptiles give new slant on conservation. *Savanna Links,* **5**: 6.

Tiffen, M., Mortimore, M. and Gichuki, F. (1994). *More people, less erosion. Environmental recovery in Kenya.* John Wiley, New York.

Tilman, D. and Wedin, D. (1991). Oscillations and chaos in the dynamics of a perennial grass. *Nature*, **353**: 653–655.

Tinley, K.L. (1982). The influence of soil moisture balance in ecosystem pattern in southern Africa. In: Huntley, B.J. and Walker, B.H. (eds), *Ecology of tropical savannas*. Springer, Berlin, pp. 175–192.

Tolsma, D.J., Ernst, W.H.O. and Verwey, R.A. (1987). Nutrients in soil and vegetation around two artificial waterpoints in eastern Botswana. *Journal of Applied Ecology*, **24**: 991–1000.

Torres, A. (1984). *Estudio de algunos aspectos de la ecofisiología de tres gramineas en la sabana estacional.* Tesis Maestría Ecología Tropical, Universidad de Los Andes, Mérida.

Tothill, J.C. (1985). American savanna ecosystems. In: Tothill, J.C. and Mott, J.J. (eds), *Ecology and management of the world's savannas*. Australian Academy of Science, Canberra, pp. 52–55.

Tothill, J.C. and Hacker, J.B. (1973). *The grasses of Southeast Queensland.* University of Queensland Press, Brisbane.

Tothill, J.C., Mott, J.J. and Gillard, P. (1982). Pasture weeds of the tropics and subtropics with special reference to Australia. In: Holzner, W. and Numata, N. (eds), *Biology and ecology of seeds*. Dr W. Junk, The Hague, pp. 403–427.

Toxopeus, A.G. (1996). *ISM, an interactive spatial and temporal modelling system as a tool in ecosystem management.* PhD Thesis, ITC Publication No. 44, ITC, Enschede, The Netherlands.

Trainor, C.R. and Woinarski, J.C.Z. (1994). Responses of lizards to three experimental fires in the savanna forests of Kakadu National Park. *Wildlife Research*, **21**(2): 131–148.

Trapnell, C.G. (1953). *The soils, vegetation and agriculture of North-Eastern Rhodesia.* Government Printer, Lusaka.

Trapnell, C.G. (1959). Ecological results of woodland burning experiments in Northern Rhodesia. *Journal of Ecology*, **47**: 129–168.

Trapnell, C.G., Friend, M.T., Chamberlain, G.T. and Birch, H.F. (1976). The effect of fire and termites on a Zambian woodland soil. *Journal of Ecology*, **64**: 577–588.

Troll, C. (1950). *Savannentypen und das problem der primarsavannen.* Proceedings of the 7th Botanical Institute Congress, Stockholm, pp. 670–674.

Trollope, W.S.W. (1984a). Fire in savanna. In: Booysen, P. de V. and Tainton, N.M. (eds), *Ecological effects of fire in south African ecosystems*. Ecological Studies 48, Springer-Verlag, Berlin, pp. 149–175.

Trollope, W.S.W. (1984b). Fire behaviour. In: Booysen, P. de V. and Tainton, N.M. (eds), *Ecological effects of fire in south African ecosystems*. Ecological Studies 48, Springer-Verlag, Berlin, pp. 199–217.

Trollope, W.S.W. (1993). Fire regime of the Kruger National Park for the period 1980–1992. *Koedoe*, **36**(2): 45–52.

Twyman, C. (1998). Rethinking community resource management: managing resources or managing people in western Botswana? *Third World Quarterly*, **19**(4): 745–770.

UNEP (1990). *Report on ad-hoc consultation meeting on assessment of global desertification: status and methodologies.* UNEP, Nairobi.

USAID (1998). *FEWS Special Report 98–2.* USAID, Washington, DC.

Vail, L. (1977). Ecology and history: the example of eastern Zambia. *Journal of Southern African Studies*, **3**: 129–155.

Van de Koppel, J. and Prins, H.H.T. (1998). The importance of herbivore interactions for the dynamics of African savanna woodlands: an hypothesis. *Journal of Tropical Ecology*, **14**: 565–576.

Van Rooyen, A.F. (1998). Combating desertification in the southern Kalahari: connecting science with community action in South Africa. *Journal of Arid Environments*, **39**: 285–297.

Van Wijngaarden, W. (1985). *Elephants–Grass–Trees–Grazers*. ITC Publications No. 4, Enschede, The Netherlands.

Van Wilgen, B.W. and Scholes, R.J. (1997). The vegetation and fire regimes of southern-hemisphere Africa. In: van Wilgen, B.W., Andreae, M.O., Goldammer, J.G. and Lindesay, J.A. (eds), *Fire in Southern African savannas. Ecological and atmospheric perspectives.* Witwatersrand University Press, Johannesburg, pp. 27–46.

Van Wilgen, B.W., Everson, B.W. and Trollope, W.S.W. (1990). Fire management in southern Africa: some examples of current objectives, practices and problems. In: Goldammer, J.G. (ed.), *Fire in the tropical biota.* Springer-Verlag, Berlin, pp. 179–215.

Van Wyk, P. (1971). Veld burning in the Kruger National Park, an interim report of some aspects of research. *Proceedings of the Tall Timbers Fire Ecology Conference*, **11**: 9–31.

Vareschi, V. (1960). Observaciones sobre la transpiración de árboles llaneros durante la época de sequía. *Boletin Sociedad Venezolana Ciencias Naturales*, **21**: 128–134.

Veenendal, E.M. (1991). *Adaptive strategies of grasses in a semi-arid savanna in Botswana.* PhD Thesis, Amsterdam Vrije Universiteit, Amsterdam.

Verdesio, J.J. (1994). Perspectivas ambientais. In: Novaes-Pinto, M. (ed.), *Cerrado, Caracterização, Ocupação e Perspectivas* (2nd edition). Editora Universidade de Brasilia, Brasilia, DF, pp. 585–605.

Vernet, J.L., Wengler, L., Solari, M.E., Ceccantini, G., Fournier, M., Ledru, M.P. and Soubies, F. (1994). Fire, climate and vegetation in central Brazil during the Holocene – data from a soil profile with charcoal (Salitre, Minas Gerais). *Comptes Rendus de l'Academie des Sciences Serie II*, **319**(11,2): 1391–1397.

Vesey-Fitzgerald, D. (1960). Grazing succession among East African game animals. *Journal of Mammalology*, **41**: 161–172.

Vetaas, O.R. (1992). Micro-site effects of trees and shrubs in dry savannas. *Journal of Vegetation Science*, **3**: 337–344.

Vidal, J. (1972). *La vegetation du Laos. I. Le milieu (conditions ecologiques)* (2nd edition). Editions Vithagna, Vientiana, Laos.

Vieira, E.M., Andrade, I. and Price, P.W. (1996). Fire effects on a *Palicourea rigida* (Rubiaceae) gall midge – a test of the plant vigour hypothesis. *Biotropica*, **28**(2): 210–217.

Viljoen, A. (1988). Long-term changes in the tree component of the vegetation in the Kruger National Park. In: Macdonald, I.A.W. and Crawford, R.J.M. (eds), *Long-term data series relating to Southern Africa's renewable natural resources.* South African National Science Progress Report 157, CSIR, Pretoria, pp. 310–315.

Villas-Boas, O. and Villas-Boas, C. (1976). *Xingu–Os indios. Seus Mitos*, Zahar Editores, Rio de Janeiro.

Vincens, A. (1991). Late Quaternary vegetation history of the south-Tanganyika basin – climatic implications in south-central Africa. *Palaeogeography, Palaeoclimatology, Palaeoecology*, **86**(3/4): 207–226.

Vincens, A., Chalie, F., Bonnefille, R., Guiot, J. and Tiercelin, J.J. (1993). Pollen-derived rainfall and temperature estimates from Lake Tanganyika and their implication for late Pleistocene water levels. *Quaternary Research*, **40**(3): 343–350.

Vogel, J.C., Fuls, A. and Danin, A. (1986). Geographical and environmental distribution of C$_3$ and C$_4$ grasses in the Sinai, Negev and Judean deserts. *Oecologia*, **70**: 258–265.

Wacharakitti, S., Chantanaparb, L. and Intrachand, P. (1971). *Study on the coppicing power and growth of some valuable tree species in dry dipterocarp forest.* Forest Research Bulletin 19, Faculty of Forestry, Kasetsart University, Bangkok.

Walker, B.H. (1981). Is succession a viable concept in African savanna ecosystems? In: West, D.C., Shugart, H.H. and Botkin, D.B. (eds), *Forest succession*, Springer-Verlag, New York, pp. 431–447.

Walker, B.H. (ed.) (1987). *Determinants of tropical savannas.* IRL Press, Oxford.

Walker, B.H. (1991). Ecological consequences of atmospheric and climatic change. *Climatic Change*, **18**: 301–316.

Walker, B.H. and Langridge, J.L. (1997). Predicting savanna vegetation structure on the basis of plant available moisture (PAM) and plant available nutrients (PAN): a case study from Australia. *Journal of Biogeography*, **24**: 813–825.

Walker, B.H. and Nix, H.A. (1993). Managing Australia's biological diversity. *Search*, **24**: 173–178.

Walker, B.H. and Noy-Meir, I. (1982). Aspects of the stability and resilience of savanna ecosystems. In: Huntley, B.J. and Walker, B.H. (eds), *Ecology of tropical savannas.* Ecological Studies 42, Springer-Verlag, Berlin, pp. 556–609.

Walker, B.H., Ludwig, D., Holling, C.S. and Peterman, R.S. (1981). Stability of semi-arid savanna grazing systems. *Journal of Ecology*, **69**: 473–498.

Walker, D. (1990). Directions and rates of tropical rainforest processes. In: Webb, L.J. and Kikkawa, J. (eds), *Australian tropical rainforests: science–values–meaning.* CSIRO, Melbourne, pp. 23–32.

Walker, J. and Gillison, A.N. (1982). Australian savannas. In: Huntley, B.J. and Walker, B.H. (eds). Ecology of tropical savannas. *Ecological Studies* **42**. Springer-Verlag, Berlin, pp. 5–24.

Walter, H. (1971). *Ecology of tropical and subtropical vegetation.* Oliver & Boyd, Edinburgh, UK.

Walter, H. (1973). *Vegetation of the earth* (2nd edition). Springer-Verlag, Berlin.

Ward, D., Ngairorue, B.T., Kathena, J., Samuels, R. and Ofran, Y. (1998). Land degradation is not a necessary outcome of communal pastoralism in arid Namibia. *Journal of Arid Environments*, **40**: 357–371.

Ward, D.E, Susott, R.A., Kauffman, J.B., Babbitt, R.E., Cummings, D.L., Dias, B., Holben, B.N., Kaufman, Y.J., Rasmussen, R.A. and Setzer, A.W. (1992). Smoke and fire characteristics for *cerrado* and deforestation burns in Brazil – Base-B experiment. *Journal of Geophysical Research – Atmospheres*, **97**(D13): 14 601–14 619.

Ward, H.K. and Cleghorn, W.B. (1970). The effects of grazing practices on tree regrowth after clearing indigenous woodland. *Rhodesian Journal of Agricultural Research*, **8**: 57–65.

Warming, E. (1892). Lagoa Santa: Et Bidrad til den biologiske plantegeografi. *Det Kongelige danske Videnskabernes selskabs Skrifter*, **6**: 153–488.

Warren, A., Sud, Y.C. and Rozanov, B. (1996). The future of deserts. *Journal of Arid Environments*, **32**: 75–89.

Watson, C.E., Fishman, J. and Reichle Jr., H.G. (1990). The significance of biomass burning as a source of carbon monoxide and ozone in the southern hemisphere tropics: a satellite analysis. *Journal of Geophysical Research*, **95**: 16 443–16 450.

Watson, J.P. (1977). The use of mounds of the termite *Macrotermes falciger* as a soil amendment. *Journal of Soil Science*, **28**: 664–764.

Weber, G.E., Jeltsch, F., VanRooyen, N. and Milton, S.J. (1998). Simulated long-term vegetation response to grazing heterogeneity in semi-arid rangelands. *Journal of Applied Ecology*, **35**(5): 687–699.

Wedin, D.A. and Tilman, D. (1990). Species effects on nitrogen cycling: a test with perennial grasses. *Oecologia*, **85**: 433–441.

Weltzin, J.F. and Coughenour, M.B. (1990). Savanna tree influence on understory vegetation and soil nutrients in northwestern Kenya. *Journal of Vegetation Science*, **1**: 325–334.

Werger, M.J.A. and Coetzee, B.J. (1978). The Sudano-Zambezian Region. In: Werger, M.J.A. (ed.), *Biogeography and ecology of southern Africa*. Junk, The Hague, pp. 301–462.

Werner, P.A. (1986). *Population dynamics and productivity of selected forest trees in Kakadu National Park*. Report to the Australian National Parks and Wildlife Service, Canberra.

Werner, P.A., Walker, B.H. and Stott, P.A. (1991). Introduction. In: Werner, P.A. (ed.), *Savanna Ecology and Management. Australian Perspectives and Intercontinental Comparisons*. Blackwell, Oxford, pp. xi–xii.

Wescott, G. and Molinski, J. (1993). Loving our parks to death: a cautionary tale. *Habitat Australia*, **21**: 14–19.

Western, D. and van Praet, C. (1973). Cyclical changes in the habitat and climate of an East African ecosystem. *Nature*, **241**: 104–106.

Westoby, M. (1991). On long-term ecological research in Australia. In: Risser, P.G. (ed.), *Long-term ecological research: an international perspective*. Wiley, New York, pp. 191–209.

Whelan, R.J. and Main, A.R. (1979). Insect grazing and post-fire plant succession in south-west Australian woodland. *Australian Journal of Ecology*, **4**: 387–398.

White, F. (1976). The underground forests of Africa: a preliminary review. *Gardens Bulletin*, **29**: 57–71.

White, F. (1983). *The vegetation of Africa*. Natural Resources Research 20, UNESCO, Paris.

White, R. (1993). *Comments*. Overseas Development Institute, Pastoral Development Network Paper 35c, pp. 15–18.

Whitmore, T.C. (1984). *Tropical rain forest of the Far East* (2nd edition). Clarendon Press, Oxford.

Whittaker, R.H. (1975). *Communities and ecosystems* (2nd edition). Macmillan Press, New York.

Whittaker, R.H., Morris, J.W. and Goodman, D. (1984). Pattern analysis in savanna woodlands at Nylsvley, South Africa. *Memoirs of the Botanical Survey of South Africa*, **49**: 1–51.

Wickham, J.D., Wu, J. and Bradford, D.F. (1995). *Stressor data sets for studying species diversity at large spatial scales*. US EPA 600/R-95/018, Office of Research and Development, US Environmental Protection Agency, Washington, DC.

Wiens, J.A. (1976). Population responses to patchy environments. *Annual Review of Ecology and Systematics*, **7**: 81–120.

Wijmstra, T.A. and Van der Hammen, T. (1966). Palynological data on the history of tropical savannas in northern South America. *Leidse Geologische Mededelingen*, **38**: 71–90.

Wilcox, D.G. and Cunningham, G.M. (1994). Economic and ecological sustainability of current land use in Australia's rangelands. In: Morton, S.R. and Price, P. (eds), *R&D for sustainable use and management of Australia's rangelands*. LWRRDC Occasional Paper Series No. 06/93, LWRRDC, Canberra, pp. 87–171.

Williams, J. (1991). Search for sustainability: agriculture and its place in the natural ecosystem. *Agricultural Science*, **4**: 32–39.

Williams, J., Day, K.J., Isbell, R.F. and Reddy, S.J. (1985). Soils and climate. In: Muchow, R.C. (ed.), *Agro-research for the semi-arid tropics: north-west Australia*. University of Queensland Press, Brisbane, pp. 31–92.

Williams, R.J. (1995). Tree mortality in relation to fire intensity in a tropical savanna of the Kakadu region, Northern Territory, Australia. *CALMScience Supplement*, **4**: 77–82.

Williams, R.J. and Douglas, M. (1995). Windthrow in a tropical savanna in Kakadu National Park, northern Australia. *Journal of Tropical Ecology*, **11**: 547–558.

Williams, R.J., Duff, G.A., Bowman, D.M.J.S. and Cook, G.D. (1996). Variation in the composition and structure of tropical savannas as a function of rainfall and soil texture along a large-scale climatic gradient in the Northern Territory, Australia. *Journal of Biogeography*, **23**: 747–757.

Williams, R.J., Myers, B.A., Muller, W.J., Duff, G.A. and Eamus, D. (1997). Leaf phenology of woody species in a northern Australian tropical savanna. *Ecology*, **78**: 2542–2558.

Williams, R.J., Cook, G.D., Gill, A.M. and Moore, P.H.R. (1999). Fire regime, fire intensity and tree survival in a tropical savanna in northern Australia. *Australian Journal of Ecology*, **24**: 50–59.

Wilson, B.A. and Bowman, D.M.J.S. (1994). Factors influencing tree growth in tropical savanna: studies of an abrupt *Eucalyptus* boundary at Yapilika, Melville Island, northern Australia. *Journal of Tropical Ecology*, **10**: 103–120.

Wilson, B.A., Brocklehurst, P.S., Clark, M.J. and Dickinson, K.J.M. (1991). *Vegetation survey of the Northern Territory: notes to accompany 1 : 1,000,000 map*. Land Resource Survey Project, Conservation Commission of the Northern Territory Technical Report No. 49, Darwin.

Wilson, B.A., Russell-Smith, J. and Williams, R.J. (1996). Terrestrial vegetation. In: Finlayson, C.M. and von Oertzen, I. (eds), *Landscape and vegetation of the Kakadu region*. Kluwer, Dordrecht, pp. 57–79.

Wilson, J.R., Ludlow, M.M., Fisher, M.J. and Schulze, E.D. (1980). Adaptation to water stress of the leaf water relation of four tropical forage species. *Australian Journal of Plant Physiology*, **7**: 207–220.

Wiman, W.D. (1990). Oil and gas developments in South America, Central America, Carribean area, and Mexico in 1989. *AAPG Bulletin – American Association of Petroleum, Geologists*, **74**(10B): 336–355.

Wiwat, K. (ed.), (1993). *Community rights: decentralisation of power to manage resources*. Local Development Institute, Bangkok (in Thai).

Woinarski, J.C.Z. (1990). Effects of fire on the bird communities of tropical woodlands and open forests in northern Australia. *Australian Journal of Ecology*, **15**: 1–22.

Woinarski, J.C.Z. and Braithwaite, R.W. (1990). Conservation foci for Australian birds and mammals. *Search*, **21**: 65–68.

Wolseley, P.A. (1997). Response of epiphytic lichens to fire in tropical forests of Thailand. *Bibliotheca Lichenologica*, **68**: 165–176.

Wolseley, P.A. and Aguirre-Hudson, B. (1997). Fire in tropical dry forests: corticolous lichens as indicators of recent ecological changes in Thailand. *Journal of Biogeography*, **24**: 345–362.

Wood, T.G. (1976). The role of termites (Isoptera) in decompositional processes. In: Anderson, J.M. and Macfadyen, A. (eds), *The role of terrestrial and aquatic organisms in decomposition processes*. Blackwell, Oxford, pp. 145–168.

Woube, M. (1998). Effect of fire on plant communities and soils in the humid tropical savannah of Gambela, Ethiopia. *Land Degradation and Development,* **9**: 275–282.

Wu, J. and Loucks, O.L. (1995). From balance of nature to hierarchical patch dynamics: a paradigm shift in ecology. *The Quarterly Review of Biology,* **70**(4): 439–466.

Yanda, P.Z. and Mung'ong'o, C.G. (1995). *Environmental changes and social development in Kasulu District, western Tanzania.* A reconnaissance report, Mimeo.

Yeaton, R.I. (1988). Porcupines, fire and the dynamics of the tree layer in *Burkea africana* savanna. *Journal of Ecology,* **76**: 1017–1029.

Yeaton, R.I., Frost, S.K. and Frost, P.G.H. (1986). A direct gradient analysis of grasses in a savanna. *South African Journal of Botany,* **82**: 482–486.

Young, M.D. (1983). Relationships between land tenure and arid zone land use. In: Messer, J. and Mosley, G. (eds), *What future for Australia's arid lands?* Australian Conservation Foundation, Melbourne.

Young, T.P. and Lindsay, W.K. (1988). Disappearance of *Acacia* woodlands: the effect of size structure. *African Journal of Ecology,* **26**: 69–72.

Young, T.P., Stubblefield, C.H. and Isbell, L.A. (1997). Ants on swollen thorn *Acacias*: species coexistence in a simple system. *Oecologia,* **109**(1): 98–107.

Zietsman, P.C., Grobbelaar, N. and van Rooyen, N. (1988). Soil nitrogenase activity of the Nylsvley Nature Reserve. *South African Journal of Botany,* **54**: 21–27.

Zimmerman, I. (1978). *The feeding ecology of Africander steers (Bos indicus) on mixed bushveld at Nylsvley Nature Reserve, Transvaal.* MSc Thesis, University of Pretoria.

Zimmerman, P.R., Greenberg, J.P., Wandiga, S.O. and Crutzen, P.J. (1982). Termites – a potentially large source of atmospheric methane, carbon-dioxide, and molecular-hydrogen. *Science,* **218**(4572): 563–565.

Index